Advanced Studies in Breakwaters and Coastal Protection

Advanced Studies in Breakwaters and Coastal Protection

Guest Editors

Maria Teresa Reis
João Alfredo Santos
Tiago Fazeres Ferradosa

Basel • Beijing • Wuhan • Barcelona • Belgrade • Novi Sad • Cluj • Manchester

Guest Editors

Maria Teresa Reis
Laboratório Nacional de Engenharia Civil (LNEC)
Lisbon
Portugal

João Alfredo Santos
Instituto Superior de Engenharia de Lisboa
Lisbon
Portugal

Tiago Fazeres Ferradosa
Faculty of Engineering
University of Porto
Porto
Portugal

Editorial Office
MDPI AG
Grosspeteranlage 5
4052 Basel, Switzerland

This is a reprint of the Special Issue, published open access by the journal *Journal of Marine Science and Engineering* (ISSN 2077-1312), freely accessible at: https://www.mdpi.com/journal/jmse/special_issues/akili_Breakwaters.

For citation purposes, cite each article independently as indicated on the article page online and as indicated below:

Lastname, A.A.; Lastname, B.B. Article Title. *Journal Name* **Year**, *Volume Number*, Page Range.

ISBN 978-3-7258-5559-9 (Hbk)
ISBN 978-3-7258-5560-5 (PDF)
https://doi.org/10.3390/books978-3-7258-5560-5

Contents

Journal of
Marine Science and Engineering

MDPI

Editorial

Advanced Studies in Breakwaters and Coastal Protection

Maria Teresa Reis [1], João Alfredo Santos [2] and Tiago Fazeres-Ferradosa [3,4,*]

1 Laboratório Nacional de Engenharia Civil (LNEC), 1700-066 Lisbon, Portugal; treis@lnec.pt
2 Instituto Superior de Engenharia de Lisboa (ISEL), 1959-007 Lisbon, Portugal; jasantos@dec.isel.ipl.pt
3 Faculty of Engineering, University of Porto, 4200-465 Porto, Portugal
4 CIIMAR/CIMAR LA, Centro Interdisciplinar de Investigação Marinha e Ambiental, Universidade do Porto, Terminal de Cruzeiros do Porto de Leixões, 4450-208 Matosinhos, Portugal
* Correspondence: tferradosa@fe.up.pt

Globally, mean sea level rise has been accelerating, populations living in coastal zones have been gradually growing, including in low-lying coasts, and the demands for deeper harbors have been quickly increasing to accommodate the future deeper-draft vessels that are being bought online by shipping companies. Specific regions around the world will also experience increasingly frequent and extreme storm events, which will inevitably lead to increased structural damage, beach/dune erosion, flooding, and saltwater intrusion, amongst other consequences. These are anticipated to have negative impacts on the economy, infrastructure, and lives. Adapted solutions that are safe and sustainable are required for the protection of coastal residents, ecosystems, coastal/harbor infrastructure, and the economy.

Breakwaters are commonly used for the sheltering of harbor basins and harbor entrances against waves, with different types being used depending on local characteristics (e.g., foundation conditions, water depths, wave climate severity), sizes and shapes of the area to be protected, amounts of large quarry stones, etc. They may be rubble-mound structures armored with rock or concrete armor units, vertical-front structures, or composite structures. Breakwaters have also been used in coastal defense schemes with the main objective of preventing shoreline erosion. In this case, detached breakwaters, reef breakwaters, and groins (similar to rubble mound breakwaters) are more commonly used.

This Special Issue on "Advanced Studies in Breakwaters and Coastal Protection" solicited original research on innovative methods and technologies for coastal and harbor protection structures, including findings on the effects of climate change and the needs for sustainable solutions, focusing on advanced studies (physical and numerical models, field observations, composite modeling, probabilistic approaches) for upgrading the reliability of existing structures in terms of safety and functionality, for the optimal design of new ones, and for improving measures to mitigate risks (e.g., warning systems and decision-support tools).

Eleven papers on these subjects were selected for this Special Issue, of which one is a case report [1], another is a technical note [2], and the remaining nine are applied research articles. The techniques used in the presented studies range from in situ measurements, in order to evaluate wave height reduction inside a port in Korea caused by the construction of a detached breakwater [1], to convolutional neural networks used to estimate overtopping discharge on several types of coastal structures [3].

Although most of the papers deal with scale model tests, there is one interesting example of a not-so-small-scale model built off the beach of Reggio Calabria [4], in Italy, at the Natural Ocean Engineering Laboratory, to study the wave impact on a vertical structure where a U_OWC was installed.

Overtopping was one of the most commonly studied phenomena in the papers, given its importance for safety matters related to breakwaters, namely, in face of climate uncertainty [2,3,5–7]. This Special Issue also features a paper that uses CFD modeling to assess the influence of the friction parameters on the overtopping response of a rubble mound breakwater under extreme wave attack [2] and another one [6] that uses a simpler numerical model for sea-wave propagation to illustrate the inclusion of such phenomena in the prototype of a forecast system.

Coastal protection against sea wave action is the subject of two papers. One paper uses scale model tests to assess the influence of seawall shape and position on a low-lying reef flat [8], whereas another paper combines scale model tests and numerical model simulations to evaluate wave transmission over rubble mound submerged breakwaters [9].

The Special Issue also contains a paper on the use of buffer blocks to dissipate the flow energy associated with tsunami events [10]. The goal of this procedure is to improve coastal protection from these events. Related to this type of energy dissipation, although at a different scale, is a paper on the hydraulic response and overtopping performance of the armor layer of a rubble mound breakwater made of single layer [5]. The shape of the artificial element, which looks like a small cube on top of a large one, promotes energy dissipation through the flow around the small cubes.

Finally, it includes a compelling paper that illustrates, using practical examples, how the potential effects of climate changes on coastal structures can be analyzed by the owners or designers of such structures [11].

We hope that readers will enjoy and learn from these papers as much as we have.

Author Contributions: Conceptualization, M.T.R., J.A.S. and T.F.-F., methodology, M.T.R., J.A.S. and T.F.-F.; validation, M.T.R., J.A.S. and T.F.-F., formal analysis, M.T.R., J.A.S. and T.F.-F.; writing—original draft preparation, J.A.S., M.T.R. and T.F.-F.; writing—review and editing, J.A.S., M.T.R. and T.F.-F.; supervision, J.A.S., M.T.R. and T.F.-F. All authors have read and agreed to the published version of the manuscript.

Conflicts of Interest: The authors declare no conflict of interest.

References

1. Ryu, K.H.; Jeong, W.M.; Oh, J.-E.; Baek, W.-D.; Chang, Y.S. Wave Height Reduction Inside Pohang New Port, Korea, Due to the Construction of a Detached Breakwater. *J. Mar. Sci. Eng.* **2022**, *10*, 1537. [CrossRef]
2. Castiglione, F.; Stagnitti, M.; Musumeci, R.E.; Foti, E. Influence of Van Gent Parameters on the Overtopping Discharge of a Rubble Mound Breakwater. *J. Mar. Sci. Eng.* **2023**, *11*, 1600. [CrossRef]
3. Tsai, Y.-T.; Tsai, C.-P. Predictions of Wave Overtopping Using Deep Learning Neural Networks. *J. Mar. Sci. Eng.* **2023**, *11*, 1925. [CrossRef]
4. Romolo, A.; Timpano, B.; Laface, V.; Fiamma, V.; Arena, F. Experimental Investigation of Wave Loads on U-OWC Breakwater. *J. Mar. Sci. Eng.* **2023**, *11*, 19. [CrossRef]
5. Safari, I.; Mouazé, D.; Aliasgary, S.; Carpentier, G.; Ropert, F. Hydraulic Response and Overtopping Performance of Single-Layer Double Cube Unit Armored Mound Breakwater. *J. Mar. Sci. Eng.* **2023**, *11*, 1382. [CrossRef]
6. Manz, A.; Zózimo, A.C.; Garzon, J.L. Application of SWASH to Compute Wave Overtopping in Ericeira Harbour for Operational Purposes. *J. Mar. Sci. Eng.* **2022**, *10*, 1881. [CrossRef]
7. Jiao, B.; Zhao, Q.; Chen, F.; Liu, C.; Fang, Q. Numerical Evaluation of Wave Dissipation on a Breakwater Slope Covered by Precast Blocks with Different Geometrical Characteristics. *J. Mar. Sci. Eng.* **2024**, *12*, 1735. [CrossRef]
8. Rossignol, G.; Sous, D. Coastal Defences on Low-Lying Reef Flats: A Laboratory Study of Seawall Shape and Position. *J. Mar. Sci. Eng.* **2022**, *10*, 1652. [CrossRef]
9. Repousis, E.; Roupas, I.; Memos, C. Wave Transmission over Rubble-Mound Submerged Breakwaters. *J. Mar. Sci. Eng.* **2023**, *11*, 1346. [CrossRef]

10. Nageswaran, P.; Oetjen, J.; Harish, S.; Sriram, V.; Sundar, V.; Schüttrumpf, H. Buffer Blocks as Wave Energy Dissipators: Flow Depth Reduction. *J. Mar. Sci. Eng.* **2024**, *12*, 1145. [CrossRef]
11. Allsop, W. Climate Change Threats to Two Breakwaters. *J. Mar. Sci. Eng.* **2022**, *10*, 1613. [CrossRef]

Journal of
Marine Science and Engineering

MDPI

Article

Numerical Evaluation of Wave Dissipation on a Breakwater Slope Covered by Precast Blocks with Different Geometrical Characteristics

Bowen Jiao [1], Qingli Zhao [1], Fang Chen [1,2], Chunhui Liu [3] and Qinghe Fang [1,2,*]

1 Research Center for Coastal Structures, School of Ocean Engineering, Harbin Institute of Technology, Weihai 264209, China; 23s130261@stu.hit.edu.cn (B.J.); zhaoqingli@hit.edu.cn (Q.Z.); fangchen@hhu.edu.cn (F.C.)
2 Jiangxi Academy of Water Science and Engineering, Nanchang 330029, China
3 School of Civil Engineering, Yantai University, Yantai 264005, China
* Correspondence: qinghefang@hit.edu.cn

Abstract: Slopes suffer damage from waves in coastal environments. Precast blocks with well-designed geometrical characteristics can benefit the construction of revetments by mitigating the issue of wave overtopping and dissipating wave energy. In this study, we numerically studied the effect of the geometrical characteristics of precast blocks on wave overtopping by carrying out a numerical simulation of wave overtopping on a slope covered with precast blocks. A total of three different types of blocks were considered in this study to determine the optimal geometric shape using a validated numerical model. Our numerical investigation demonstrated that the roughness of the precast block plays an important role in lessening the height of the wave run-up. Concave and embedded regular hexagons could reduce the wave run-up height by 44.6% compared with smooth slopes within a 2 s wave period. Herein, we evaluate and discuss the influence of the geometrical characteristics of a given precast block, such as thickness, aperture, and wave dissipation notch, on wave run-up. We also present an empirical formula for predicting wave run-up on a slope covered by a concave and embedded regular hexagon-type prefabricated block. This study provides valuable insights into the design of prefabricated revetment blocks.

Keywords: slope; precast block; numerical simulation; geometrical characteristic; wave run-up

1. Introduction

With rising sea levels, waterways and their neighboring infrastructure face a heightened risk of wave overtopping [1]. It is essential to protect the coast and the banks of inland waterways, such as rivers, canals, or lakes, from wave erosion [2,3]. Slope armoring is an economical and practical approach through which to defend the coast and banks from the damaging effects of wave action [4]. Relative to smooth slopes, precast block slopes effectively reduce wave overtopping due to their different geometrical characteristics, which act as rough elements. This roughness significantly enhances their ability to lessen the wave run-up height. The wave run-up height is an important parameter to consider within the engineering and design of bank protection facilities, as highlighted in [5]. Moreover, precast blocks can be manufactured on site, and their individual weights can be adjusted according to various design specifications [6]. These blocks also help to reduce costs, making them particularly suitable for special environments such as steep slopes [7,8]. Their modular nature further enables easier transportation and installation. It should be noted that the wave run-up on the slopes covered by prefabricated blocks is a complex fluid–structure interaction issue. Therefore, run-up on the exact shape of the block surface should be of concern in engineering practice [9].

When attempting to optimize existing designs, an in-depth physical and process-based understanding of how waves interact is highly beneficial [10]. The wave run-up process is crucial in designing the crest of seawalls, and the distribution of the wave run-up height is a valuable factor in planning elevated structures. The wave run-up height is also a useful metric through which we can assess how effectively waves are dissipated and is derived from numerical models simulating the interaction between waves and structures. This coefficient integrates the roughness of the precast blocks, reflecting the performance of different block shapes under the impact of waves. Roughness affects the fluid flow, while permeability describes the ability of a fluid to pass through the material. Therefore, the roughness permeability coefficient combines these two aspects to more accurately simulate the ability of precast blocks to protect land from waves.

Although some research has been conducted on the ability of slope structures to dissipate waves [11–13], there are few published articles on reducing the wave run-up through changing the shape of precast blocks [13]. Mainstream slope protection methods include vegetation and concrete precast blocks, with vegetation offering wave dissipation [14,15]. However, considering the local hydrological conditions and maintenance costs, precast concrete blocks present a more practical solution [16]. In this study, we used precast blocks due to their ease of construction and adaptability. Through physical experiments and numerical simulations of slopes made up of precast blocks, we were able to demonstrate that precast block slopes have a sufficient capacity to dissipate waves [17] while also being cost-effective in terms of production, transportation, installation, and maintenance [18,19]. Precast blocks are also suitable for various environments and offer longer lifespans with lower maintenance costs, partly due to their enhanced erosion resistance [18,20].

Precast blocks are usually designed with a regular rectangular shape to cover the banks of inland waterways—such as rivers, canals, and lakes—from erosional damage. [21–23]. Thanks to their cost-effectiveness and practicability, precast blocks are commonly used in China and other countries. However, the safety of various forms of bank protection built using precast blocks is significantly influenced by overtopping [24]. It is evident that the geometrical characteristics of precast blocks have a degree of influence on such overtopping [25]. To address this issue, we proposed three innovative slope protection structures and evaluated their ability to dissipate waves in regular conditions. These three types of precast blocks were concave and embedded regular hexagon-type prefabricated blocks, Z-type prefabricated concrete blocks, and embedded quadrangle-type prefabricated hollow blocks, referred to hereafter as B1, B2, and B3, respectively.

In this research, both physical experiments and numerical simulations were integrated to advance coastal engineering practices by evaluating the performance of three specially shaped precast block designs as sloping surfaces for embankments. The novelty of this study is reflected in two main aspects. First, the wave run-up reduction efficiency of these block shapes, which had previously been overlooked as they were traditionally applied for ornamental purposes in China's inland lakes, was systematically assessed. Second, through the optimization of the precast block designs based on their wave dissipation effectiveness, adjustments were made to various shape parameters including aperture, thickness, and wave sill size to determine the most suitable configuration. This research contributes to identifying the most effective design and improving wave dissipation performance in embankment construction [26]. However, it is important to note that in shallow or intermediate waters, it is common practice to use spectral waves in the form of irregular wave trains. This represents a limitation of the current study, which primarily focused on regular waves. Future research should include spectral waves to provide a more comprehensive analysis of wave interactions and their effects on coastal structures.

The innovation of this study lies in its integration of numerical simulations and physical experiments to derive an empirical formula through which we can predict the wave run-up on different slope designs. This approach enhances the accuracy of such predictions and provides a reliable tool for coastal engineers. Furthermore, the derived formula can be applied in future projects to predict the wave run-up more accurately, thereby contributing to the safety and durability of coastal infrastructure. This combination of numerical and experimental methods also indicates potential opportunities for further research in coastal engineering, particularly as it relates to the interactions between different wave types and slope designs. The rest of this paper is organized as follows.

Section 2 presents our methodology, detailing the physical model testing and a comparison between numerical and experimental results. The numerical setup is also described including the mesh design, initial conditions, wave generation, and turbulence modeling. Section 3 focuses on the validation of the model, featuring discussions on mesh convergence and validation of the numerical model using free surface elevation and cross-sectional water surface profile data. Section 4 presents the model results, and finally, Section 5 concludes the paper with key findings and potential future research directions.

2. Numerical Methodology

2.1. Numerical Setup

ANSYS FLUENT 2021, a widely used CFD software, employs the finite volume method to solve fluid flow problems. In the process of analysis, variables are laid out on a structured or unstructured grid to effectively capture complex flow dynamics. The solver ensures the accurate conservation of mass, momentum, and energy across the control volumes, providing reliable simulation results. In this study, we employed numerical simulation to not only validate but also enhance the accuracy of the computational process by replicating real-world conditions. Specifically, the numerical result represented the performance of several prefabricated wave protection blocks in terms of their efficiency in both slope protection and wave energy dissipation in conditions similar to those encountered in real coastal environments. Navier–Stokes equations were solved for the two-phase flow in this study, and the interface of water and air was captured using the volume of fluid (VOF) method.

In this study, turbulence was modeled using the stabilized RNG k-ε turbulence model. The initial turbulent kinetic energy (k) was set to 0.4 m^2/s^2, and the turbulent dissipation rate (ε) was set to 0.04 m^2/s^3, ensuring an accurate representation of the turbulent flow. These values were based on experimental data and were carefully calibrated to align with the observed turbulence levels.

A sensitivity analysis was conducted on the k factor and the dissipation rate (ε). The results indicated that varying the k value from 0.35 m^2/s^2 to 0.45 m^2/s^2 did not significantly alter the wave overtopping discharge, thus confirming the reliability of the initially selected values. Similarly, the ε parameter was varied between 0.035 m^2/s^3 and 0.045 m^2/s^3, and the dissipation rates remained consistent with our experimental observations. This sensitivity analysis confirmed that the chosen values of k and ε were within a stable and accurate range for this model. Based on these findings, the model was adjusted to ensure realistic turbulence levels.

In this study, we employed finite volume discretization to handle complex geometries and interfaces in fluid dynamics, allowing for a detailed analysis of flow separation, recirculation zones, and pressure gradients, all of which are critical in predicting wave overtopping and energy dissipation. This approach effectively manages irregular meshes while maintaining geometric precision. To ensure the accuracy of our model, various mesh densities and refinement techniques were utilized during mesh generation to accurately capture the behavior of waves. Specifically, non-uniform mesh refinement was applied to the water's surface area in order to enhance the resolution of wave details, particularly near the wave crest and trough regions wherein steep gradients occur. Additional mesh refinement around the block's surface ensured accurate predictions of boundary layer development and localized turbulence.

2.2. Numerical Wave Flume and Slope

Subsequently, four types of geometrically distinct precast blocks were imported into the software for analysis. The 3D numerical wave flume, as depicted in Figure 1, required boundary conditions that replicate realistic air–water interactions. In the numerical simulation, a total length of 42.5 m was established, of which the slope length was 17.5 m.

Additionally, the wave height was 0.5 m at a water depth of 3 m. Constant pressure at the atmospheric boundary allowed for natural air inflow and outflow, while relaxation zones (sponge layers) were implemented at both the inlet and outlet boundaries to mitigate wave reflections and ensure an accurate representation of wave dissipation. A piston-type wave generator on the left side was used to create regular waves with specified heights and periods, thus mimicking typical inland or coastal conditions. Numerical damping within the sponge layers efficiently absorbed reflected waves, preventing interference in our downstream wave measurements. These configurations ensured accurate simulation of wave–block interactions, providing not only reliable but also realistic data for the analysis of prefabricated block performance under various wave conditions. It should be noted that the results of the wave run-up on the slope would be significantly influenced by the scale effect. Unfortunately, there is not a well-rounded method to analyze the scale effect quantitatively. A feasible method to avoid the effect of scale is using a large-scale model as much as possible [27]. In this study, a full-size prefabricated block was set up to investigate the wave run-up, which was influenced by scale effect limitedly.

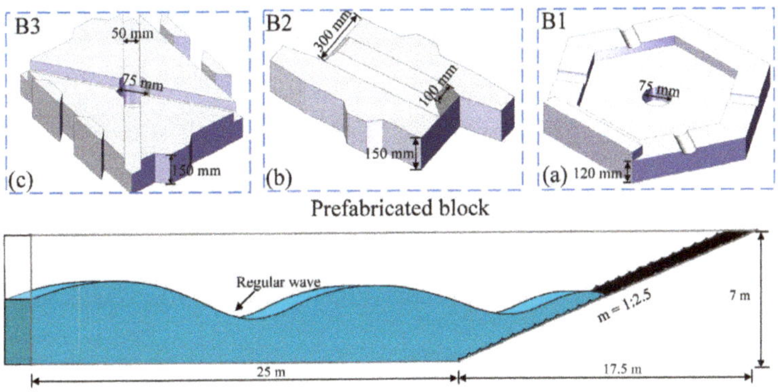

Figure 1. Three-dimensional panorama and details of the individual model. (**a**) B1: concave and embedded regular hexagon-type prefabricated blocks; (**b**) B2: Z-type prefabricated concrete blocks; (**c**) B3: embedded quadrangle-type prefabricated hollow blocks.

For the four types of slopes, the numerical mesh was created with blockMesh and SnappyHexMesh in OpenFOAM 2021. The mesh from the inlet boundary to the toe of the dike was orthogonal and conformal, with a cell size of 0.02 m both in the X-direction (horizontal) and in the Z-direction (vertical). Overtopping discharge is closely related to wave run-up, and [28] suggested that quadrilateral grids parallel to the slope surface may improve the accuracy with which the wave run-up on a slope can be simulated. Thus, in this study, we utilized this type of mesh near the surface, as shown in Figure 2. The mesh around the protrusions was a refinement of the base mesh achieved using SnappyHexMesh with a refined grid cell size of 0.002 m in the direction perpendicular to the slope and 0.003 m in the direction parallel to the slope. The influence of the grid size around the protrusions was analyzed for the overtopping discharge, as presented in [29].

With a constant single refinement level, the grid around the free surface was refined using SnappyHexMesh. The aspect ratio of the computational cells in all simulations was kept close to 1, following the recommendation of [30] for simulating wave propagation and breaking.

The physical boundary conditions were configured as follows. Gravitational acceleration in the Z-direction was set to $-9.81\ \mathrm{m/s^2}$. The horizontal boundary X_{Max} was defined as a velocity inlet, and the boundary X_{Min} was set as a wall boundary. The initial water level was set at 3 m, with a wave height of 0.5 m and a period of 2 s. The top boundary Z_{Max} was configured with a pressure outlet boundary condition representing air, and the bottom boundary was set as a wall boundary. The vertical boundaries Y_{Max} and Y_{Min} were also set as wall boundaries.

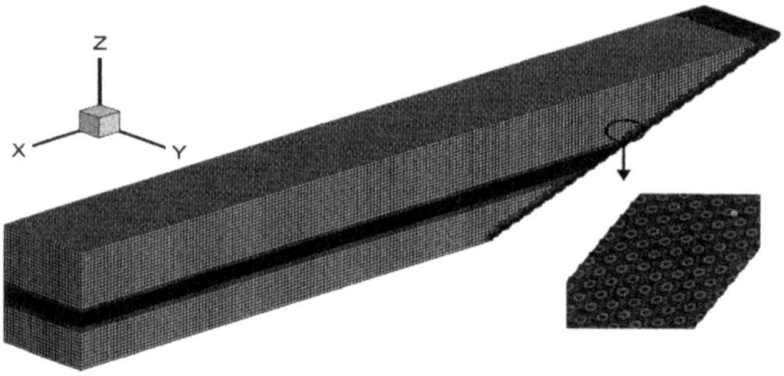

Figure 2. Global and local view of the mesh.

Additionally, the wave characteristics in this setup were analyzed using several dimensionless parameters. The Ursell number was calculated to evaluate the nonlinearity of the waves, and the Irribaren number was used to assess the influence of the wave slope. For this configuration, the Ursell number was determined to be 2.16, indicating that the waves were linear in the present conditions. The Irribaren number was 1.49, which suggests a moderate probability of wave breaking.

As mentioned, a wave tank is needed for wave generation. Hence, the boundary conditions of the wave tank were as follows. The left-hand side of the tank was regarded as the velocity inlet. In the upper part, we applied atmospheric pressure. The right-hand side and bottom of the wave flume were considered wall boundaries, where no slip was enforced at the walls (i.e., the normal velocity component was set to zero). The river channel boundary conditions were set to open channel wave boundary conditions (open channel wave BC).

When scaling a real-life problem down to a numerical wave flume, the Froude number and Reynolds number are crucial parameters. In this study, the Froude number was found to equal 1.0, ensuring dynamic similarity between the model and real-life scenarios. The Reynolds number was 590,000, indicating turbulent flow. We validated the use of this configuration using these dimensionless numbers, proving that the model accurately represented the intended physical phenomena. In our simulations, the Courant number was kept below 0.5 to ensure the stability and accuracy of our numerical solution. This configuration allowed us to accurately capture the wave dynamics while preventing numerical instability.

The total simulation time was set to 20 s, with a time step size of 0.01 s, to ensure numerical stability and accuracy, resulting in a Courant number of 1.084. While this was slightly above the conventional threshold of 1, it was still acceptable in this context. However, the elevated Courant number indicates a potential limitation, which could affect the numerical stability and accuracy. Future work should consider refining the time step to enhance the simulation results. A wave generator was placed on the left side of the wave flume to match the target wave conditions, and a wave absorber was placed on the right side to prevent wave reflection from affecting the results. The standard k-ε turbulence model was selected for this study to accurately simulate the turbulent interactions between the waves and the prefabricated blocks. This model effectively captures the turbulent characteristics of fluid flow and is suitable for most engineering applications. The initial turbulent kinetic energy (k) was set to 0.4 m^2/s^2, and the turbulent dissipation rate (ε) was set to 0.04 m^2/s^3, ensuring accurate representation of the turbulent flow. To identify the height and extent of the wave run-up in the model, a wetting and drying algorithm was used within ANSYS FLUENT. This algorithm accurately tracked the movement of the water's surface as it interacted with the structure. By monitoring changes in the free surface elevation over time, the maximum wave run-up height could be determined. The algorithm ensured that cells smoothly transitioned between wet and dry states, thereby indicating the dynamic interface between water and air and providing reliable data for subsequent analyses of wave run-up.

2.3. Wave Generation

ANSYS FLUENT, a widely used form of computational fluid dynamics (CFD) software for simulating fluid flow and wave dynamics, was employed to generate a consistent time series of regular waves in line with the experimental data. The accuracy of this software in modeling wave generation and propagation in coastal engineering applications has been validated [31]. In this study, we generated regular waves in the form of push plate waves in order to mirror the physical experiment. According to linear wave theory, also known as Airy wave theory, the free surface profile of the wave is given by Equation (1).

$$\eta = \frac{H}{2}\cos(kx - \sigma t) \tag{1}$$

where H is the wave height from crest to trough [m], h is the average water depth [m], x is the distance along the longitudinal direction [m], and t is time [s]. Additionally, σ is the wave frequency [rad s^{-1}], determined with Equation (2), and k is the wave number.

$$\sigma = \sqrt{gk\tanh(kh)} \tag{2}$$

In this study, we aimed to recreate a virtual environment that closely matched our experimental conditions. The generation of regular waves through linear wave theory can improve the utility of physical experiments. Moreover, FLUENT 2021 simulation operates in tandem with a structural analysis solver when evaluating the interaction between waves and any physical structures located within the domain. The results of the numerical simulations need to be verified by experimental results to ensure the accuracy of the former. It should be noted that the incident wave was simplified to a regular wave in this study to reduce the computing complexity, which means that studies with complex wave conditions are still lacking.

3. Convergence and Validation

3.1. Convergence of Numerical Model

To simulate the effect of four types of wave protection precast blocks on wave run-up, our numerical simulation focused on B1 as a case study. Given the arbitrary fabrication of these physical models, B1 was simplified and designed as a three-dimensional geometric model. To simulate a realistic environment despite the limits imposed by physical conditions, we scaled up the model to fit a more realistic physical situation.

Mesh convergence tests for the prefabricated block slope B1 were performed with different mesh densities. The details of the mesh convergence test are presented in Table 1. Δx and Δz represent the minimum mesh size in the direction of wave propagation and the wave height, respectively [32].

Table 1. Details of the mesh convergence test.

Case	Mesh	Total Cell Count	L (Wavelength)/Δx	H (Wave Height)/Δz
B1	Mesh 1	1.06×10^6	20	10
	Mesh 2	2.65×10^6	40	10
	Mesh 3	4.96×10^6	60	10
	Mesh 4	2.79×10^6	40	15
	Mesh 5	2.79×10^6	40	20

$L/\Delta x$ refers to refinement of the grid size in the wavelength direction. $H/\Delta z$ refers to refinement of the grid size in the wave height direction.

The computational mesh was set up with a horizontal length of 42.5 m, ranging from $X_{Min} = 0$ to $X_{Max} = 42.5$ (a height of 7 m and a width of 4.75 m). To accurately capture the wave dynamics, a gradient mesh was employed within the water surface area. The mesh was scaled in four distinct zones: from the boundary at $Z = 0$ to $Z = 2.5$, the mesh height was 0.1 m; from $Z = 2.5$ to $Z = 3.5$, the mesh height remained at 0.05 m, with 100 divisions in the horizontal direction; and from $Z = 3.5$ to $Z = 7$, the mesh height was 0.2 m. Comparisons of the free surface elevation under the wavelength L are presented in Figure 3. The results showed that different mesh densities had a minimal effect on the wave height. To better present the details of the protection provided by prefabricated rough block slopes, we used the mesh number from Mesh 3 for the following numerical simulation.

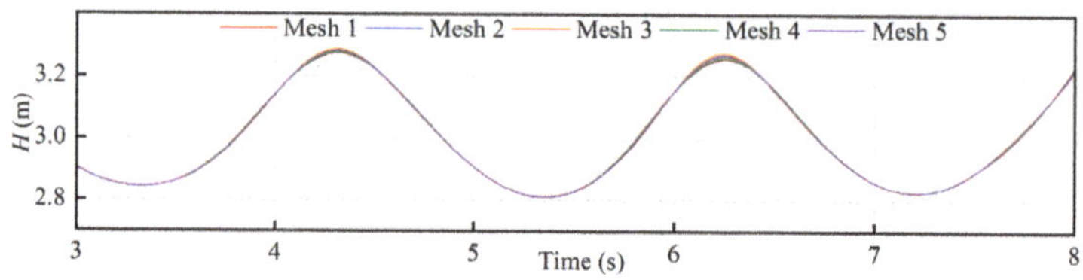

Figure 3. Comparison of the numerical results of free surface elevation.

Following the simulation, six horizontal points of contact between the water surface line and the embankment were selected at various moments: *a* (33.5, 0, 3), *b* (33.5, 1, 3), *c* (33.5, 2, 3), *d* (33.5, 3, 3), *e* (33.5, 4, 3), and *f* (33.5, 4.75, 3). The defined values in parentheses refer to x, y, and z in the spatial domain. We intend to study the variation over time in terms of hydrodynamic characteristics at these points. Additionally, we analyzed the hydrodynamic characteristics at different moments by selecting transverse sections A1-A1 and A2-A2 [33] alongside vertical section points. The wave height recorders R1, R2, and R3 were set for each wave segment to determine whether the waves dissipated in real-time. The specific distribution of the sections and points can be found in the accompanying Figure 4.

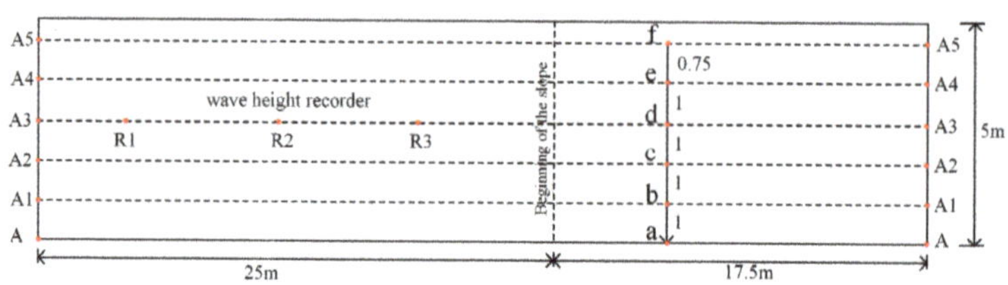

Figure 4. The sections and points in the model floor plan.

3.2. Validation

The results of an experimental wave were employed to validate the present numerical model. The experiment was conducted in the water flume located at Jiangxi Academy of Water Sciences, China, which measures 8 m in length, 0.4 m in width, and 0.5 m in height. Figure 5 shows a diagram of the experimental setup. Waves with two different wave periods and wave heights were generated in the wave flume using a flap-type wave generator. The wave generator was hinged with the bottom of the groove, and the rotation angle could be adjusted to generate waves of different heights. By adjusting the wavemaker, waves of different periods could be generated. The wave height was set to 0.05 m in the physical experiment.

The experiment involved two wave periods, 2.5 s and 1.875 s, at depths of 23.6 cm and 30.6 cm, respectively. To accurately evaluate the data, we designated three points on the cross-section. We observed the average maximum wave run-up height at these three points. Wave run-up on a given slope was determined via a combination of slope wave traces and photography. We extracted twenty cycles of wave run-up images from the captured video and then found concurrent wave run-up values on the cross-section of these three points.

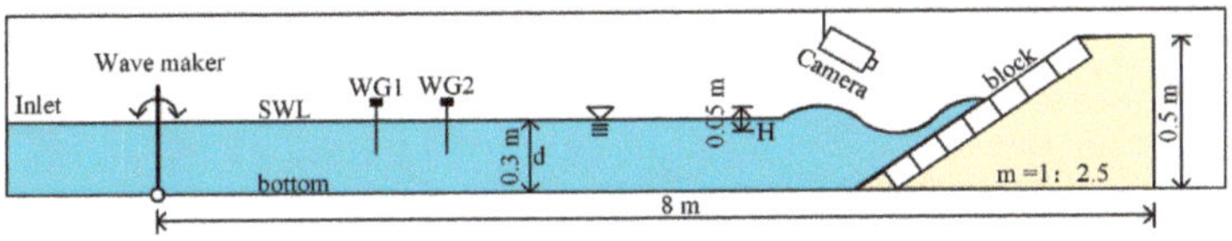

Figure 5. Layout diagram of the sink and model.

Convergence analysis was carried out to show that numerical simulation could effectively reflect the results of the physical experiments. In this section, numerical simulations were carried out in the same conditions as the physical experiments. The wave generation and turbulence modeling methods were consistent with those in Section 2. During the numerical simulation, three points were set, and the wave run-up heights of B0, B1, B2, and B3 were selected. As mentioned earlier, the three chosen types of precast blocks were concave and embedded regular hexagon-type prefabricated blocks, Z-type prefabricated concrete blocks, and embedded quadrangle-type prefabricated hollow blocks, referred to as B1, B2, and B3, respectively. Table 2 shows an exact comparison of the numerical and experimental results. It is evident that at no point did the error exceed 5%. The boundary condition settings and turbulence model were therefore proven to be suitable for simulation. To address turbulence, the RNG k-ε model was selected for its ability to capture small-scale turbulence structures, which are critical in simulating the interactions between breaking waves and dike surfaces. This model accounts for microscale turbulence effects, providing a precise calculation of turbulent viscosity, particularly in high-shear regions.

Table 2. Comparison of the wave run-up heights between the numerical simulations and experimental measurements.

Time/Horizontal Distance	B1 (cm)		B2 (cm)		B3 (cm)		Pearson Correlation	*NRMSE* (%)	*NError* (%)
	Exp	Num	Exp	Num	Exp	Num			
1.875 s/23.6 cm	5.07	5.23	5.55	5.55	6.50	6.81	0.95	2.97	3.15
1.875 s/30.6 cm	14.72	15.26	14.76	15.32	14.99	15.71	0.96	3.44	3.67
2.5 s/30.6 cm	7.17	7.34	7.29	7.64	7.43	7.79	0.92	2.85	2.37

Note: The table shows a comparison of wave run-up heights and experimental measurements under different conditions. The Pearson correlation coefficient was calculated based on 20 data points. Due to space constraints and for clarity, only a subset of the data points is shown in the table. The *NRMSE* and *NError* values are expressed as percentages (%) relative to the experimental measurements.

Turbulence impacts flow velocity and the way in which we predict discharge from wave overtopping. Calibration of the model parameters, specifically the turbulent kinetic energy k and dissipation rate ε, thus ensured alignment with the experimental data, reducing potential errors in predictions of flow. The resulting error margin of less than 5% highlighted the model's efficacy.

In addition to wave run-up heights, our analysis also included velocities and shear stresses, ensuring a comprehensive evaluation of the model's performance. Comparison of the experimental and numerical velocity results indicated a high degree of correlation, with Pearson correlation coefficients exceeding 0.9 in different conditions. This suggests that the model accurately captures flow dynamics and the interactions of waves with structures. Similarly, the shear stress measurements indicated that the RNG k-ε model provides reliable predictions of the distribution of stress across the slopes' surfaces. The discrepancies observed between the experimental and simulated shear stresses were minimal, with *RMSE* values remaining below the acceptable thresholds.

In this study, to facilitate the comparison between the numerical simulation results and experimental measurements, we normalized the RMSE and Error. The purpose of this normalization was to adjust the data to a common scale, enabling the relative comparison of errors across different magnitudes.

Specifically, the *RMSE* and Error were normalized based on the average experimental values:

$$NRMSE = \frac{\sqrt{\frac{1}{n}\sum_{i=1}^{n}\left(x_{sim,i} - x_{\exp,i}\right)^2}}{\overline{x}_{\exp}} \times 100\% \quad (3)$$

where $\overline{x}_{\exp}$ is the average of the experimental values.

$$NError = \frac{\sum_{i=1}^{n}\left(x_{sim,i} - x_{\exp,i}\right)}{n \cdot \overline{x}_{\exp}} \times 100\% \quad (4)$$

4. Wave Run-Up on the Slope

4.1. Free Surface Elevation

The graph illustrates the variation in free surface elevation at points *a*, *b*, *c*, *d*, *e*, and *f* in Figure 6. The data extracted from the selected measurement points indicate that the water depth at the embankment increased as the waves climbed, reaching the maximum water level at the crest of the wave. Subsequently, as the water level on the embankment began to recede, the water surface line started to descend back to its initial position. Due to the interaction between incoming and retreating waves, there was a period in which the waves were unable to reach that location, resulting in a stable water level that remained at the height of the embankment surface [34]. After four to five wave cycles, the waves on the embankment began to climb again, with greater energy, and within the period of a given wavelength, they began to rise and reach their maximum elevation at the end of the cycle. As shown in the graph, all six points experienced the highest wave run-up during this period. The parameter measured here was the free surface elevation (η), which indicates the vertical height of the wave run-up relative to a fixed cross-section on the slope. The patterns illustrated in Figure 6 were highly intermittent and did not follow regular repetitions for each wave period. This irregularity can be attributed to several factors, the first of which is wave–solid boundary interactions. Water near the lateral boundaries slows down and dissipates due to drag from the lateral solid boundaries. This boundary-induced dissipation of wave energy leads to variations in the recorded run-up heights, preventing uniform wave behavior across the slope and causing fluctuations in the measured surface elevation. The second such factor is the measurement limitation; the setup of the free surface elevation (η) measurement does not capture the full wave cycle when the water level drops below set detection points, especially during the trough of the wave. As a result, there are moments when the water level is below the detection point, creating the impression of intermittent data. The last factor is localized flow dynamics; small changes in local flow conditions such as shear stresses, flow separation, and turbulence near the boundaries can introduce additional variability in the free surface elevation (η). This sensitivity to localized hydrodynamics contributes to the irregular patterns in the data.

As the water ascended to the embankment, it passed through various scattered points and moved higher up the slope. When the water level at these points remained unchanged, we confirmed that the water had reached surface height at that location. In analyzing the free liquid surface elevation (η) at these six scattered points over a total computational time of 20 s, we were able to discern whether the water level had risen to that height. The wave run-up on the embankment clearly exhibited a certain periodicity, with the water level remaining unchanged at the embankment surface even across several time intervals. This periodic behavior provided insights into the dynamics of the waves' interactions with the embankment surface, offering valuable information on the influence that wave characteristics—such as frequency and amplitude—have on the wave run-up dynamics.

During the first wave cycle, from 0 s to 2 s, all scattered points indicated an increase in water level, briefly reaching a peak value before descending back to the surface, with rapid fluctuations in the water level. The variation in the water level at the 1.4 s position aligned with this pattern, suggesting that the water level quickly decreased after reaching this point, causing a swift decrease in the free liquid surface elevation. These fluctuations are indicative of the complex interactions between waves and slopes and the dissipation of energy that occurs during wave breaking. Variability in the wave-induced velocities and shear stresses also plays a crucial role in shaping these dynamics.

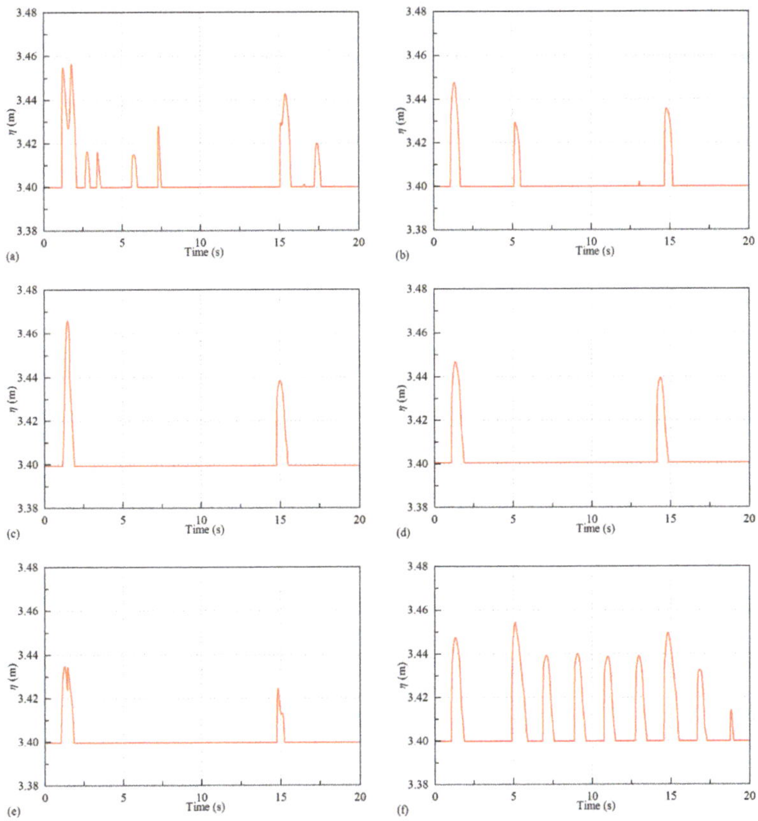

Figure 6. Free surface elevation (η) over time for the B1 prefabricated block during regular wave interaction. **(a–f)** The free surface elevation of points a, b, c, d, e and f from Figure 4.

Due to the varying positions of the scattered points and the randomness of the surface morphology, the water level fluctuations at these points can also vary. Point f, for instance, experienced multiple changes in free liquid surface elevation within 20 s, with a frequency closely matching the wave frequency. The free surface elevation at point f correlated strongly with the local velocity field, which suggests that areas of high velocity correspond to an increased wave run-up height. Moreover, point c reached the highest elevation of 3.46 m among all of the scattered points, indicating that the water level at the center of the embankment was able to reach its maximum height during ascent. The water level at point c rose to this level within the first wave cycle and then decreased rapidly. Other scattered points also showed changes in elevation during this time, but their maximum elevations were relatively lower, failing to match the height reached at point c. This suggests that the center of the embankment receives significant impact from the wave, highlighting the importance of considering spatial variability when assessing the risk of overtopping.

After the first harmonic wave cycle, the water level at point c decreased, and due to the interaction between the waves and the embankment slope, the water level crossed the waterline again at 13 s and 15 s. The reoccurrence of these peaks emphasizes the role of slope geometry and wave reflection in dictating wave run-up patterns. Upon analyzing the wave–structure interactions at key points, we found that regions likely experiencing higher shear stresses aligned with the observed wave run-up peaks, providing insights into the mechanics of wave overtopping. The free liquid surface elevation remained unchanged for several cycles before returning to 3.40 m and then its initial position. The water level changes at points a, b, d, and e were also concentrated at approximately 1.4 s and 15 s, respectively. These observations underscore the need to consider both temporal and spatial variations in wave run-up studies, without which we cannot accurately predict overtopping scenarios and inform coastal defense strategies.

Figure 7a–d depicts three-dimensional free surface elevation maps for B1 at specific times: t = 1.4 s, 6.6 s, 12.8 s, and 17.2 s. At t = 1.4 s, the wave on the embankment surface rose, and due to the influence of the precast blocks, the liquid surface exhibited irregular patterns. The fluid experienced localized breaking as it climbed, and asymmetry in the embankment slope resulted in variation in the free liquid surface at different scattered points. At t = 6.6 s, the free liquid surface was more intensely affected by the precast blocks, leading to more pronounced wave breaking on the embankment surface. The unevenness of the slope surface also caused a small amount of water to linger on the embankment slope. At t = 12.8 s, during the wave run-up, the free liquid surface elevation on the left side of the embankment rose more sharply than that on the middle side, as there were no protrusions on the left side of the slope surface to obstruct the upward movement of the water. This further illustrates the impact of slope treatment on wave run-up. At t = 17.2 s, the liquid surface tended to level out as the waves receded.

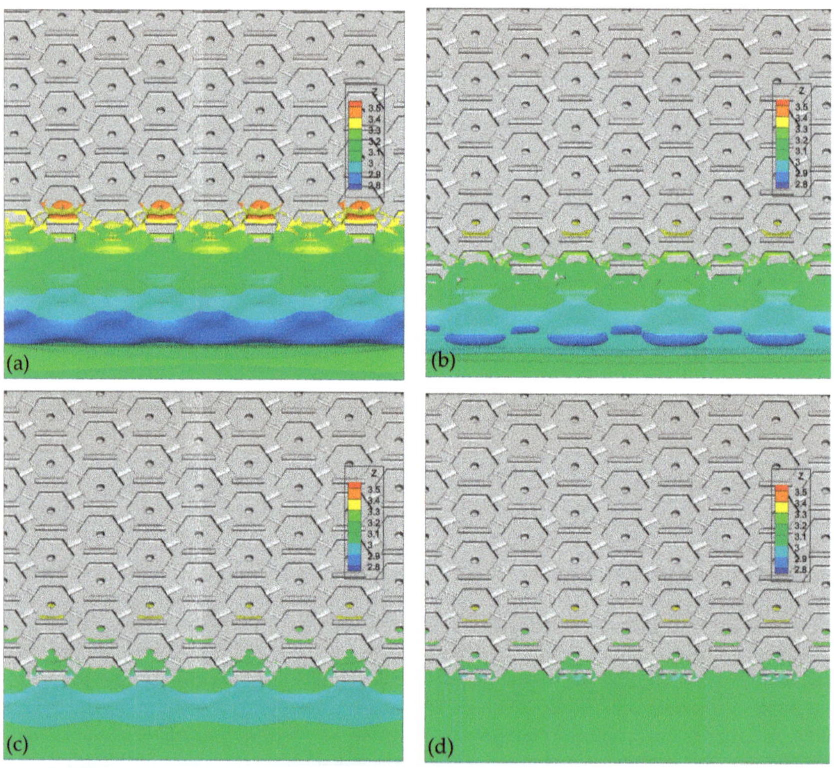

Figure 7. Three-dimensional map depicting the local water surface profile of the embedded hexagonal model. (**a**) t = 1.4 s; (**b**) t = 6.6 s; (**c**) t = 12.8 s; (**d**) t = 17.2 s.

4.2. Distribution of the Cross-Sectional Water Surface Profile

In comparing the free surface elevation at scattered points, we could identify the sections in which the maximum wave run-up occurred, and the free surface and water depth distribution data could be gleaned in sections A-A, A1-A1, A2-A2, A3-A3, A4-A4, and A5-A5 during these time intervals. The water depth decreased as the wave run-up height increased, reaching zero at the point at which the wave reached its maximum height on the embankment surface. The water depth remained zero above this point, indicating an absence of liquid on the embankment surface. The water depth here refers to the depth starting from the slope. The maximum wave run-up height was defined as the vertical distance from the still water level to the highest point reached by the wave on the embankment surface, measured when the wave stopped advancing and began to recede. We statistically analyzed the maximum wave run-up conditions at these cross-sections by processing the maximum wave run-up values at each time.

Figure 8 shows the fluctuations in the wave run-up height for B1 in sections A-A, A1-A1, A2-A2, A3-A3, A4-A4, and A5-A5. The waves reached the surface of the embankment within 5 s. Initially, within the first 1.5 s, the liquid surface elevation in each section increased, peaking at approximately 3.43 m around the 1.7 s mark. Subsequently, troughs were observed in all profiles at 3.1 s, and the overall trend showed a continuous increase from 3.1 s to 3.8 s, indicating synchronization between the peak liquid level rise and the wave period. Notably, at section A-A, the liquid level bottomed out at 3.11 m at 3.1 s, followed by a swift ascent to 3.2 m, which then escalated rapidly to 3.40 m within 3 s. Across the six sections, the waves showed nonlinear progression marked by distinct steep slopes and flattened troughs, reflecting the complex interplay of wave dynamics. Such abrupt shoaling is the main reason for wave breaking.

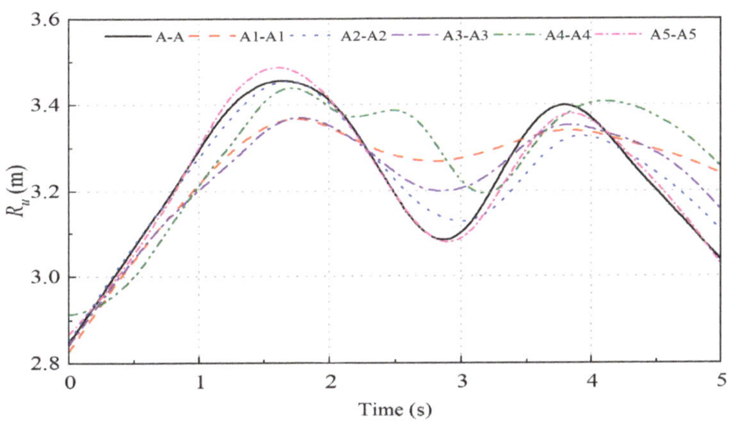

Figure 8. Cross-sectional wave run-up height (*Ru* typically refers to the vertical distance from the still water level (or mean sea level) to the highest point reached by the waves on a slope.).

The divergence from the general trend identified for section A4 can provide us with greater insight. Specifically, the wave run-up height at A4 showed a noticeable deviation from the other sections between 2 s and 4 s. This deviation is likely due to localized hydrodynamic effects. Complex flow dynamics including flow separation and reattachment around the embankment might have been more pronounced in section A4, contributing to the observed anomalies in wave run-up height.

The situation at the embankment corner was characterized by trailing waves pushing against the preceding waves, causing fluctuations in the water surface and creating several free liquid surface elevations of varying heights at that location. Additionally, after the waves on the embankment climbed and then receded to the corner, they collided with the following waves, resulting in breaking and further fluctuations. This interaction affected the subsequent ability of the wave to climb the embankment, preventing the free liquid surface elevation from consistently increasing with the arrival of each new wave. Consequently, the wave run-up on the embankment surface decreased for a period following this interaction.

5. Results and Discussion

5.1. Effect of Different Prefabricated Block Slopes on Wave Run-Up Height and Wave Dissipation

In the model tests, we observed that the maximum wave run-up on each prefabricated block slope occurred within the first few cycles. Consequently, we measured the wave run-up values for the first 20 s of testing each precast block; we also identified the maximum wave run-up for a comparison of wave dissipation. The maximum recorded water run-up heights for the four different conditions are presented in Table 3 and Figure 9. B0 refers to a smooth slope form of protection, which was used as the control group to demonstrate the wave-dissipating effects of the other block slopes.

Table 3. Maximum wave run-up heights of the prefabricated block slopes.

Period (s)	B0 (m)	B1 (m)	B2 (m)	B3 (m)
1.5	0.660	0.362	0.397	0.432
2.0	0.672	0.432	0.443	0.453
2.5	0.725	0.478	0.530	0.546
3.0	1.044	0.524	0.644	0.649

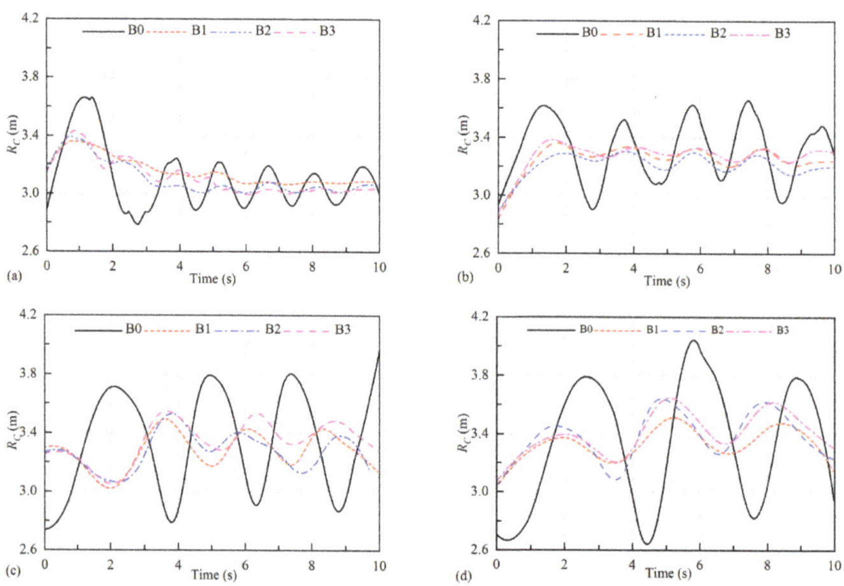

Figure 9. Maximum water height on the prefabricated block slopes. (**a**) T = 1.5 s; (**b**) T = 2.0 s; (**c**) T = 2.5 s; (**d**) T = 3.0 s.

We designated the wave run-up on smooth concrete prefabricated blocks as our control group and then calculated the degree to which the wave run-up was attenuated by the three newly engineered precast block designs relative to the baseline.

The metric of the wave run-up reduction efficiency was introduced to better illustrate the efficacy of slope protection structures in reducing the wave run-up height. The formula for its calculation is as follows:

$$D_i = \frac{R_i}{R_g} \times 100\% \quad (5)$$

The above formula obtains the rate (D_i) of the reduction in wave run-up for three novel prefabricated block designs in comparison to a conventional smooth concrete prefabricated block [35]. R_i is the wave run-up height for the new prefabricated block designs, and R_g is the corresponding wave run-up height for the smooth concrete prefabricated block.

Such calculations, as described in Table 4, assess the degree of wave dissipation achieved by the three precast block designs relative to smooth concrete blocks. For a wave period of 3.0 s, the embankment constructed with B1 demonstrated the most significant wave run-up attenuation at 50.7%. B2 showed 38.3% attenuation, and B3 showed a 37.8% reduction. These results indicate a consistent trend in wave dissipation performance across various wave periods.

Table 4. Wave dissipation effect of three precast blocks compared with that of the smooth concrete prefabricated blocks (as shown in the calculation of the wave run-up reduction efficiency in Section 5.1.).

Period (s)	B1 (%)	B2 (%)	B3 (%)
1.5	45.2	39.8	34.5
2.0	44.6	41.0	37.6
2.5	34.1	26.9	24.7
3.0	50.7	38.3	37.8

After determining the optimal wave dissipation performance of the B1 prefabricated block slope, we carried out further investigation into B1's effect on the wave run-up height when its geometrical characteristics changed. The roughness of B1 encompassed its thickness, aperture, and wave dissipation notch, as shown in Figure 10. A prefabricated B1 block with a thickness of 120 mm, aperture of 100 mm, and wave dissipation notch width of 50 mm was selected [35]. In the numerical simulation process, the wave run-up reduction efficiency D_i was used to reflect the rate at which the wave run-up height was reduced when different geometrical characteristics were present. Although the EurOtop Manual provides comprehensive guidelines for analyzing roughness in coastal and hydraulic engineering contexts, it does not specifically address methods for observing concrete blocks with unique geometrical features such as those depicted in our study.

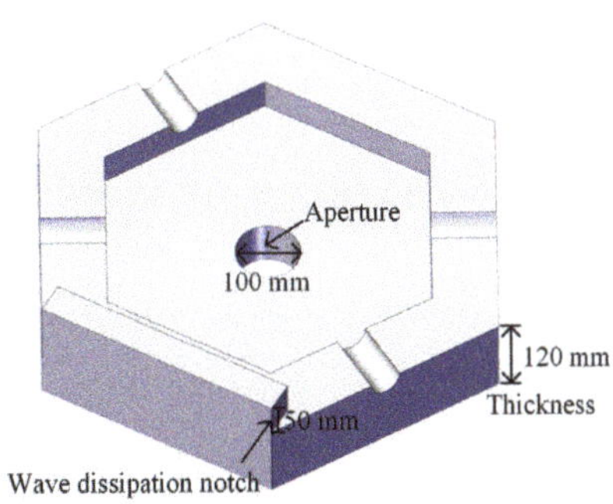

Figure 10. Different geometrical characteristic conditions of B1.

The complexity and uniqueness of the block designs, particularly the presence of apertures and wave dissipation notches, necessitated a tailored approach to evaluating their impact on roughness. To this end, a method of normalizing the roughness factor, γ_f, was proposed:

$$\gamma_f = \frac{x - \min(x)}{\max(x) - \min(x)} \tag{6}$$

Here, x represents a set of key dimensional parameters of the blocks including aperture, thickness, and the dimensions of the wave dissipation notch. By normalizing these values, the varying sizes and configurations across different block designs can be standardized, facilitating a clear and consistent analysis of how these dimensions influence the waves' interaction with block structures.

This normalization allowed us to quantify the specific effects of each dimensional parameter on the hydraulic performance and wave dissipation efficiency of the blocks. These findings are presented in Figure 11, which demonstrates the impact of varying sizes and shapes on the roughness factor, and consequently on the wave run-up and dissipation characteristics.

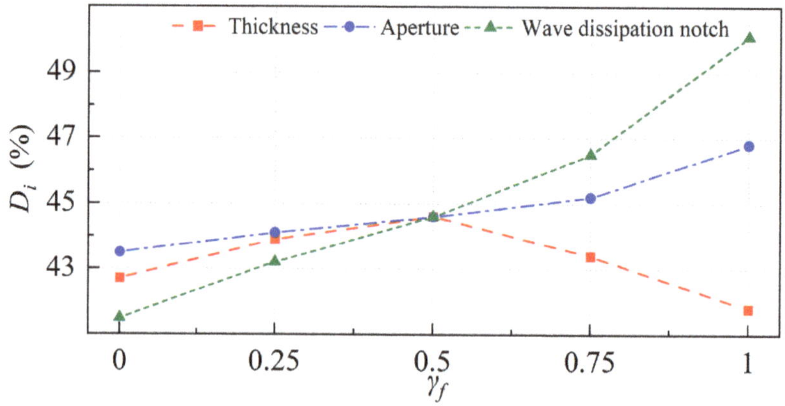

Figure 11. The impact of the three roughness changes on the wave run-up reduction efficiency.

Our approach, while not explicitly outlined in the EurOtop Manual, aligns with the mission of rigorous empirical analysis advocated by the manual. It ensured that our findings were both scientifically sound and applicable to the practical design and implementation of coastal protection structures.

This method, therefore, not only adheres to established engineering practices but also contributes to the field by providing a detailed and adaptable framework for evaluating novel block designs, which are increasingly relevant in contemporary coastal engineering projects.

In this study, precast blocks with unique geometrical features (as shown in Figure 10) were utilized to study the dissipation of wave energy. To ensure that the geometrical characteristics of the blocks were appropriately scaled alongside the wave properties, the following dimensionless parameters were introduced for a comprehensive analysis and comparison. To compare the dimensions of the precast block with the wave properties, the following dimensionless geometric ratios were defined.

Block height ratio (H_r):

$$H_r = \frac{h_{block}}{H_{wave}} \tag{7}$$

where h_{block} is the thickness of the block, and H_{wave} is the wave height.

Aperture ratio (A_r):

$$A_r = \frac{d_{aperture}}{L_{wave}} \tag{8}$$

where $d_{aperture}$ is the diameter of the aperture, and L_{wave} is the wavelength.

Notch ratio (N_r):

$$N_r = \frac{d_{notch}}{H_{wave}} \tag{9}$$

where d_{notch} is the depth of the wave dissipation notch, and H_{wave} is the wave height.

Using the defined dimensionless parameters, these ratios were calculated under various wave conditions, and an analysis was conducted using the following representative wave properties: wave height (H_{wave}), wavelength (L_{wave}), and wave speed (v) (or each wave's condition). The Froude number, Reynolds number, and the geometric ratios (H_r, A_r and N_r) were also calculated. The results are summarized in Table 5.

Table 5. Dimensionless analysis of the geometrical characteristics of the precast block and wave.

Wave Condition	*Fr*	*Re*	H_r	A_r	N_r
H= 0.5 m, L = 10 m, v = 0.5 m/s	0.50	5×10^5	0.24	0.01	0.10
H = 1 m, L = 15 m, v = 0.5 m/s	0.41	7.5×10^5	0.12	0.007	0.05
H = 1.5 m, L = 20 m, v = 0.5 m/s	0.35	1×10^6	0.08	0.005	0.03

As shown in the table, the geometric ratios between the precast block and wave properties varied significantly with increasing wave height and wavelength, reflecting the block design's adaptability to different wave conditions. In conditions with higher wave heights and longer wavelengths, the block height ratio and notch ratio were relatively small, indicating that the block's effectiveness in dissipating wave energy may have been limited under these conditions.

We conducted the numerical simulation with a maximum wave height of 0.5 m and a period of 2.0 s. From our research, we can conclude that the maximum wave run-up height occurred during the period between 1.5 s and 2.0 s. We were able to measure the wave run-up time within 10 s, and the specific data obtained were analyzed.

Figure 11 shows that the impact of the three different roughness levels on the wave run-up reduction efficiency did differ. An increase in roughness exerted a certain effect on the attenuation of waves, with the most significant effect being caused by an increase in the aperture and wave dissipation notch. The effect of the aperture was minimal. Thickness had a more pronounced effect on wave dissipation in the first half of the experiment and showed a clear downward trend in the second half. Upon analysis, when the wave moved upward, a layer of water formed when the previous wave moved downward. This residual layer of water reduced the roughness of the prefabricated block slope. Moreover, due to a single increase in thickness, the height of the wave dissipation notch decreased, resulting in a decrease in wave dissipation efficiency. However, relative to smooth slopes, prefabricated block slope protection nonetheless proved able to reduce wave run-up and dissipate more wave energy. That said, we can conclude that the most effective dissipation of wave run-up was achieved by altering the height of the wave dissipation notch.

5.2. Empirical Formula of the Wave-Absorbing Effect of Prefabricated Blocks

It is clear from the above conclusion that B1 had the best wave dissipation effect. To estimate the wave run-up height, the following formula [36] was used:

$$R_u = K_\Delta R_1 H \tag{10}$$

where R_u is the wave run-up height [m]. Starting from a calm water surface, the upward direction was positive. K_Δ is the roughness permeability coefficient related to the type of slope protection structure. In this study, values ranging from 0.50 to 0.55 were used.

$$R_1 = K_1 \tanh(0.432M) + [(R_1)_m - K_2]R(M) \tag{11}$$

R_1 is the wave run-up height when K_Δ = 1 and H = 1. H is the wave height.

$$M = \frac{1}{m}\left(\frac{L}{H}\right)^{1/2}\left(\tanh\frac{2\pi d}{L}\right)^{-1/2} \tag{12}$$

M is a function related to the slope m. m is the slope gradient coefficient, with a slope gradient of 1:m. L is the wavelength [m], and d (m) is the water depth in front of the embankment. To produce the wave run-up formula, we referred to empirical formulas provided in the EurOtop Manual. This manual offers detailed design methods and parameters for wave–structure interactions. Specifically, the roughness coefficient Ks, water depth d, and wavelength L were used in our calculations:

$$(R_1)_m = \frac{K_3}{2}\tanh\frac{2\pi d}{L}\left(1 + \frac{\frac{4\pi d}{L}}{\sinh\frac{4\pi d}{L}}\right) \tag{13}$$

where $(R_1)_m$ is the maximum wave run-up value corresponding to d/L [m].

$$R(M) = 1.09M^{3.32}\exp(-1.25M) \tag{14}$$

where $R(M)$ is the run-up function. Empirical data and design guidelines from the EurOtop Manual were integrated to ensure the scientific accuracy of our methodology. A dimensionless analysis, incorporating geometric and hydraulic characteristics, was also conducted to comprehensively analyze the wave behavior. Wave parameters such as the roughness coefficient (Ks), water depth (d), and wavelength (L) were directly taken from the EurOtop formulas. We made minor adjustments such as modifying the roughness coefficient by a factor of 2 and introducing a new term to better capture the wave–structure interactions, thereby most accurately reflecting the specific hydrodynamic and structural characteristics of our prefabricated block design. K_1, K_2, and K_3 can be obtained by combining Equations (9)–(12) with the maximum wave run-up from the numerical simulation, meaning that in engineering projects, the wave run-up on prefabricated blocks such as B1 can be predicted in advance.

The coefficient K_3 was calculated from the wave run-up height with a wave period of 2 s using Formula (13). The error observed in predictions of wave height for other periods using K_3 did not exceed 5%, indicating strong prediction ability. K_1 and K_2 were calculated from the wave run-up height using Formulas (10)–(12). K_1, K_2, and K_3 were found to be 1.448, −0.86, and 1.313, respectively. The error observed in predictions of wave height for other periods using K_3 did not exceed 10%. There was a notable discrepancy in the initial prediction for a wave period of 3.0 s. A comprehensive fluid dynamics analysis identified several factors contributing to the larger error in calculating coefficient K_3 for a wave period of 3.0 s. Its longer wave period led to increased wavelength, enhancing wave–slope interactions, resonance effects, and the concentration of energy. Nonlinear effects like wave breaking and higher harmonics further affected the wave run-up height, meaning that deviation from linear predictions may have occurred. These discrepancies, likely due to detection errors resulting from wave splashing, suggest the need for model refinement. By incorporating roughness permeability and slope gradient coefficients (m), our empirical formula may improve the accuracy with which wave run-up is predicted for structures like precast blocks (allowing for a better integration of numerical simulations and theoretical models) and can enhance reliability through mutual verification and the correction of errors.

6. Conclusions

In this study, we utilized numerical simulations to measure the wave run-up (and the extent to which it could be attenuated) on three different types of prefabricated blocks. The major conclusions drawn from this study can be summarized as follows.

(1) Prefabricated block slopes provide effective protection for inland waters in China, particularly along lakes and riverbanks. These structures are designed to dissipate wave run-up, making them ideal for environments with mild and regular wave activity. Their cost-effectiveness and aesthetic appeal make them suitable for urban and recreational slope protection, offering both economic and visual benefits. The

paper used a numerical method to evaluate these blocks, demonstrating their practical value for inland water protection.

(2) Among the three prefabricated block types tested, block B1 exhibited superior performance, with a wave run-up reduction efficiency of 44.6% at T = 2.0 s and 50.7% at T = 3.0 s. The optimized aperture, thickness, and wave dissipation notch of B1 were key to its higher performance. The wave dissipation notch was crucial in disrupting wave energy, making B1 more effective. This study highlights the innovative role of this notch in reducing wave run-up, offering design insights for artificial blocks. Adjusting its size impacts the run-up height, providing a novel approach to optimizing wave attenuation structures.

(3) Empirical formulas were used to predict the wave run-up height when using B1 in specific engineering scenarios. These formulas were based on a limited dataset and primarily validated through numerical simulations, demonstrating an error margin of no more than 10% within the scope of the study. It is important to note that these formulas are intended for preliminary analysis and should not be used to draw definitive conclusions about real-world dike performance or actual wave conditions. Further validation through broader experimental data and physical testing is necessary if we are to improve the accuracy and applicability of these formulas. As such, this study serves as an initial step in exploring the behavior of prefabricated blocks in controlled-wave conditions rather than offering conclusive predictions for use within large-scale coastal infrastructure. With additional refinement and testing, these formulas could enhance the reliability of wave run-up forecasts, contributing to more effective design and implementation of coastal protection in the future.

In this study, we numerically investigated the influence of the geometry of prefabricated blocks on wave run-up. However, a general run-up model is still lacking; such a model could be developed in further studies and physical tests. Regular waves are employed as incident waves in this study to ensure the efficiency and stability of the present numerical model. It should be noted that real-world waves, either in oceans or inland waters, cannot be represented by regular waves. Further studies on irregular waves should be carried out in future.

This study highlights the significance of the roughness in wave run-up reduction, particularly for small-amplitude wave conditions. It should be noted that the impact of roughness should be considered for small-amplitude wave run-up on the slope. However, for larger-amplitude wave run-up, one should refer to EurOtop, as the stability of blocks under wave attack and on the permeability of the top layer of the placed blocks are more important topics.

Author Contributions: B.J.: Methodology, Data curation, Writing—original draft, Visualization. Data curation, Writing—review and editing. Q.Z.: Writing—original draft, Visualization. C.L.: Methodology. F.C.: Writing—review and editing. Q.F.: Project administration, Supervision. All authors have read and agreed to the published version of the manuscript.

Funding: This research was funded by the National Key R&D Program of China (2023YFB2604400), the National Natural Science Foundation of China (52178128), the Shandong Provincial Higher Education Institutions Youth Plan Team (2022KJ081), Jiangxi Provincial Project of Key Discipline Leaders (20232BCJ23105), and the Open Research Fund Projects of Jiangxi Academy of Water Science and Engineering (2020GCZX05, 2021SKSG07).

Institutional Review Board Statement: Not applicable.

Informed Consent Statement: Not applicable.

Data Availability Statement: The data presented in this study are available from the corresponding author upon request.

Conflicts of Interest: The authors declare no conflicts of interest.

References

1. Samaras, A.G.; Karambas, T.V. Modelling the impact of climate change on coastal flooding: Implications for coastal structures design. *J. Mar. Sci. Eng.* **2021**, *9*, 1008. [CrossRef]
2. Galiatsatou, P.; Makris, C.; Prinos, P. Optimized Reliability Based Upgrading of Rubble Mound Breakwaters in a Changing Climate. *J. Mar. Sci.* **2018**, *6*, 92. [CrossRef]
3. De-León, D.; Loza, L. Reliability-based analysis of Lázaro Cárdenas breakwater including the economical impact of the port activity. *Int. J. Disaster Risk Reduct.* **2019**, *40*, 101276. [CrossRef]
4. Koutrouveli, T.I.; Dimas, A.A. Wave and hydrodynamic processes in the vicinity of a rubble-mound, permeable, zero-freeboard breakwater. *J. Mar. Sci. Eng.* **2020**, *8*, 206. [CrossRef]
5. Magdalena, I.; Erwina, N. An efficient two-layer non-hydrostatic model for investigating wave run-up phenomena. *Computation* **2019**, *8*, 1. [CrossRef]
6. Baker, R.J.W. Precast Concrete Blocks for High-Velocity Flow Applications. *Water Environ. J.* **1990**, *4*, 552–558. [CrossRef]
7. Safari, I.; Mouazé, D.; Aliasgary, S.; Carpentier, G.; Ropert, F. Hydraulic Response and Overtopping Performance of Single-Layer Double Cube Unit Armored Mound Breakwater. *J. Mar. Sci. Eng.* **2023**, *11*, 1382. [CrossRef]
8. Zhao, H.; Ding, F.; Ye, J.; Jiang, H.; Chen, W.; Gu, W.; Yu, G.; Li, Q. Physical Experimental Study on the Wave Reflection and Run-Up of a New Ecological Hollow Cube. *J. Mar. Sci. Eng.* **2024**, *12*, 664. [CrossRef]
9. Breteler, M.K.; Pilarczyk, K. Stability of artificial roughness elements and run-up reduction. In *Coastal Engineering 1996*; ASCE Library: Reston, VA, USA, 1996; pp. 1556–1568. [CrossRef]
10. Stagnitti, M.; Lara, J.L.; Musumeci, R.E.; Foti, E. Assessment of the variation of failure probability of upgraded rubble-mound breakwaters due to climate change. *Front. Mar. Sci.* **2022**, *9*, 986993. [CrossRef]
11. Kreyenschulte, M.; Schürenkamp, D.; Bratz, B.; Schüttrumpf, H.; Goseberg, N. Wave Run-Up on Mortar-Grouted Riprap Revetments. *Water* **2020**, *12*, 3396. [CrossRef]
12. Zhao, E.; Mu, L.; Hu, Z.; Wang, X.; Sun, J.; Zhang, Z. Physical and Numerical Investigations on Wave Run-Up and Dissipation under Breakwater with Fence Revetment. *J. Mar. Sci. Eng.* **2021**, *9*, 1355. [CrossRef]
13. De Waal, J.; Van der Meer, J. Wave runup and overtopping on coastal structures. In *Coastal Engineering 1992*; ASCE Library: Reston, VA, USA, 1993; pp. 1758–1771. [CrossRef]
14. Tang, W.; Mohseni, E.; Wang, Z. Development of vegetation concrete technology for slope protection and greening. *Constr. Build. Mater.* **2018**, *179*, 605–613. [CrossRef]
15. Cao, W.; Omran, B.A.; Lei, Y.; Zhao, X.; Yang, X.; Chen, Q.; Tian, G. Studying early stage slope protection effects of vegetation communities for Xinnan Highway in China. *Ecol. Eng.* **2018**, *110*, 87–98. [CrossRef]
16. Dang, B.-L.; Nguyen-Ngoc, H.; Hoang, T.D.; Nguyen-Xuan, H.; Abdel Wahab, M. Numerical investigation of novel prefabricated hollow concrete blocks for stepped-type seawall structures. *Eng. Struct.* **2019**, *198*, 109558. [CrossRef]
17. Hashim, A.M.; Noor, N. Performance evaluation of innovative concrete armor unit for coastal protection structure. In Proceedings of the 2013 IEEE Business Engineering and Industrial Applications Colloquium (BEIAC), Langkawi, Malaysia, 7–9 April 2013; pp. 673–678.
18. Asamoah, R.O.; Ankrah, J.S.; Offei-Nyako, K.; Tutu, E.O. Cost analysis of precast and cast-in-place concrete construction for selected public buildings in Ghana. *J. Constr. Eng.* **2016**, *2016*, 8785129. [CrossRef]
19. Denagama, T.; Jayasooriya, J. Comparative analysis of precast construction and conventional construction of small-scale concrete building in terms of cost. *Int. J. Res. Appl. Sci. Eng. Technol.* **2023**, *11*, 573–578. [CrossRef]
20. Chan, T.K. Comparison of precast construction costs–case studies in Australia and Malaysia. In Proceedings of the 27th Annual ARCOM Conference, Bristol, UK, 5–7 September 2011; pp. 5–7.
21. Attri, G.K.; Gupta, R.C.; Shrivastava, S. Sustainable precast concrete blocks incorporating recycled concrete aggregate, stone crusher, and silica dust. *J. Clean. Prod.* **2022**, *362*, 132354. [CrossRef]
22. Zhao, Z.; Courard, L.; Groslambert, S.; Jehin, T.; LÉOnard, A.; Xiao, J. Use of recycled concrete aggregates from precast block for the production of new building blocks: An industrial scale study. *Resour. Conserv. Recycl.* **2020**, *157*, 104786. [CrossRef]
23. Lima, P.R.; Barros, J.A.; Roque, A.B.; Fontes, C.M.; Lima, J.M. Short sisal fiber reinforced recycled concrete block for one-way precast concrete slabs. *Constr. Build. Mater.* **2018**, *187*, 620–634. [CrossRef]
24. Stagnitti, M.; Lara, J.; Musumeci, R.; Foti, E. Numerical modeling of wave overtopping of damaged and upgraded rubble-mound breakwaters. *Ocean Eng.* **2023**, *280*, 114798. [CrossRef]
25. Shounda, J.; Barman, K.; Debnath, K. Effect of wave-current flow on double-averaged turbulence properties over rough bed of hemispherical obstacles. *Coast. Eng. J.* **2023**, *65*, 572–596. [CrossRef]
26. Magdalena, I.; La'lang, R.; Mendoza, R. Quantification of wave attenuation in mangroves in manila bay using nonlinear shallow water equations. *J. Results Appl. Math.* **2021**, *12*, 100191. [CrossRef]
27. Andersen, T.L.; Burcharth, H.F.; Gironella, X. Comparison of new large and small scale overtopping tests for rubble mound breakwaters. *Coast. Eng.* **2011**, *58*, 351–373. [CrossRef]
28. Wroniszewski, P.A.; Verschaeve, J.C.; Pedersen, G.K. Benchmarking of Navier–Stokes codes for free surface simulations by means of a solitary wave. *Coast. Eng.* **2014**, *91*, 1–17. [CrossRef]
29. Chen, W.; Warmink, J.; Van Gent, M.; Hulscher, S. Numerical modelling of wave overtopping at dikes using OpenFOAM®. *Coast. Eng.* **2021**, *166*, 103890. [CrossRef]

30. Jacobsen, N.G.; Fuhrman, D.R.; Fredsøe, J. A wave generation toolbox for the open-source CFD library: OpenFoam®. *J. Numer. Methods Fluids* **2012**, *70*, 1073–1088. [CrossRef]
31. Ferziger, J.H.; Perić, M. *Computational Methods for Fluid Dynamics*; Springer: Berlin/Heidelberg, Germany, 2002.
32. Gao, Y.; Zhu, J.; Qu, X.; Wang, L.; Wang, B. Numerical simulation of wave run-up on three cylinders in an equilateral-triangular arrangement. *Ocean Eng.* **2021**, *233*, 109176. [CrossRef]
33. Molin, B.; Kimmoun, O.; Liu, Y.; Remy, F.; Bingham, H.B. Experimental and numerical study of the wave run-up along a vertical plate. *J. Fluid Mech.* **2010**, *654*, 363–386. [CrossRef]
34. Craik, A.D. *Wave Interactions and Fluid Flows*; Cambridge University Press: Cambridge, UK, 1988.
35. Clavero, M.; Díaz-Carrasco, P.; Losada, M.Á. Bulk wave dissipation in the armor layer of slope rock and cube armored breakwaters. *J. Mar. Sci. Eng.* **2020**, *8*, 152. [CrossRef]
36. EurOtop Manual. Manual on Wave Overtopping of Sea Defences and Related Structures: An Overtopping Manual Largely Based on European Research, Home—Overtopping Manual. 2018. Available online: http://www.overtopping-manual.com (accessed on 5 March 2024).

Journal of
Marine Science and Engineering

MDPI

Article

Buffer Blocks as Wave Energy Dissipators: Flow Depth Reduction

Praba Nageswaran [1], Jan Oetjen [1,2], S. Harish [1,*], V. Sriram [3], V. Sundar [3] and Holger Schüttrumpf [1]

1 Institute of Hydraulic Engineering and Water Resources Management, RWTH Aachen University, 52074 Aachen, Germany; nageswaran@iww.rwth-aachen.de (P.N.); jan.oetjen@brd.nrw.de (J.O.); schuettrumpf@iww.rwth-aachen.de (H.S.)
2 Bezirksregierung Düsseldorf, 40474 Düsseldorf, Germany
3 Department of Ocean Engineering, IIT Madras, Chennai 600036, India; vsriram@iitm.ac.in (V.S.); vallamsundar@gmail.com (V.S.)
* Correspondence: selvam@iww.rwth-aachen.de

Abstract: High-energy events such as tsunamis pose significant threats to coastal infrastructure and buildings, necessitating effective mitigation strategies to minimize damage. Compared to massive construction measures, buffer blocks are increasingly recognized as a potential solution for reducing the impact of such events. Understanding their effectiveness and optimizing their placement configurations is crucial for enhancing coastal resilience. The present study aims to experimentally evaluate the influence of buffer blocks on the depth of tsunami inundation. By comparing different configurations of buffer blocks, this study seeks to identify the most effective one for reducing the inundating flow depth. The laboratory tests revealed that the buffer blocks significantly influence flow depth and its characteristics. For ratios of flow depth to block height (R) below 1.5, the buffer blocks exhibited a delay in the arrival of the flow and effectively reduced the flow depths. However, for R values greater than 1.5, the effectiveness reduced as the flow began to overtop the obstructions. The increase in the flow depth at the buffer blocks was the least for a single-row configuration, while the multiple-row configurations with wider spacing offered higher flow resistance and increased the flow depths in front of the blocks. These findings suggest that strategically configured buffer blocks can substantially enhance coastal protection against high-energy flow events during natural coastal hazards. This study provides critical insights into the design and optimization of buffer block configurations, contributing to improved coastal infrastructure resilience and disaster mitigation strategies.

Keywords: buffer blocks; tsunami flow depth reduction; tsunami energy reduction; coastal protection

1. Introduction

Coastal areas are of crucial importance for the settlements and industries. Neumann et al. [1] showed that in the year 2000, the average population density in low-elevation coastal zones, which are coastal areas less than 10 m above sea level [2], was more than five times higher compared to the global mean population density. Predictions indicate an increase in the population in low-elevation coastal zones from 680 million people in 2019 to over one billion by 2050 [3].

This rapid population growth involves a corresponding expansion of infrastructure, change in land use, and, eventually, the emergence of economic agglomerations [4,5]. As a result, the exposure of coastal inhabitants and economic assets to coastal hazards has increased, thereby leading to higher vulnerability and, thus, risk of extreme events along the coast, such as storm surges or tsunamis [1,6].

Tsunamis, in particular, can cause devastating damage due to their destructive nature, as are two striking examples over the past two decades, resulting in huge human and economic losses. The Indian Ocean tsunami, IOT of 2004, cost the lives of nearly

230,000 people in 14 countries around the Indian Ocean and caused total economic damage of about USD 9.9 billion [7], making it the deadliest tsunami hazard recorded in history. The Tohokuku tsunami, TOT of 2011, has been reported as yet another dreadful natural disaster, resulting in total economic damage of more than USD 200 billion [8] and approximately 20,000 deaths [9].

These catastrophic events have spurred research efforts in tsunami risk management, mitigation, and protection, as many coasts exposed to this risk, especially in structurally weak regions, do not have adequate protective measures against extreme coastal hazards. To mitigate tsunami damages, hard (e.g., sea walls) or soft (e.g., early warning systems) countermeasures can be implemented, with each of the measures having its specific benefits and shortcomings [10].

In light of the above, developing advanced countermeasures, mitigation strategies, early warning systems, and emergency response plans tailored to the needs of coastal communities is crucial for future coastal management. This paper aims to conduct a comprehensive quantification of the performance and effectiveness of buffer blocks (which are traditionally used for reducing wave run-up), describes the possibility of using buffer blocks as energy dissipators to mitigate the impact and resulting damage, and highlights the need for further research.

2. Tsunami Mitigation Techniques

2.1. Shortcomings of Traditional Approaches

Japan, which is the country with the highest susceptibility to tsunamis, has acquired considerable knowledge and experience in the field of tsunami risk mitigation [11]. In response to a significant tsunami event in 1993, the Council on Earthquake Disaster Prevention (CEDP) of Japan proposed ten countermeasures for mitigating the impact of future tsunamis [11,12]. These proposals were later supplemented by proposals from the National Oceanic and Atmospheric Administration of the USA (NOAA [13]) and UNESCO [14], as depicted in Figure 1.

Depending on the expected intensity of tsunami hazards, the establishment of these (partial) barriers can be achieved using four distinct strategies, as outlined by NOAA [13]. The traditional protection strategy includes blocking structures (e.g., massive sea walls) that act as a permeable or non-permeable barrier to withstand the incoming flow due to the extreme event along the coast. Another strategy employs steering structures (e.g., angled walls) to divert the tsunami force away from vulnerable buildings and coastal communities. A less conventional technique, the slowing strategy, aims to reduce the wave impact by implementing macro-roughness elements, which can be easily integrated into densely developed areas. Alternatively, in line with the philosophy of accommodating, buildings can be constructed on higher ground or elevated structures such as piers or podiums or far enough from the shore to avoid the inundation area [13].

The two most commonly applied measures are sea walls (blocking) and breakwaters (blocking and slowing) [10]. Although these measures have the potential to significantly reduce the damage caused by tsunami hazards to coastal communities and could easily be implemented in developed areas, there are crucial shortcomings that need to be considered. Table 1 presents the primary findings from the comprehensive review of structural tsunami countermeasures by Oetjen et al. [10].

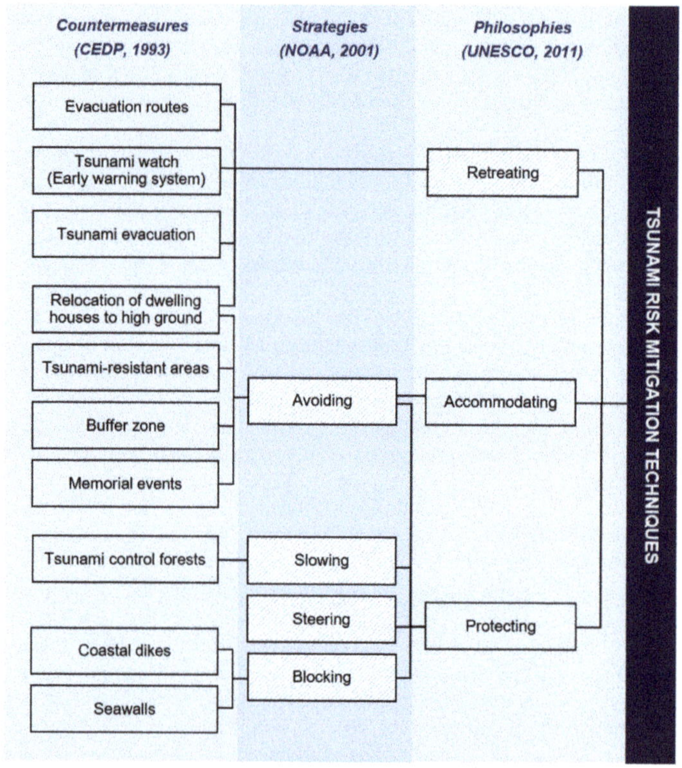

Figure 1. Tsunami mitigation techniques following CEDP [12], NOAA [13], UNESCO [14].

Table 1. Overview of shortcomings of tsunami mitigation strategies following Oetjen et al. [10] and references therein.

Strategy	Targeted Tsunami Level	Financial Effort	Area of Application	Shortcomings
Blocking	Level 1 *	Very high	Developed area	Increased hazard by failure of structure, wave overtopping, wave reflection, wave amplifications, or redirection of wave energy into unintended directions Intervention into water ecology Disconnection of settlements from sea (fisheries, tourism, etc.) False sense of security
Steering	Level 1		Not developed area	Increased hazard for subsequent buildings Increased flood velocities due to neighboring steering structures
Slowing	Level 1	Reasonable	Developed area	Insufficient knowledge No design guidelines exist
Avoiding	Level 1	High	Not developed area/in development	Subsequent enforcement of coastal areas impossible
Retreating	Level 1, Level 2		Not developed area/recently affected areas	Enormous intervention for local communities

* Level 1 = tsunamis with a return period of 50–160 years with inundation depths below 10 m; Level 2 = tsunamis with a return period of hundreds to thousands of years with inundation depths above 10 m.

2.2. Macro-Roughness Elements

In light of the aforementioned shortcomings of current tsunami mitigation strategies, there is a clear need to revise existing design guidelines based on recent tsunami hazards and to develop advanced mitigation techniques. The complex interaction between tsunamis and countermeasures requires further research and understanding [10]. To contribute to filling this research gap, this study supports the proposal of Percy et al. [15] of a space-saving approach based on the idea of inducing energy dissipation and flow deceleration.

The concept of using buffer blocks as a tsunami mitigation measure is based on the installation of salient structural elements along the coastline in front of vulnerable structures and areas. These elements, which function as macro-roughness elements, are designed to induce wave energy dissipation by creating additional turbulences and reflections in the flow, ultimately reducing the incoming inundation distance and height and, subsequently, the potential damage (Figures 2 and 3). The key idea is to create disturbances in the streamline and inhibit the flow using its own energy. Furthermore, the buffer blocks should be considered multi-functional coastal structures since tsunamis are rare events. These blocks can serve as relaxing spaces for beachgoers without obstructing the aesthetics of the coastal view. Hence, the number of rows and the height of the buffer blocks should be optimal to enhance the aesthetics of the beach, as seen on the coast along the Island of Nordeney (Figure 4).

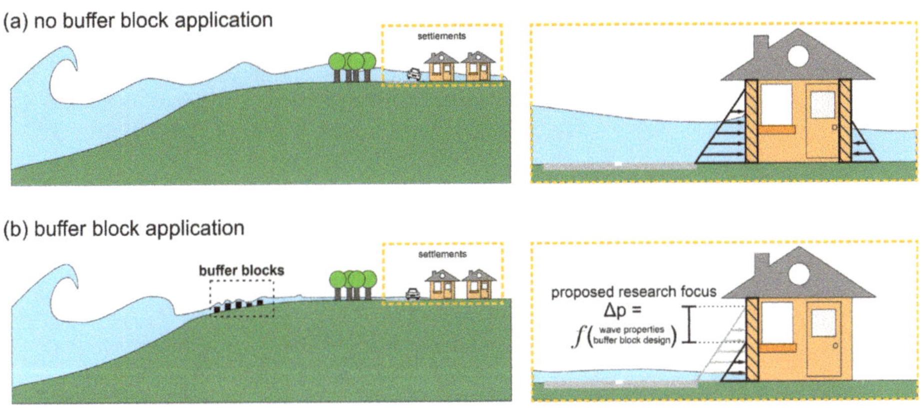

Figure 2. Intended functionality of buffer blocks as mitigation measure (conceptual): (**a**) Bore impact without buffer blocks: the hydraulic bore reaches the subsequent structure with only slightly reduced energy, stressing it further with forces from backwash. (**b**) Bore impact with buffer blocks: additional drag and turbulences are introduced to the flow, hampering flow height and velocity before it reaches the structure of interest. Note: The buffer blocks are initially subaerial and overflow only during the wave impact.

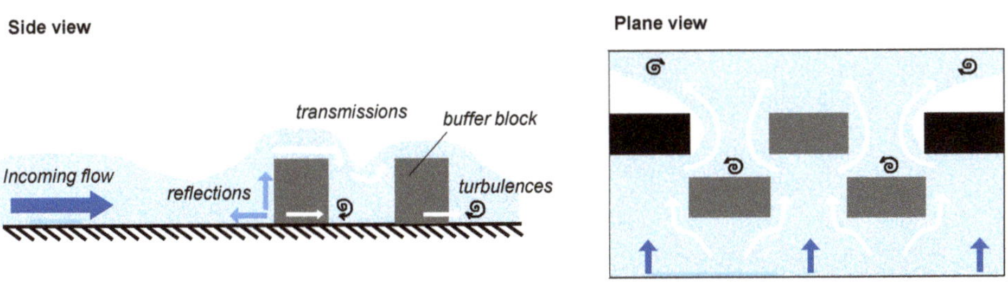

Figure 3. Additional reflections and turbulences introduced by buffer blocks in side view and plane view. Permeable design allows propagation of flow through and over buffer blocks.

Figure 4. Buffer blocks on Norderney, Germany (North Sea): (**a**) Straight and curved staggered roughness elements serve as an onshore wave attenuation measure. (**b**) Smaller roughness elements are integrated into the dike revetment. (© Jan Oetjen 2023).

Unlike massive and continuous coastal protection structures, buffer blocks do not have to withstand the full impact energy as the flow is allowed to propagate through and over the blocks. This permeable design requires less material and space, making buffer blocks comparably economical and avoiding critical shortcomings of other conventional measures.

The buffer block concept shares similarities with the performance of vegetation belts as a mitigation measure due to their attenuation effect, which has been observed during the IOT 2004 and TOT 2011 and has been investigated in several studies [16–27].

Buffer blocks serve as an artificial low-maintenance substitute for vegetation in areas where vegetation zones are neither feasible nor desirable. In areas where vegetation is beneficial as a mitigation measure, buffer blocks can act as a reliable reinforcement and a secondary defense line behind vegetation belts or vice versa.

The concept of buffer blocks as a tsunami protection measure and also to serve as a multi-functional coastal structure is a new approach, and its development is in an early stage of research. To date, buffer blocks have only been implemented as roughness elements on dikes and as space-saving reinforcement measures to existing dikes in the context of storm surges (as shown in Figure 4), allowing the dissipation of wave energy during overflow [28]. Along the North Sea Island of Norderney, straight elements measuring 5.5 m long and curved elements measuring 7.7 m long, both with a height of 1.3 m, successfully endured multiple storm surges over several years without suffering any damage [29].

The application of buffer blocks in the context of tsunamis has been discussed in general by several authors [15,29–35], but empirical relationships between wave impact and buffer block configuration have not yet been published.

Oumeraci [29] suggested that the concept of discontinuous damping elements could be modified by wider and more stable buffer blocks that prevent ecological and social interference with the coastal ecosystem and allow significant attenuation of the incoming tsunami flow.

In his experimental studies, Goseberg [4,30] investigated the wave run-up reduction in non-breaking long waves due to the interaction with macro-roughness elements, representing coastal urban development that is not fully submerged during the run-up. The experiments were carried out in a closed-circuit wave flume with a sloped beach inclined at 1:40. Based on his experiments, Goseberg [4,30] concluded that the run-up height is mainly influenced by the element configuration (arrangement and direction of wave impact) and the long-shore obstruction ratio (distance between the blocks), whereas the cross-shore obstruction ratio (number of rows) only have a minor influence.

Similar investigations were conducted by Giridhar and Reddy [36] regarding the effect of differently shaped buffer blocks (rectangular/semi-circular/trapezoidal) on wave-run up and wave reflection. For trapezoidal, rectangular, and semi-circular blocks, a trend in decreasing wave run-up was reported. Both Goseberg [4,30] and Giridhar and Reddy [36] did not consider the flow depth and force reduction behind the macro-roughness elements.

In contrast, the research carried out by Rahman et al. [31] focused on the attenuation of wave forces on coastal buildings fronted by continuous sea wall setups of different heights and a perforated sea wall setup. The results indicated a significant decrease in both flow heights and forces exerted on the building model with and without a sea wall. Notably, the case without sea wall protection showed greater increases in wave heights and resulting forces. A maximum of 41% force reduction was observed with the higher wall configuration, compared to 27% with the smaller wall, suggesting that an increase in wall height can provide more substantial attenuation of forces. Walls closer to the model were more effective in reducing forces, indicating that the proximity of the sea wall to the onshore structure is a highly relevant factor. The performance characteristics of both solid and perforated walls are similar, Rahman et al. [31] recommended the use of perforated sea walls in view of the fact that the overtopped water is enabled to flow back to the sea.

The experimental study by Thomas et al. [32] is one of the first three-dimensional works investigating the influence of buffer blocks on the wave run-up and induced forces on coastal structures. Thomas et al. [32] observed that smaller wake clearance angles generally reduced forces (up to 75%), whereas larger angles increased them (up to more than 100%) due to wave deflection and focusing. Angles between 20° and 35° tended to increase the forces on the structures. Angles less than 15° led to substantial decreases in wave run up and velocity. The clearance angle is defined as the one between a line perpendicular to the wave propagation direction and centered on the instrumented structure, and a line connecting the front center of the instrumented structure to the front inside corner of the nearest buffer block.

In line with the idea of flow disturbance, Ono and Hiraishi [37] concluded through physical modeling that increasing obstacle heights has a significant effect on the impact of force reduction and demonstrated that obstacles that were both higher and less permeable were more effective in mitigating wave forces.

Similar to the investigations by Goseberg [4,30], the experimental study by Moon et al. [34] examined the influence of nearby buildings (macro-roughness elements) on the impact force of tsunamis on buildings (structure of interest), providing insights into the hydrodynamic interactions in between the buffer blocks. Compared to a non-protected scenario, Moon et al. [34] pointed out that the presence of seaward blocks reduced the forces and pressures on the front face of the structure of interest up to 38% and 30% on the back face. Landward blocks generally increased the maximum back-face pressure, emphasizing the potential negative effects of improperly positioned buffer blocks.

Ishii et al. [35] also pointed out the significance of the block placement, examining their shielding effects on onshore structures during wave run-up. Both the three-dimensional experimental and numerical studies involved cuboid elements, representing coastal buildings and differing in their blockage ratio (single structure/structures parallel/perpendicular to shoreline) and spacing. They concluded that onshore obstruction elements have a significant influence on the flow characteristics, as the surface velocity was reduced by 40% in an area extending a minimum of three times the structure's width. The study reported that blocks placed directly behind others offered significant protection against the impact, suggesting potential strategies for urban planning in tsunami-prone areas [35].

Most of the studies above and listed in Table 2 refer to building structures and their interactions with tsunami flows, which can broadly be considered as macro-roughness elements, indicating the potential of buffer blocks as tsunami mitigation measures. However, detailed knowledge regarding the effectiveness of these buffer blocks—which are eventually smaller and thus interact differently with flows, particularly in terms of their overtopping, reflection, and dissipation characteristics—is still limited.

Table 2. Overview of existing studies on the interaction between roughness elements and tsunami-like flows.

Study	Type of Buffer Blocks	Characteristics Varied in Buffer Block Configurations	Wave Generation	Scale	Bed Type and Condition
Goseberg [4,30]	Cubic elements 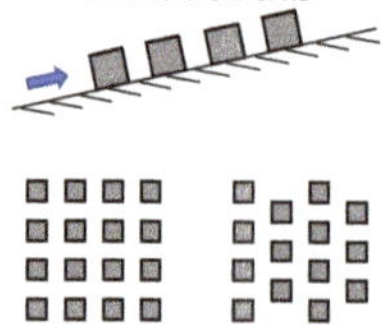	Arrangement (staggered/aligned) Orientation (perpendicular/45° oblique) Obstruction ratio (cross-shore/long-shore)	Single sinusoidal waves (pump system)	Not specified	Sloped (1:40), Dry
Giridhar and Reddy [36]	Rectangular/semi-circular/trapezoidal elements 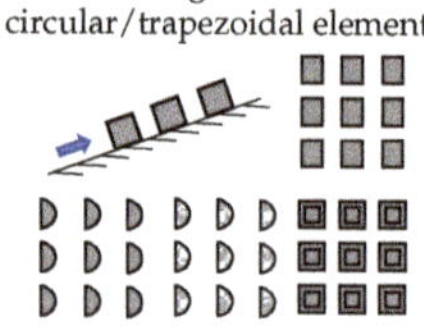	Distance between blocks Porosity	Monochromatic waves (wave paddle)	Not specified	Sloped (1:35), Dry
Rahman et al. [31]	Continuous and perforated elements 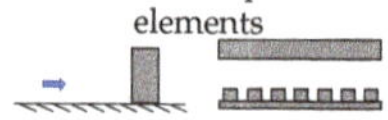	Height Position Permeability (26% perforated/solid)	Tsunami-like waves (dam-break setup)	1:35	Flat, Dry
Thomas et al. [32]	Cubic elements 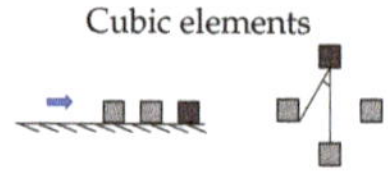	Wake clearance angle	Long-period waves (wavemaker)	1:25	Flat, Dry
Ono and Hiraishi [37]	Continuous elements 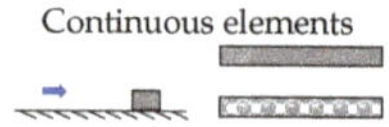	Height Material (wood/artificial raisin)	Solitary waves (piston-type wavemaker)	1:40	Flat, Dry
Moon et al. [34]	Cuboid elements 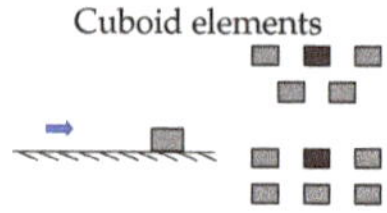	Arrangement (staggered/aligned) Presence of seaward/landward roughness	Solitary waves (dam-break setup)	1:100	Flat, Dry
Ishii et al. [35]	Cuboid elements 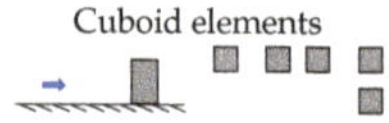	Blockage ratio (single/parallel/perpendicular)	Tsunami-like waves (pump-driven dam- break setup)	1:80	Flat, Dry
Percy et al. [15]	Cuboid elements 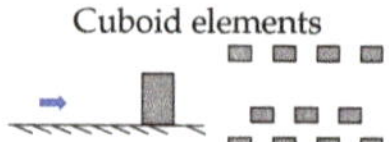	Number of rows Distance between rows	Hydraulic bores (dam-break setup)	1:8	Flat, Dry
Present study	Cuboid elements 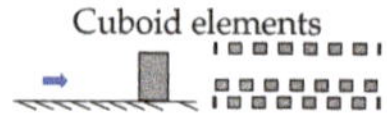	Number of rows Distance between rows	Hydraulic bores (pump system)	1:16	Flat, Dry

Therefore, the study by Percy et al. [15] focused on roughness elements of smaller heights, providing additional time for emergency evacuation with decreased forces on coastal buildings. Using OpenFOAM, the study numerically examined the relationship between buffer block characteristics and their attenuation effect on the flow depth and velocity in a 1 m wide flume. The work involved buffer blocks placed in a staggered arrangement with differing numbers of rows and spacings between the rows and a dam-break scenario generating tsunami-like flows. Percy et al. [15] identified that a single row of buffer blocks reduced the momentum flux by 5 to 40% and the three-row configuration by 15 to 65%, depending on the flow characteristics. Moreover, it was stated that increasing the number of rows leads to a significant reduction in flow depth behind the blocks, compared to the single-row configuration (10 to 40%). As more rows relate to more obstructions in the flow, the flow depth in front of the buffer blocks increases due to the build-up of water.

Wider inter-row spacings lead to less pressure on the rear side of the blocks, less turbulence, and slower flow reflection than closer spacings. Percy et al. [15] stated that the performance of the blocks depends on the incoming flow and the resulting flow characteristics. In case of flow depths smaller than the block height, the blocks were reported to be more effective in reducing the flow depth behind them.

By extending the examination of smaller-height roughness elements, this study initiates preliminary physical experiments aimed at understanding the flow characteristics in the presence of the buffer blocks. The present study focuses on five buffer block configurations that vary by the number of rows and the spacing between rows. Unlike the dam-break scenario used in Percy et al.'s [15] research, the present study utilizes a pump-based method for generating the bores, which can simulate tsunami inundation in the long term. Additionally, this study includes an experimental examination of the single-row configuration explored by Percy et al. [15] and a continuous block setup analogous to a sea wall. This approach enables a systematic comparison in terms of both flow generation techniques and scale effects. This research targets more extreme events, focusing on pure overtopping scenarios on the buffer blocks where the ratio of flow depth to block height (R) is greater than or equal to 1, compared to the conditions examined by Percy et al. [15], where $0.7 \leq R \leq 1.5$.

3. Materials and Methods

To evaluate how and to what extent buffer blocks reduced the tsunami flow energy, a comprehensive study employing both physical and numerical modeling is being conducted as part of a collaborative research project (buffer blocks as wave energy dissipators: BB-WEnDis) between the Institute of Hydraulic Engineering and Water Resources Management (IWW) at RWTH Aachen University in Germany and the Department of Ocean Engineering (DOE) at the Indian Institute of Technology Madras in India. The reproduction of similar experimental and numerical investigations at both laboratories (IIT Madras and RWTH Aachen) involved in the present research provides the basis for this and enables a technical, methodical, and well-controlled comparison.

3.1. Experimental Facility

The physical experiments for the present study were conducted in the large tilting flume of the laboratory at the IWW, utilizing a pump-based bore generation method (Figure 5). The flume, measuring 33 m in length, 1 m in width, and 1 m in height, was used for numerous tsunami-related studies [38–42]. Similar to the previous studies, the tsunami-like flow was simulated as a broken bore. Two remote-controlled pumps, which are switched on and off with a time interval of 2 s, extract water from an underground water reservoir, which is then directed upwards through a metal perforated plate and a flow straightener consisting of a stack of pipes arranged in parallel (Figure 6). The perforated plate is intended to minimize water splashing out of the flume, while the flow straightener ensures the homogenization of the flow and thus creates two-dimensional flow characteristics. After the homogenization, the flow encounters the edge of the 20 cm high false bottom (i.e., a pedestal; see Figures 5 and 6) and, finally, flows in the experimental flume. The broken bore then propagates across the 24.5 m long false bottom and impacts the buffer blocks 14 m from the start of the false bottom. Finally, the water flows back into the underground reservoir via the outlet. For the generation of bores varying in heights, lengths, and velocities, the pumping time (duration) and the percentage of valve openings were adjusted. This allowed the generation of well-controlled bores.

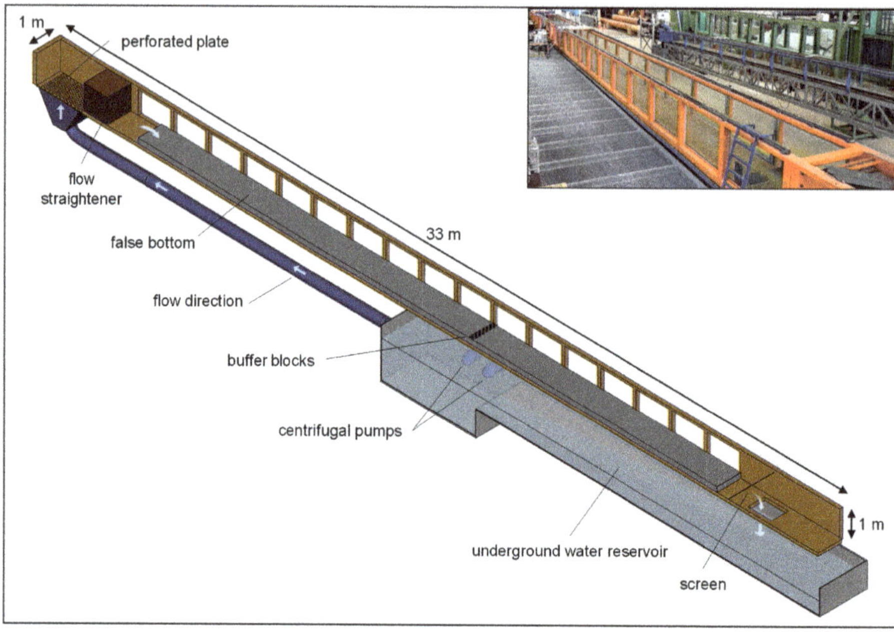

Figure 5. Three-dimensional view of the experimental facility with a built-in false bottom. Arrows indicate the flow direction.

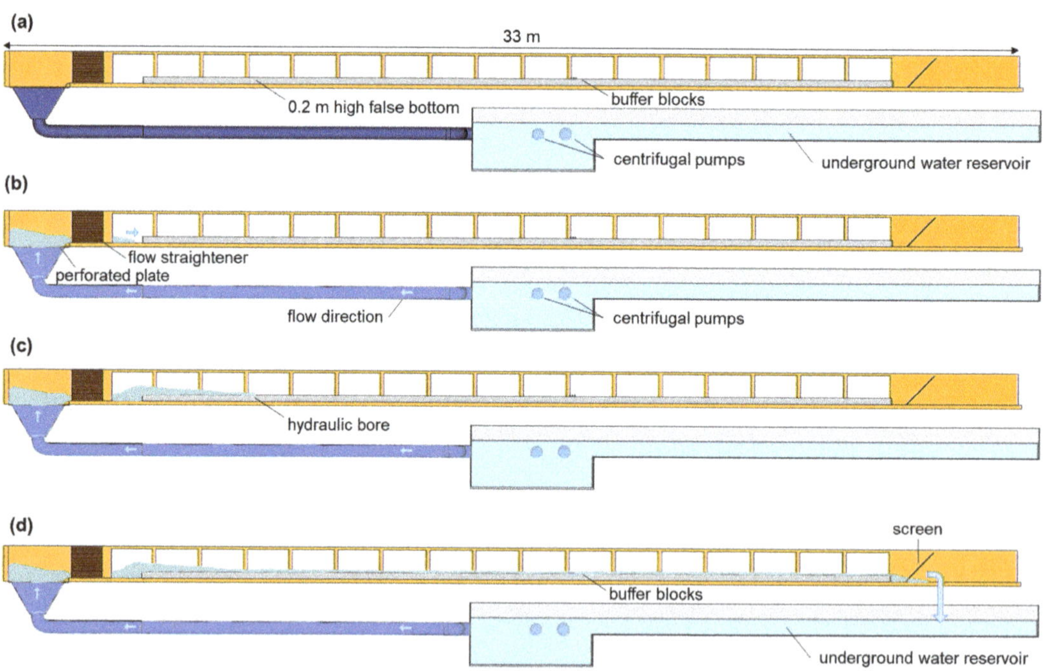

Figure 6. Bore generation in the large tilting flume: (**a**) initial state; (**b**) pumping phase; (**c**) bore development; (**d**) bore impact.

3.2. Model Setup

This study aims to investigate the interaction between extreme flow, such as that due to a tsunami, and buffer blocks and to establish correlations between different buffer block configurations and flow characteristics. The prototype buffer block used in this study measures 0.8 m in height, 1.2 m in width, and 0.4 m in length, dimensions slightly smaller than those along the East Frisian Island of Nordeney of the North Sea coast of Germany. The height of the buffer blocks (h) was chosen to ensure access to the sea while also providing a functional seating area on the beach under normal conditions and to integrate well with the beach landscape. The dimensions, with the width being 1.5 times the height and the length half the height, were derived following the USBR guidelines, as previously specified by Percy et al. [15].

The first phase of experiments, conducted on a scale of 1:16, involved buffer blocks of PVC of the dimensions (7.5 × 2.5 × 5 cm). For a systematical investigation of the dissipation capacity of buffer blocks, five different configurations (C1–C5) in a staggered arrangement were investigated. These configurations varied in number of rows (one, two, or three), and the distance between the rows was investigated (see Figure 7). The first and third rows comprise six complete blocks and two blocks partially intersected by the flume wall. Furthermore, a reference scenario without any buffer blocks was also tested for comparison. The five configurations of buffer blocks are depicted in Figure 7. It is important to note that the distance between the rows was defined from the front of one block to the front of the next and was set to be equal to or double the height of the buffer blocks (h), as illustrated in Figure 8. In order to limit the range of parameters investigated to a reasonable number, the model setups were restricted to a flat shore model. It should, therefore, be noted that real applications in practice may partially deviate from this "idealization" of a flat shore and that corresponding hydraulic effects are neglected in the present study. Nevertheless, after the beach slope, the ground would be mostly flat. We are more focused on the flat region where the tsunami flow already overflowed the beach slope.

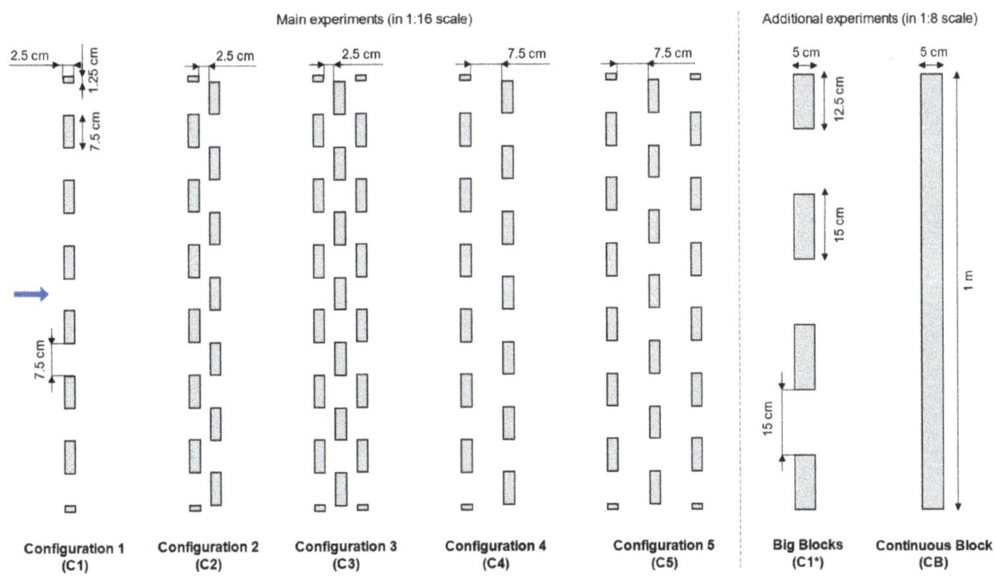

Figure 7. Test setups varying in size, number of elements, number of rows, and distance between rows. Setup C1* describes a variant of the C1 setup (1:8 scale).

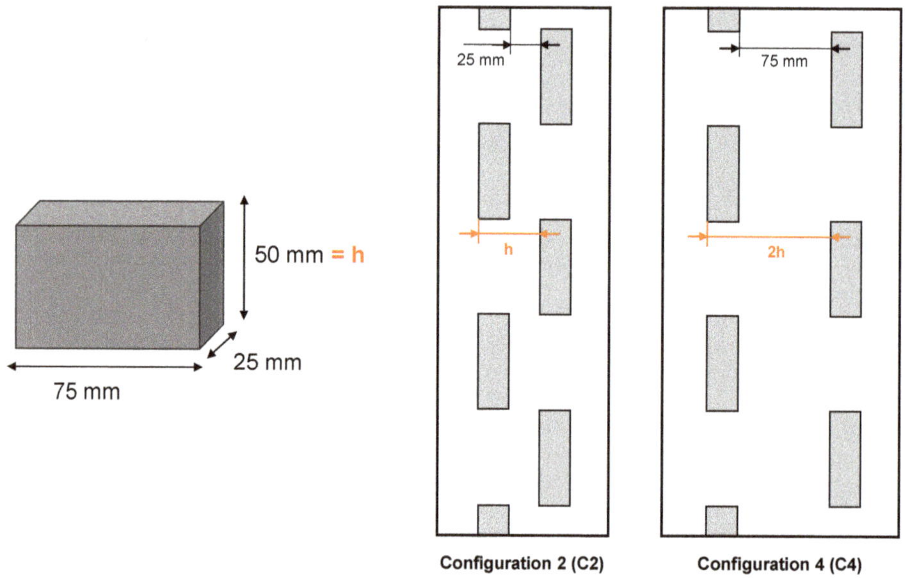

Figure 8. Dimensions of one buffer block (**left**). The distances between the rows (front to front) equal the height of the buffer block (C2 and C3) and twice the height of the buffer block (C4 and C5) (**right**).

The research plan at the IITM (scale 1:8) includes the execution of physical experiments using a dam-break setup within a flume spanning 2 m in width, subsequent to the preliminary numerical analyses. Since the tilting flume at IWW is only half the width, the scale was set to be half of those to be conducted at IITM. This way, the same long-shore obstruction rate was maintained, allowing a 1:1 comparison of the results. The long-shore obstruction rate was calculated as the sum of the widths of buffer blocks (b) divided by the width of the flume (w):

$$\frac{b_{IWW}}{w_{IWW}} = \frac{2 \times 12.5\ \text{mm} + 6 \times 75\ \text{mm}}{1000\ \text{mm}} = \frac{475\ \text{mm}}{1000\ \text{mm}} = 0.475$$

$$\frac{b_{IITM}}{w_{IITM}} = \frac{2 \times 25\ \text{mm} + 6 \times 150\ \text{mm}}{2000\ \text{mm}} = \frac{950\ \text{mm}}{2000\ \text{mm}} = 0.475$$

Previously, Percy et al. [15] investigated tsunami-attenuation characteristics of the buffer blocks by carrying out numerical simulations with the dam-break flow (Table 2). Taking advantage of the fact that the width of the flumes in both studies was consistent at 1 m, additional experiments in a 1:8 scale were conducted using buffer blocks (test setup C1*) with dimensions identical to those used by Percy et al. [15] in their numerical study (35 × 5 × 10 cm). This setup allowed the investigation of the flow characteristics and the comparison of the outcomes from the physical model tests with those obtained from the numerical model. The C1* blocks were arranged in a single row, symmetrically to the center of the flume cross-section and perpendicular to the flow direction (Figure 7). Similarly, to the prior experiments, the lateral spacing between the blocks was set equal to their width (15 cm), thereby maintaining a 1:1 ratio of the lateral gap to the block width. To investigate the impact of permeable structures in comparison to a massive sea wall, a 10 cm high continuous block (CB) extending across the full width of the flume was also tested in comparison to the buffer blocks.

3.3. Test Program

Prior to the experiments, free-flow tests with different input settings were conducted. These initial tests facilitated the identification of four distinct design flows. The preliminary tests revealed that with a pumping time shorter than 10 s, the water level did not rise above the false bottom. Hence, the pumping time period was always varied between 15 s and 20 s. Table 3 provides an overview of the parameters considered for the experiments with a systematic variation in the flow parameters, as well as test setups. Each experiment was repeated three times, resulting in 96 experimental runs.

Table 3. Selected flow depth to buffer block height ratios (block configuration in Figure 7).

Free-Flow Depth at Buffer Block Position H [cm]	Free-Flow Depth to Block Height Ratio R [−]	
	For Setups C1 to C5 (1:16 Scale)	For Setups C1* and CB (1:8 Scale)
5	1	0.5
7.5	1.5	0.75
10	2	1
15	3	1.5

A ratio of the flow depth to the buffer block height (R) below 1 indicates that the buffer block has emerged, while a ratio greater than 1 signifies a submerged condition. The selection of the four design flows was based on their profiles and specific flow depths at the buffer block position under free-flow conditions, as shown in Figure 9 and Table 3. Given this study's focus on more extreme events than those considered by Percy et al. [15], the design flows were chosen to ensure their height exceeds the block height (5 cm), leading to selected flow depth to block height ratios of 1, 1.5, 2, and 3, corresponding to flow depths of 5, 7.5, 10, and 15 cm respectively. For the additional experiments to investigate it at a 1:8 scale (Figure 7) (i.e., C1* and CB configuration), the exact same design flows were employed (Table 3), as they correspond to R values ranging from 0.5 to 1.5, which align with the ratios explored in the study by Percy et al. [15].

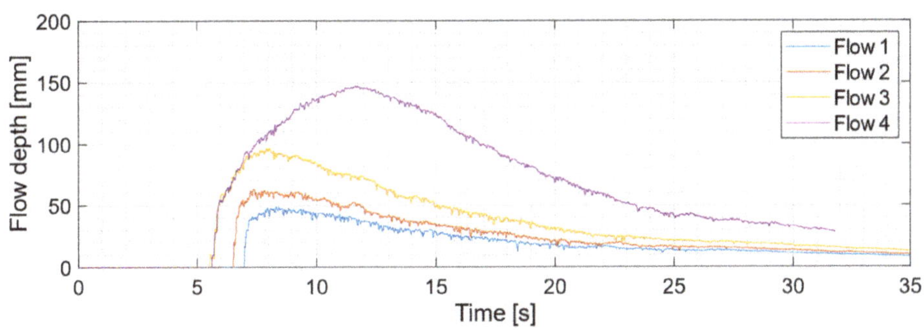

Figure 9. Flow profiles of selected design flows at buffer block position during free-flow. Flow 1 to Flow 4 represent the free-flow depth presented in Table 3 in the increasing order of the flow depth.

3.4. Instrumentation

As mentioned earlier, the experiments were carried out with and without buffer blocks to derive their influence on flow depth reduction. In this first set of experiments, the focus lay on investigating the flow properties in front and behind the buffer blocks. Therefore, only flow depth was measured. Six ultrasonic depth sensors (WG) (6 micro-sonic pico+35/I) were placed every 3.5 m from the start of the false bottom to measure the variation in flow depth across the entire flow path. WG 1 served as the reference point for the time (t = 0 s). Figure 10 shows the positions of the structures and instruments.

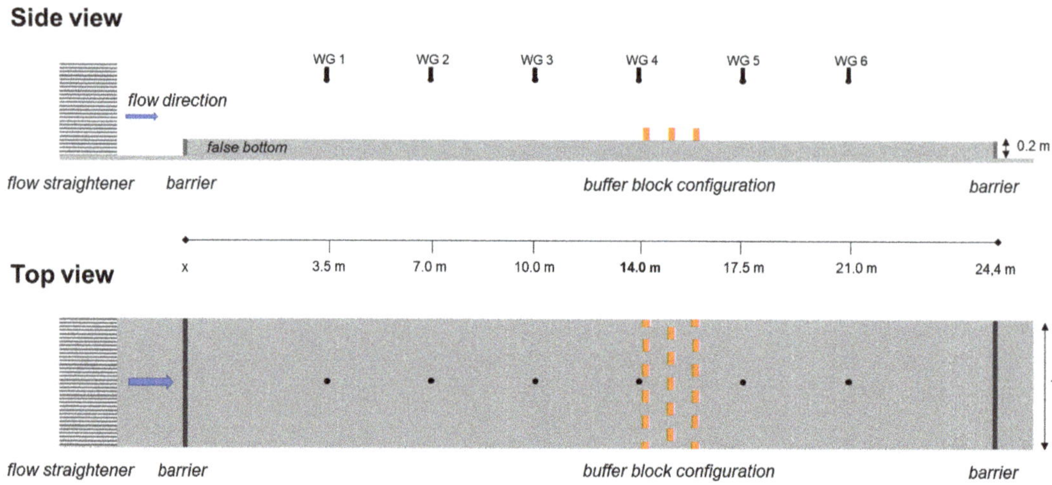

Figure 10. Simplified sketch of experimental setup (not to scale).

4. Results and Discussion

4.1. Experiments with Configurations C1 to C5

The flow depth profiles over time produced in the presence of the buffer blocks are depicted in Figure 11. For R = 1 and R = 1.5, the buffer blocks delayed the arrival of the flow at WG 6 by about 1 s compared to the free flow, corresponding to the results from the first set of experiments. In the case of R = 2 and R = 3, the delay in flow arrival is hardly noticeable, suggesting that the buffer blocks might be less effective for flow depths that are at least double the height of the block. Overall, the flow depth profiles with buffer blocks are comparable to those of the free flow, with no significant difference detectable from the different configurations. Only for the least flow tested (R = 1) does the single row configuration appear to have the least reduction in flow depth, while all other configurations show a slightly greater reduction.

Additionally, oscillations in the measured flow depth behind the buffer blocks could be detected, indicating an interaction between the flow and the buffer blocks. For $R \leq 1.5$, these oscillations are slightly below the free flow curve, whereas for $R \geq 2$, they oscillate slightly above it. A fast Fourier transform analysis was performed to understand whether the frequencies of the flows relate to the buffer block configurations. However, due to the irregular nature and low amplitude of the oscillations, it proved difficult to establish a well-defined relationship. Visual observations suggest that the oscillations might stem from the interference of streams deflected through the buffer blocks, as depicted in Figure 12. This phenomenon appears to be a local phenomenon, possibly due to the proximity to the buffer blocks.

A closer examination of the flow depth profiles in front of the buffer blocks reveals a consistent alignment of flow peaks across all test cases, proving reliable reproduction and comparability of the incoming flows. Additionally, the reflected flows depict similar profiles across different flow depth to block height ratios. Notably, the two-row configuration with a 2 h gap (C4) results in the fastest and highest reflected flow. This is followed, in decreasing order of flow reflection, by the three-row configuration with 2 h gap (C5), the three-row configuration with 1 h gap (C3), the two-row configuration with 1 h gap (C2), and finally, the single-row configuration (C1). To analyze and quantify the influence of buffer blocks on flow depths, the dimensionless parameter relative flow depth ($\frac{H_b}{H_{ff}}$) defined as the ratio of the maximum flow depth with buffer blocks (H_b) to the maximum flow depth in the free flow scenario (H_{ff}) is introduced (Figure 13). It is observed that the presence of buffer blocks consistently increased the flow depth in front of the blocks, ranging from 1.3 to 2.25 times higher than in the free flow case. The lowest increase in flow depth was noted for the single-row configuration (C1), whereas the highest increase was observed for the multiple-row configuration with wider spacing (C4 and C5). This could be because the single-row configuration offers less obstruction to the flow, resulting in minimal disruption and lower flow depth increase. In contrast, the multiple-row configurations with wider spacing create more significant obstructions, causing greater flow resistance and a subsequent increase in flow depth. The wider spacing between the rows of buffer blocks allows for more turbulent flow interactions and energy dissipation, leading to greater accumulation of water and, subsequently, higher flow depths. Additionally, for most of the buffer block configurations, the relative flow depths appear to increase until $R < 1.5$ and decrease for $R > 1.5$. This trend suggests that as the incoming flow depth rises, the influence of buffer blocks diminishes as the flow begins to overtop the blocks. The data for $R = 3$, which is less than for $R = 1$, also supports this observation.

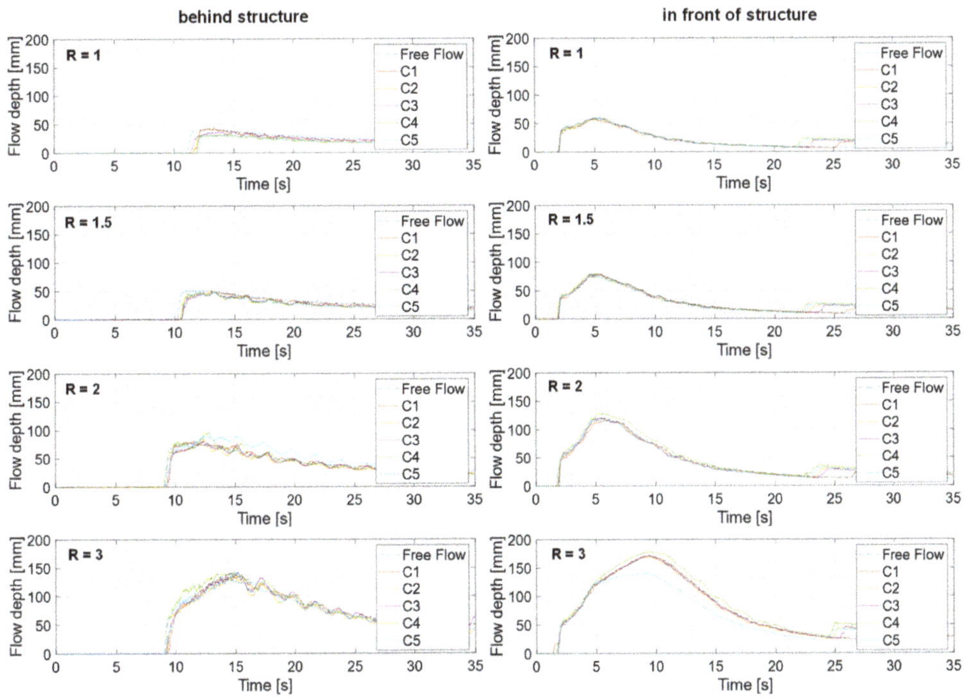

Figure 11. Comparison of flow depths due to the buffer block configurations C1 to C5 with the free flow. These plots correspond to WG 6 (x = 21 m) (**left**) and WG 2 (x = 7 m) (**right**) for varying free flow depth to block height ratios.

Behind the buffer blocks, the relative flow depths are found to be in the range of 0.78 to 1.34. The values at WG 5 are generally found to be higher than those at WG 6, indicating that the relative flow depths decrease with increasing distance from the blocks. This is due to the rooster tail formed behind such structures, as explained by Harish et al. [39]. However, in most cases, the relative flow depths are greater than 1, signifying higher flow depths compared to the scenario without blocks. Only for the $R \leq 1.5$ cases do the flow depths decrease. This observation suggests that the influence of the buffer blocks is strongly dependent on the ratio of flow depth to block height. Further, it could also be realized from Figure 13 that the C5 configuration performs better in reducing the flow depth behind the buffer blocks compared to the other configuration.

Figure 12. Interference of deflected streams through buffer blocks: (**a**) Initial bore approach and flow deflection at gaps between blocks. (**b**) Bore impact on buffer blocks: splash and overtopping.

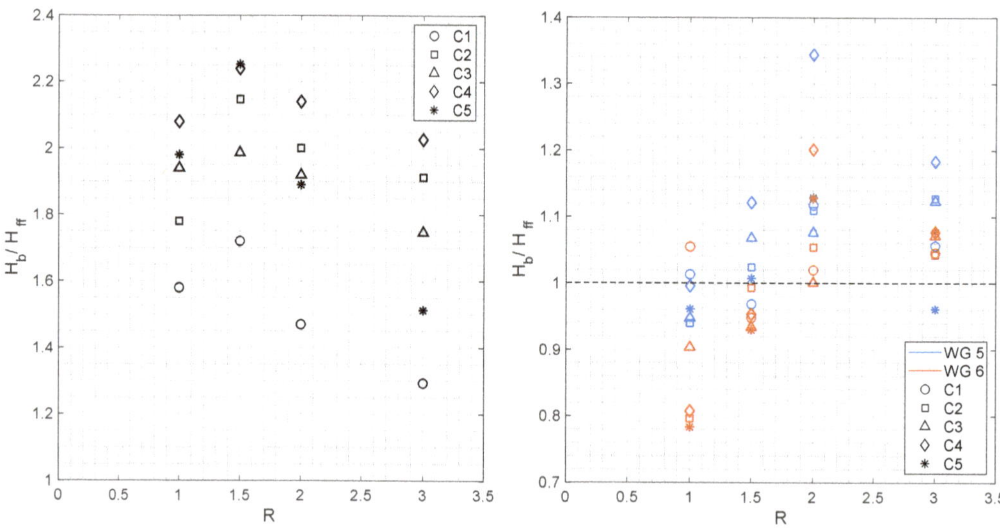

Figure 13. Comparison of the relative flow depths due to the buffer block configurations C1 to C5. These plots correspond to the relative flow depths at WG 4 (x = 14 m) (**left**), WG 5 (x = 17.5 m), and WG 6 (x = 21 m) (**right**) for varying free flow depth to block height ratios.

To quantify and compare the flow depth reduction in relation to the incoming flow and the position, the reduction in flow depth is similarly plotted against the flow depth to block height ratio R for WG 4, 5, and 6 (Figure 14). The reduction in flow depth due to the presence of the buffer blocks was calculated as $\frac{dH}{H_{ff}} = \frac{H_{ff}-H_b}{H_{ff}}$ [−]. Herein, '*H*' indicates the flow depth, '*ff*' depicts the free flow case, and '*b*' is the case in the presence of the blocks. In all the test scenarios, the results display negative values for the flow depth reduction in front of the buffer block position. Confirming with the observations above, these negative values signify an increase in water level, which can be attributed to the obstruction caused by the buffer blocks, leading to the accumulation of water in front of them. The increase in water levels ranges from 125% for the multiple-row configurations with wider spacing (C4, C5) to 30% for the single-row configuration (C1). Similar to the profile of the relative flow depth, the data for the flow depth reductions show different behavior for R > 1.5. This aligns with the earlier finding that for R values above 1, the flow predominantly overtops rather than passes through the blocks. Consequently, this results in the bore rolling over the blocks and, thus, inducing a relatively lower accumulation of water in front of the structure. The experiments with configurations C1 to C5 indicate that the single-row configuration (C1) results in the least increase in flow depth. As the number of rows and the gaps between them increase, so does the magnitude of the flow depth.

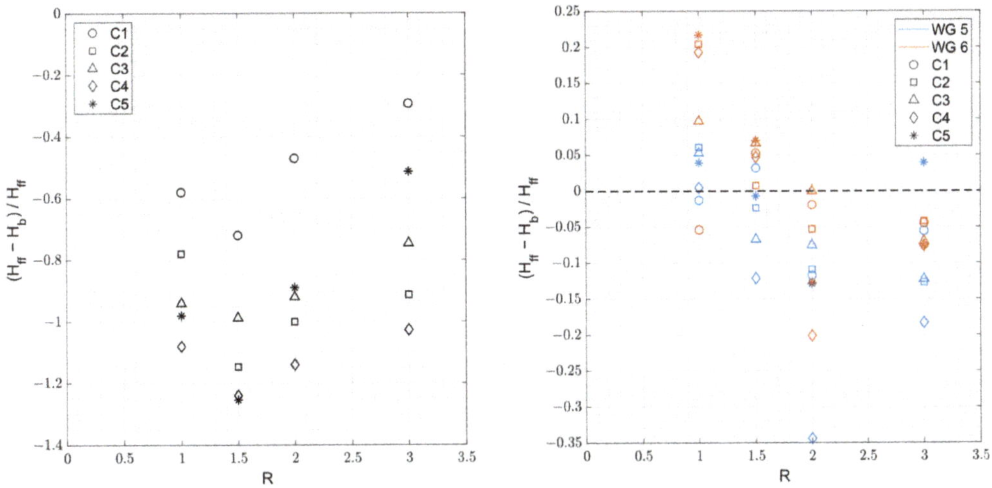

Figure 14. Comparison of the flow depth reductions due to the buffer block configurations C1 to C5. These plots correspond to the relative flow depths at WG 4 (x = 14 m) (**left**), WG 5 (x = 17.5 m), and WG 6 (x = 21 m) (**right**) for varying free flow depth to block height ratios.

Leeside of the blocks, the flow depth reduction becomes greater with an increase in the distance to the blocks. Overall, the buffer blocks appear to have a negligible impact on the flow depth, with a slight increase in flow depth compared to the free flow observed downstream. The values at WG 6 range from −20% to +22%. This arbitrariness of the data may be attributed to the previously discussed local phenomenon of oscillations and turbulences influencing the WG 6 measurements. Therefore, conducting measurements far away from the buffer blocks might offer a clearer understanding of the flow regime unaffected by this localized phenomenon.

In order to enhance the understanding of how the rows and their spacing influence flow depth reduction, the reduction values are plotted against the number of rows in Figure 15. The flow depth reduction is calculated in comparison to the free flow depth at position 6. A pattern emerges wherein increasing bore heights correlate with a diminished reduction in flow depth, while an increased number of rows tends to enhance the reduction slightly. Regarding the row spacing, observations from the experiments indicate that for configurations with two rows, a gap of 2 h yields a greater reduction than a gap of 1 h at low R values.

To assess whether the size of the prototype buffer block (1.2 × 0.4 × 0.8 m) should potentially be doubled, the performances of the C1* and C1 configurations were compared, as shown in Figure 16. In a single-row configuration and with the same incoming flow parameters, there is no significant difference in flow depth reduction whether small (C1) or larger buffer blocks (C1*) are implemented. However, the reflection of flow caused by larger blocks is higher and faster (−5 s), as can be observed on the right-hand side of Figure 16. The larger buffer block height (C1) serves as the reference point for calculating the ratio R in this comparison.

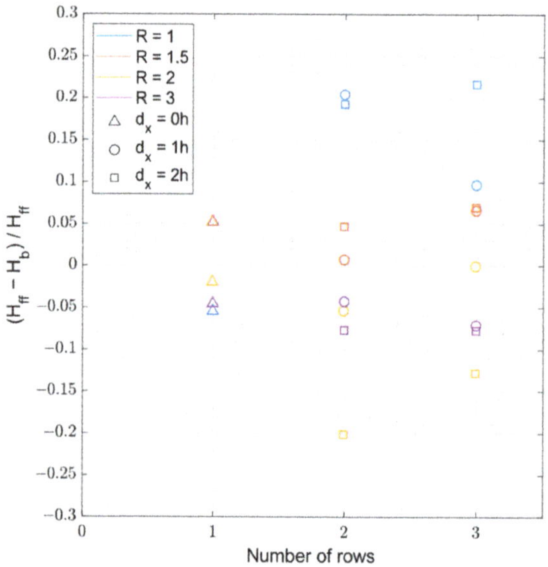

Figure 15. Reduction in flow depth relative to the number of rows and their spacing dx. Flow depth reduction in comparison to the free flow depth at WG 6 (x = 21 m).

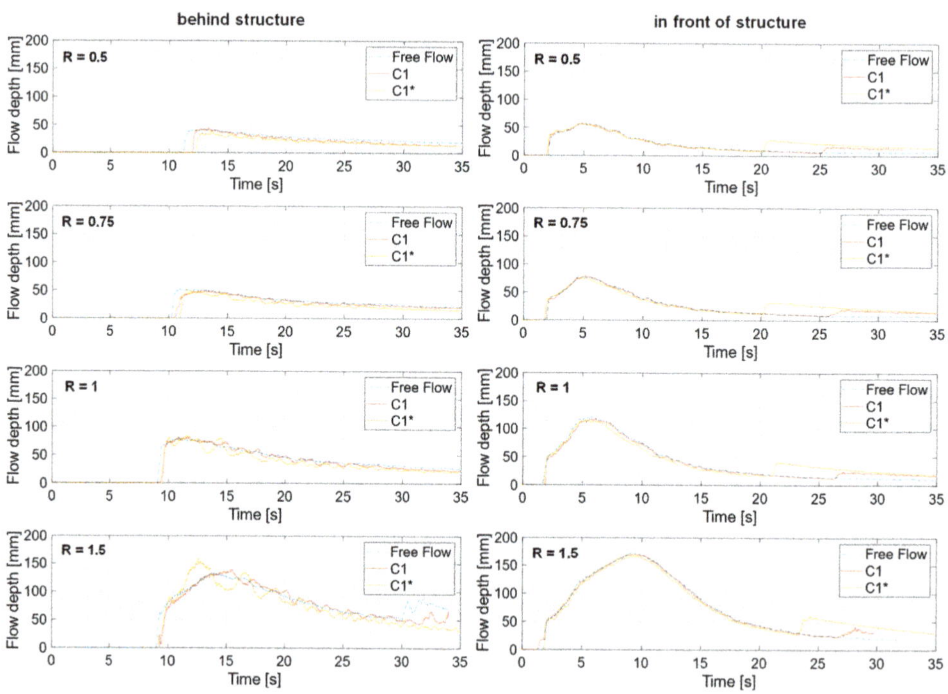

Figure 16. Comparison of flow depths due to the small buffer blocks (C1) and larger buffer blocks (C1*) with the free flow. These plots correspond to WG 6 (x = 21 m) (**left**) and WG 2 (x = 7 m) (**right**) for varying free flow depth to block height ratios. The ratio R is calculated using the height of the smaller buffer blocks.

4.2. Additional Experiments with Configuration C1* and CB

The flow characteristics over time provide an initial insight into the influence of permeable structures such as buffer blocks compared to continuous structures such as a sea wall. Figure 17 compares the flow depths over time behind the structure for the different incoming flow characteristics due to buffer blocks or continuous blocks, as opposed to the scenario without any structure (free flow). Both the continuous block and the buffer blocks delay the arrival of the flow at WG 6 by approx. 1 s (Figure 17; left-hand side), in comparison to the free flow. Across all ratios of R, the flow depths resulting from the continuous blocks were consistently lower than those generated by the buffer blocks and for the free flow case. This indicates that in the case of a continuous block, most of the flow is reflected, and a lesser part of the energy is transmitted due to the impermeable nature of the structure. Oscillations in the time history can be observed in the downstream flow profile due to the presence of the buffer blocks. These oscillations tend to be greater for the submerged condition ($R \geq 1$) than for the emerged scenarios ($R = 0.5$ and $R = 0.75$). For ratios above 1, the flow mainly propagates over the blocks, which could lead to more turbulence than the flow propagation along the block.

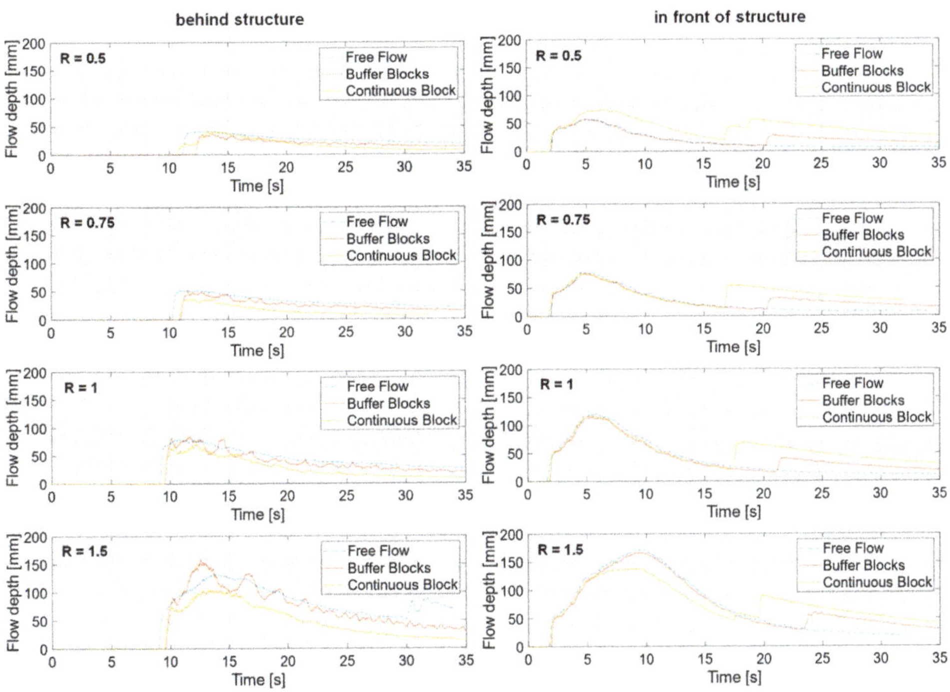

Figure 17. Comparison of flow depths due to the buffer blocks and continuous block with the free flow. These plots correspond to WG 6 (x = 21 m) (**left**) and WG 2 (x = 7 m) (**right**) for varying free flow depth to block height ratios.

Focusing on the reflection component in front of the buffer blocks, it is to be noted that the flow reflection at a continuous block has a larger bore height than the one at a buffer block. Based on the distribution of the flow peaks, it can also be seen that the reflection of flow induced by a continuous block propagates approx. 3 s faster than the flow induced by the buffer blocks. This is in line with the results mentioned above. Since more water is reflected in the case of a continuous block, the reflecting bore height is greater than in the case of buffer blocks.

The numerical study by Percy et al. [15] presents flow profiles at the buffer block position, depicted in Figure 18, that bear a notable resemblance to those observed in the IWW experiments. In the experimental data, there is a pronounced initial sharp peak, which is then followed by a subtler, more gently rounded peak. On the other hand, the flow profiles generated by the numerical model exhibit a smoother, more prolonged ascent in water level. The sharp peak in the flow depth might be a splash captured by the ultrasonic sensor, which is not being captured in the numerical model due to the numerical gauge extraction method; however, such a splash was noticed in the numerical model results [15].

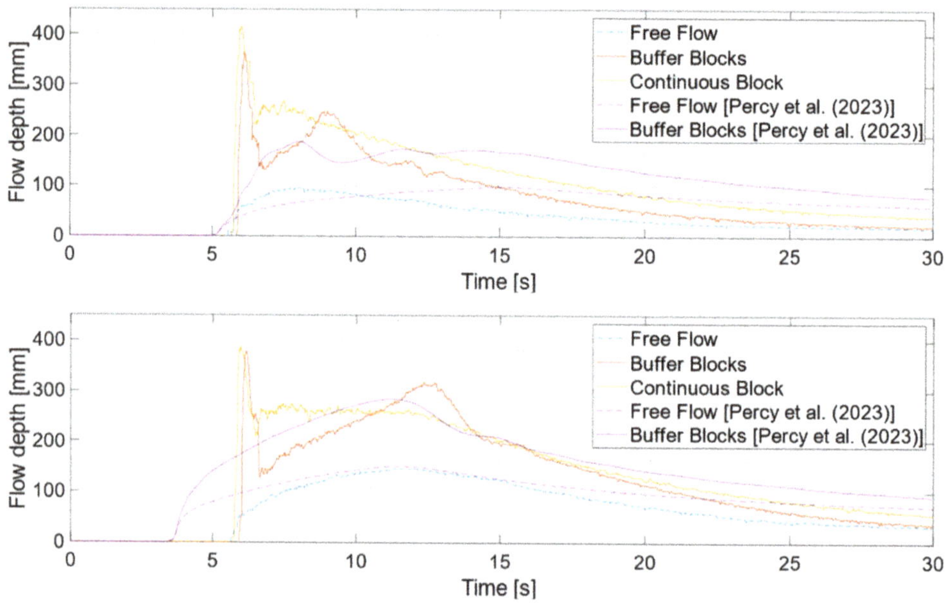

Figure 18. Comparison of physical results of the present study with numerical results of Percy et al. [15] for the flow depths for R = 1 (**top**) and R = 1.5 (**bottom**) at buffer block position. Plots correspond to the single-row configuration and continuous block structure.

Examining the relative flow depth in front of the buffer block (Figure 19), a trend similar to that observed in C1 to C5 configurations is evident. For the continuous block, a substantial increase in water level is observed, with a rise of 250% for R = 0.5 and 150% for R = 0.75 (Figure 20). For higher ratios of flow depth to block height, such as R = 1.5, the increase is comparatively lower at 70%. This is due to significant flow overtopping. In the case of the buffer block configuration (C1*), the increase in water levels noted is less, with a rise of 140% for R = 0.5 and 115% for R = 0.75. The study by Percy et al. [15] reported even smaller increases (80%) than those measured in these experiments. This discrepancy could be attributed to the differences in extreme flow generation methods. In the present study, the flows were produced by pumps, whereas Percy et al. [15] used a finite-length dam break setup. The finite-length dam break setup could lead to differences in flow characteristics, such as energy and velocity, which might result in varying impacts on flow depth increases. For the continuous block configuration, the reduction in flow depth behind the block is also found to be higher with an increasing distance to the blocks for all investigated ratios R. The most significant reduction of 24.5% was observed at 7 m behind the block (WG 6) for R = 0.75, while the reduction due to buffer blocks (C1*) is minimal with 0.06%.

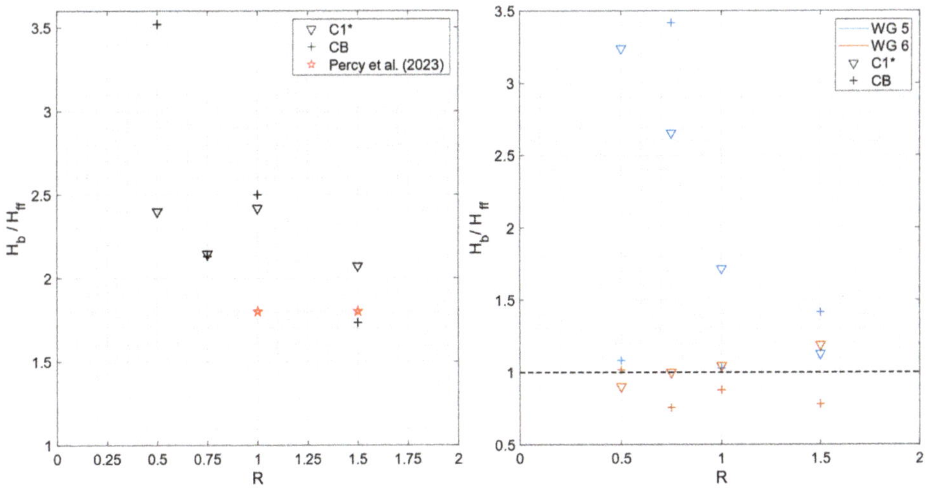

Figure 19. Comparison of the relative flow depths due to the C1* and CB configurations and the results of Percy et al. [15]. These plots correspond to the relative flow depths at WG 4 (x = 14 m) (**left**), WG 5 (x = 17.5 m), and WG 6 (x = 21 m) (**right**) for varying free flow depth to block height ratios.

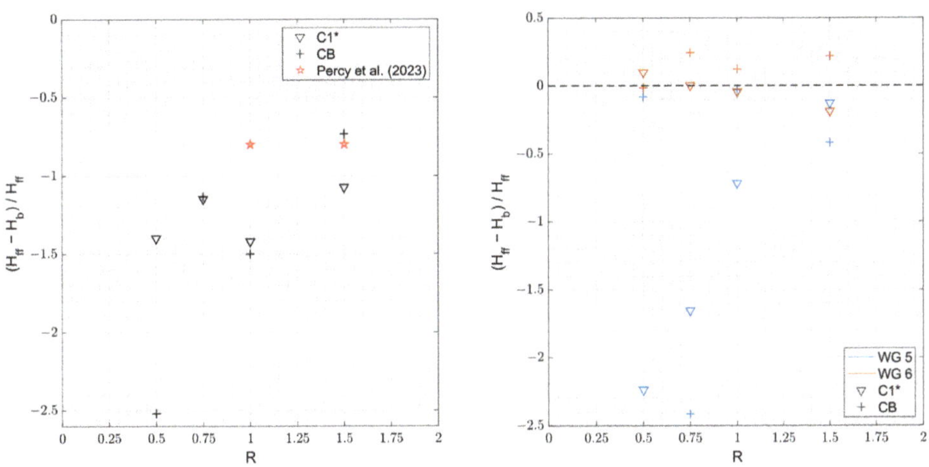

Figure 20. Comparison of the flow depth reductions due to the C1* and CB configurations and the results of Percey et al. [15]. These plots correspond to the relative flow depths at WG 4 (x = 14 m) (**left**), WG 5 (x = 17.5 m), and WG 6 (x = 21 m) (**right**) for varying free flow depth to block height ratios.

5. Conclusions and Future Research

The rapid population growth in coastal areas has resulted in increased infrastructure and economic concentrations, making coastal inhabitants and assets more vulnerable to hazards such as tsunamis. The study shows that buffer blocks present a promising solution for mitigating tsunami impacts by dissipating flow energy and overcoming the limitations of conventional measures. Different configurations of buffer blocks were tested at a 1:16 scale by varying the number of rows and spacing between them. The following vital conclusions arrived:

- Buffer blocks effectively delay flow arrival and reduce flow depths for flow depth to block height ratios (R) below 1.5. However, their effectiveness reduces for higher R as

the flow overtops the blocks. The flow overtopping, in most cases, resulted in increased flow depth immediately behind the buffer blocks and, hence, behaves negatively.
- Increasing the number of rows enhanced the flow depth reduction behind the buffer blocks and increased the flow depth in the vicinity of the buffer blocks on its upstream side due to the increase in the obstruction.
- Multiple-row configurations with wider spacing achieved the best flow depth reduction behind the buffer blocks, hence necessitating sufficient space between the buffer blocks' rows. C5 configuration yielded the best results out of the tested configuration when $R \leq 1.5$.
- Comparisons with continuous block structures revealed that buffer blocks perform better in flow depth reduction by inducing more turbulence and oscillations. Thus, buffer blocks can be realized as an alternative to the conventional sea walls of low heights.

Despite the present study identifying that increasing the spacing between the rows and increasing the number of buffer block rows can yield a better reduction in flow depth, the tested configuration was limited considering the bi-functional nature (i.e., both tsunami-resistant and beach aesthetics) of these buffer blocks, as argued in Section 2.2.

While flow depth measurements provide valuable insights, they offer a limited analysis of energy dissipation. Nevertheless, the assessment of flow depth information is more critical as it forms the integral component of flow energy and momentum evaluation, which are necessary for tsunami hazard and risk mapping. Further, such information is necessary to establish the minimum vertical evacuation height required for evacuating the people. The research highlights the complexity of accurately predicting the performance of such protection measures under varying conditions. To address these limitations, future studies should include detailed measurements of flow velocity and incorporate force sensors to quantify the forces exerted on buffer blocks and the structures behind them. These data are crucial for designing buffer blocks that can withstand the impact forces of high-energy events. Future experimental or numerical experiments should also investigate the beach slope, which may have a significant impact on the buffer block functionality.

Future research should focus on refining the understanding of buffer blocks' performance under different flow conditions and integrating the measurement of velocity and forces into improved design guidelines and planning strategies. This will help ensure that coastal communities worldwide are better protected against the increasing threat of tsunamis and other extreme flow events. By advancing the knowledge and application of buffer blocks, we can enhance coastal resilience and safeguard vulnerable populations and infrastructure.

Author Contributions: Conceptualization, P.N., J.O., S.H. and H.S.; methodology, P.N., J.O. and S.H.; investigation, P.N.; resources, H.S., V.S. (V. Sriram) and V.S. (V. Sundar); writing—original draft preparation, P.N.; writing—review and editing, S.H., J.O., V.S. (V. Sriram), V.S. (V. Sundar) and H.S.; supervision, V.S. (V. Sriram), V.S. (V. Sundar) and H.S.; project administration, S.H.; funding acquisition, H.S. All authors have read and agreed to the published version of the manuscript.

Funding: This research was funded by the German Research Foundation (Deutsche Forschungsgemeinschaft; DFG) for the project "Buffer blocks as wave energy dissipators (BB-WEnDis)" grant number 453313678.

Data Availability Statement: The data presented in this study are available from the corresponding author upon reasonable request.

Acknowledgments: The authors would like to express their gratitude to the German Science Foundation, Deutsche Forschungsgemeinschaft e.V. (DFG), for sponsoring this joint research project between the Institute of Hydraulic Engineering and Water Resources Management at RWTH Aachen University, Aachen, Germany, and the Department of Ocean Engineering at IIT Madras, Chennai, India, through grant number 453313678. The first author gratefully acknowledges the financial support for the research stay at DOE, IIT Madras, provided by the Global Water and Climate Adaptation Centre—Aachen, Bangkok, Chennai, Dresden (ABCD-Centre). Additionally, the first author would like to thank Kristian Brodersen, Head of the Instrumentation Department, and Arne Harz, Head of the Model Construction Department, along with Christian Vogelgesang, Master Bricklayer, and Industrial Mechanics Lars König, Malte Schmitz, and Peter Kläser, for their contributions in building the experimental setup. Special thanks are extended to Golda Percy, IIT Madras, for providing the numerical results for this study.

Conflicts of Interest: The authors declare no conflict of interest.

References

1. Neumann, B.; Vafeidis, A.T.; Zimmermann, J.; Nicholls, R.J. Future coastal population growth and exposure to sea-level rise and coastal flooding—A global assessment. *PLoS ONE* **2015**, *10*, e0118571. [CrossRef] [PubMed]
2. McGranahan, G.; Balk, D.; Anderson, B. The rising tide: Assessing the risks of climate change and human settlements in low elevation coastal zones. *Environ. Urban.* **2007**, *19*, 17–37. [CrossRef]
3. IPCC. Summary for Policymakers. In *IPCC Special Report on the Ocean and Cryosphere in a Changing Climate*; Cambridge University Press: Cambridge, UK; New York, NY, USA, 2019. [CrossRef]
4. Goseberg, N. Reduction of maximum tsunami run-up due to the interaction with beachfront development—Application of single sinusoidal waves. *Nat. Hazards Earth Syst. Sci.* **2013**, *13*, 2991–3010. [CrossRef]
5. Mehvar, S.; Filatova, T.; Dastgheib, A.; van Ruyter Steveninck, E.; Ranasinghe, R. Quantifying Economic Value of Coastal Ecosystem Services: A Review. *JMSE* **2018**, *6*, 5. [CrossRef]
6. UNEP. *Marine and Coastal Ecosystems and Human Well-Being: A Synthesis Report Based on the Findings of the Millennium Ecosystems Assessment*; UNEP: Cambridge, UK, 2006; Available online: https://www.unep.org/resources/report/marine-and-coastal-ecosystems-and-human-well-being-synthesis-report-based-findings (accessed on 8 May 2024).
7. Telford, J.; Cosgrave, J.; Houghton, R. *Joint Evaluation of the International Response to the Indian Ocean Tsunami: Synthesis Report*; Edita: October City, Egypt, 2006.
8. Kajitani, Y.; Chang, S.E.; Tatano, H. Economic Impacts of the 2011 Tohoku-Oki Earthquake and Tsunami. *Earthq. Spectra.* **2013**, *29*, 457–478. [CrossRef]
9. Goto, K.; Ishizawa, T.; Ebina, Y.; Imamura, F.; Sato, S.; Udo, K. Ten years after the 2011 Tohoku-oki earthquake and tsunami: Geological and environmental effects and implications for disaster policy changes. *Earth Sci. Rev.* **2021**, *212*, 103417. [CrossRef]
10. Oetjen, J.; Sundar, V.; Venkatachalam, S.; Reicherter, K.; Engel, M.; Schüttrumpf, H.; Sannasiraj, S.A. A comprehensive review on structural tsunami countermeasures. *Nat. Hazards* **2022**, *113*, 1419–1449. [CrossRef]
11. Shuto, N.; Fujima, K. A short history of tsunami research and countermeasures in Japan. *Proc. Jpn. Acad. Ser. B* **2009**, *85*, 267–275. [CrossRef]
12. CEDP. Note on prevention against tsunamis, Sanshu-sha, Tokyo. Council on Earthquake Disaster Prevention, Ministry of Education (in Japanese) **1993**, 1–14.
13. NOAA (Ed.) *Designing for Tsunamis: Seven Principles for Planning and Designing for Tsunami Hazards*; National Tsunami Hazard Mitigation Program (NTHMP): Galveston, TX, USA, 2001.
14. UNESCO; IOC. *Reducing and Managing the Risk of Tsunamis: Manuals and Guides*; UNESCO: Paris, France, 2011; Volume 57.
15. Golda Percy, V.P.; Sriram, V.; Sundar, V.; Schüttrumpf, H. Effect of the buffer blocks in attenuating a tsunami-like flow. *Ocean Eng.* **2023**, *286*, 115489. [CrossRef]
16. Kathiresan, K.; Rajendran, N. Coastal mangrove forests mitigated tsunami. *Estuar. Coast. Shelf Sci.* **2005**, *65*, 601–606. [CrossRef]
17. Olwig, M.F.; Sørensen, M.K.; Rasmussen, M.S.; Danielsen, F.; Selvam, V.; Hansen, L.B.; Nyborg, L.; Vestergaard, K.B.; Parish, F.; Karunagaran, V.M. Using remote sensing to assess the protective role of coastal woody vegetation against tsunami waves. *Int. J. Remote Sens.* **2007**, *28*, 3153–3169. [CrossRef]
18. Iverson, L.R.; Prasad, A.M. Modeling tsunami damage in Aceh: A reply. *Landsc. Ecol.* **2008**, *23*, 7–10. [CrossRef]
19. Tanaka, N.; Sasaki, Y.; Mowjood, M.I.M.; Jinadasa, K.B.S.N.; Homchuen, S. Coastal vegetation structures and their functions in tsunami protection: Experience of the recent Indian Ocean tsunami. *Landsc. Ecol. Eng.* **2007**, *3*, 33–45. [CrossRef]
20. Tanaka, S.; Istiyanto, D.C.; Kuribayashi, D. *Planning and Design of Tsunami-Mitigative Coastal Vegetation Belts*; International Centre for Water Hazard and Risk Management under the Auspices of UNESCO, Public Works Research Institute: Tsukuba, Japan, 2010.
21. Baird, A.H.; Kerr, A.M. Landscape analysis and tsunami damage in Aceh: Comment on Iverson and Prasad (2007). *Landsc. Ecol.* **2008**, *23*, 3–5. [CrossRef]
22. Yanagisawa, H.; Koshimura, S.; Goto, K.; Miyagi, T.; Imamura, F.; Ruangrassamee, A.; Tanavud, C. The reduction effects of mangrove forest on a tsunami based on field surveys at Pakarang Cape, Thailand and numerical analysis. *Estuar. Coast. Shelf Sci.* **2009**, *81*, 27–37. [CrossRef]
23. Sundar, V.; Murali, K.; Noarayanan, L. Effect of vegetation on run-up and wall pressures due to cnoidal waves. *J. Hydraul. Res.* **2011**, *49*, 562–567. [CrossRef]
24. Noarayanan, L.; Murali, K.; Sundar, V. Role of Vegetation on Beach Run-up due to Regular and Cnoidal Waves. *J. Coast. Res.* **2012**, *278*, 123–130. [CrossRef]
25. Noarayanan, L.; Murali, K.; Sundar, V. Manning's 'n' for staggered flexible emergent vegetation. *J. Earthq. Tsunami* **2013**, *7*, 1250029. [CrossRef]

26. Strusińska-Correia, A.; Husrin, S.; Oumeraci, H. Tsunami damping by mangrove forest: A laboratory study using parameterized trees. *Nat. Hazards Earth Syst. Sci.* **2013**, *13*, 483–503. [CrossRef]
27. Nateghi, R.; Bricker, J.D.; Guikema, S.D.; Bessho, A. Statistical Analysis of the Effectiveness of Seawalls and Coastal Forests in Mitigating Tsunami Impacts in Iwate and Miyagi Prefectures. *PLoS ONE* **2016**, *11*, e0158375. [CrossRef]
28. EurOtop. Manual on Wave Overtopping of Sea Defences and Related Structures. An Overtopping Manual Largely Based on European Research, But for Worldwide Application. Available online: www.overtopping-manual.com (accessed on 7 April 2023).
29. Oumeraci, H. Nonconventional Wave Damping Structures. In *Handbook of Coastal and Ocean Engineering*; World Scientific: Singapore, 2009; pp. 287–315. ISBN 978-981-281-929-1.
30. Goseberg, N. The Run-Up of Long Waves: Laboratory-Scaled Geophysical Reproduction and Onshore Interaction with Macroroughness Elements: Mitteilungen Heft 97. Ph.D. Thesis, Leibniz Universität Hannover, Hannover, Germany, 2011.
31. Rahman, S.; Akib, S.; Khan, M.T.R.; Shirazi, S.M. Experimental study on tsunami risk reduction on coastal building fronted by sea wall. *Sci. World J.* **2014**, *2014*, 729357. [CrossRef] [PubMed]
32. Thomas, S.; Killian, J.; Bridges, K. Influence of Macroroughness on Tsunami Loading of Coastal Structures. *J. Waterw. Port Coast. Ocean Eng.* **2015**, *141*, 04014028. [CrossRef]
33. Capel, A. Wave run-up and overtopping reduction by block revetments with enhanced roughness. *Coast. Eng.* **2015**, *104*, 76–92. [CrossRef]
34. Moon, W.C.; Chiew, L.Q.; Cheong, K.W.; Tee, Y.C.; Chun, J.B.; Lau, T.L. An experimental study for estimating tsunami wave forces acting on building with seaward and landward macroroughness. *Ocean Eng.* **2019**, *186*, 106116. [CrossRef]
35. Ishii, H.; Takabatake, T.; Esteban, M.; Stolle, J.; Shibayama, T. Experimental and numerical investigation on tsunami run-up flow around coastal buildings. *Coast. Eng.* **2021**, *63*, 485–503. [CrossRef]
36. Giridhar, G.; Muni Reddy, M.G. Hydrodynamic Study of Energy Dissipation Blocks on Reduction of Wave Run-up and Wave Reflection. *Aquat. Procedia* **2015**, *4*, 281–286. [CrossRef]
37. Ono, S.; Hiraishi, T. Tsunami Force Reduction due to Obstacle in Front of Coastal Dike and Evaluation of Collision Force by Driftage. *JEPE* **2017**, *11*, 300–310. [CrossRef]
38. Oetjen, J.; Engel, M.; Pudasaini, S.P.; Schuettrumpf, H. Significance of boulder shape, shoreline configuration and pre-transport setting for the transport of boulders by tsunamis. *Earth Surf. Process. Landf.* **2020**, *45*, 2118–2133. [CrossRef]
39. Harish, S.; Sriram, V.; Schüttrumpf, H.; Sannasiraj, S.A. Tsunami-like flow induced force on the structure: Prediction formulae for the horizontal force in quasi-steady flow phase. *Coast. Eng.* **2021**, *168*, 103938. [CrossRef]
40. Harish, S.; Sriram, V.; Schüttrumpf, H.; Sannasiraj, S.A. Flow-structure interference effects with the surrounding structure in the choked quasi-steady condition of tsunami: Comparison with traditional obstruction approach. *Appl. Ocean Res.* **2022**, *126*, 103255. [CrossRef]
41. Harish, S.; Sriram, V.; Schüttrumpf, H.; Sannasiraj, S.A. Tsunami-like flow induced forces on the structure: Dependence of the hydrodynamic force coefficients on Froude number and flow channel width in quasi-steady flow phase. *Coast. Eng.* **2022**, *172*, 104078. [CrossRef]
42. Harish, S.; Sriram, V.; Schüttrumpf, H.; Sannasiraj, S.A. Tsunami-like flow induced forces on the building with openings and orientation in the quasi-steady flow phase. *Ocean Eng.* **2024**, *301*, 117337. [CrossRef]

Journal of
Marine Science and Engineering

MDPI

Article

Predictions of Wave Overtopping Using Deep Learning Neural Networks

Yu-Ting Tsai [1,*] and Ching-Piao Tsai [2,*]

1 Bachelor's Program of Precision System Design, Feng Chia University, Taichung 407, Taiwan
2 Department of Civil Engineering, National Chung Hsing University, Taichung 402, Taiwan
* Correspondence: yuttsai@fcu.edu.tw (Y.-T.T.); cptsai@nchu.edu.tw (C.-P.T.)

Abstract: Deep learning techniques have revolutionized the field of artificial intelligence by enabling accurate predictions of complex natural scenarios. This paper proposes a novel convolutional neural network (CNN) model that involves deep learning technologies, such as the bottleneck residual block, layer normalization, and dropout layer, to predict wave overtopping at coastal structures under a wide range of conditions. To optimize the performance of the CNN model, the hyperparameter tuning process via Bayesian optimization is used. The results of validation demonstrate that the proposed CNN model is highly accurate in estimating wave overtopping discharge from hydraulic and structural parameters. The testing accuracy of the overtopping predictions using a prototype dataset shows that the proposed CNN model outperforms those existing machine learning models. An example application of the CNN model is presented for predicting prototype overtopping considering various crest freeboards of coastal structures.

Keywords: wave overtopping; deep learning technique; convolutional neural network; bottleneck residual block; convolution layer; layer normalization; dropout layer; hyperparameter optimal tuning process; Bayesian optimization; bootstrap resampling

1. Introduction

The accurate assessment of wave overtopping discharges is crucial for the appropriate design of coastal structures that can prevent severe coastal flooding. Wave overtopping at coastal structures has been studied extensively through various methods, including field investigations, scale model measurements, and numerical simulations [1–7]. The CLASH (Crest Level Assessment of Coastal Structures) project [8] conducted numerous physical modeling and prototype investigations of wave overtopping discharge at various types of coastal structures. Various empirical formulas are included in the EurOtop manual [9] for the estimation of wave overtopping discharge. On the other hand, robust artificial neural network (ANN) models, utilizing hydraulic and structural parameters sourced from either the new EurOtop database [9] or the original CLASH database [10] as input variables, have demonstrated efficient performance in predicting wave overtopping discharge at coastal structures [4,7,11–14]. Each test in the EurOtop database is characterized by a reliability factor (RF) and complexity factor (CF) [10], which are used to select reliable data for training and improving ANN models.

Previous studies have proposed a variety of multilayer architectures for ANN models in the field of machine learning. Verhaeghe et al. [7] proposed a two-phase neural network model, including a classifier–quantifier scheme, to distinguish negligible from significant overtopping and avoid large overpredictions in areas of low overtopping. However, Zanuttigh et al. [11] argued that classifier–quantifier schemes do not truly improve ANN performance and instead increase the complexity of the ANN architecture while creating undesirable discontinuities in predictions. den Bieman et al. [12] presented a new model using an advanced machine learning technique called XGBoost (XGB) associated with feature engineering, which generally outperforms conventional ANN models and empirical formulas in terms of prediction accuracy. Bootstrap resampling techniques were employed in these machine learning models to evaluate the uncertainties associated with their predictions.

Accurately predicting wave overtopping discharge in complex scenarios remains a challenging task. As an alternative to machine learning artificial neural networks (ANNs), deep convolutional neural networks (CNNs) present a promising avenue to tackle this issue. Originally developed for identifying the characteristics of individual images from large datasets, CNNs have found applications in big data analysis and decision making [13,14]. Utilizing deep convolutional neural networks for the prediction of wave overtopping discharge represents a relatively recent research area, yet it has already exhibited significant potential. Through convolutional processing for feature extraction, CNNs can effectively capture the distinctive attributes of data representations and employ filters to accentuate the respective importance of each output task. As a result, CNNs may offer significant improvements in the prediction accuracy of wave overtopping discharge.

This article presents a CNN model using a serial bottleneck residual block architecture that incorporates convolution layers, layer normalization, and a dropout layer. The model is designed to determine the relevance of hydraulic and coastal structure inputs for accurately predicting wave overtopping discharge. This paper is organized as follows. Section 2 describes the new EurOtop database which is applied to train the CNN model. Then, we explain the model expression, network layer architecture, and hyperparameter optimization process used for training the CNN model. As applied in previous ANNs, the bootstrap resampling technique is also introduced in training the CNN model. Next, we discuss the accuracy of the CNN model predictions and compare them with those of existing ANN models. An example application of the model for predicting prototype overtopping is also presented. Finally, the conclusions of the study are addressed.

2. EurOtop Database

The proposed CNN model was trained using data from the new EurOtop database [9], which is an extensive dataset containing over 17,000 tests. This dataset includes nearly 13,500 records specifically related to wave overtopping discharge. This new EurOtop database is an extension of the original CLASH database, which was compiled from approximately 10,000 schematized tests conducted worldwide. By utilizing this comprehensive dataset, the CNN model can learn from a wide range of test scenarios and improve its predictive capabilities for wave overtopping discharge. Figure 1 depicts the general geometric and relevant hydraulic parameters of coastal structures. Full descriptions of the parameters, wave conditions, and structural cross sections in the database are provided in the literature [9,11,12,15,16]. The overtopping experiments in the database are categorized into seven groups based on the structure type and oblique wave attack conditions [9]; they are rock permeable straight slopes (group A), rock impermeable straight slopes (group B), armor units with straight slopes (group C), smooth and straight slopes (group D), structures with combined slopes and berms (group E), vertical walls (group F), and oblique wave attacks (group G).

The parameters selected for training the CNN model are listed in Table 1 and comprise three hydraulic parameters of waves and thirteen structural parameters as inputs and one parameter (wave overtopping discharge) as the output. The training dataset of the neural network model consists of data derived from laboratory tests conducted at various scales. To ensure the applicability of the model to prototype scenarios, the use of dimensionless parameters is advantageous. Following [9,11,16], the significant wave height and water depth at the structure toe are represented as the wave steepness $H_{m0,t}/L_{m-1,0,t}$ and the relative water depth $h/L_{m-1,0,t}$, respectively, by dividing the wavelength $L_{m-1,0,t}$, in which $L_{m-1,0,t}$ is calculated by $1.56T^2{}_{m-1,0,t}$, and $T_{m-1,0,t}$ is the spectral wave period at the structure toe. The parameters B_t, G_c, and B describing horizontal measures of structure widths are normalized by the wavelength $L_{m-1,0,t}$ and the parameters h_t, D_u, D_d, A_c, R_c, and h_b

describing vertical measures of structure heights are normalized by the wave height $H_{m0,t}$. By incorporating dimensionless parameters, the neural network model can effectively capture the underlying physics and behavior of wave overtopping, regardless of the specific scale of the experiments.

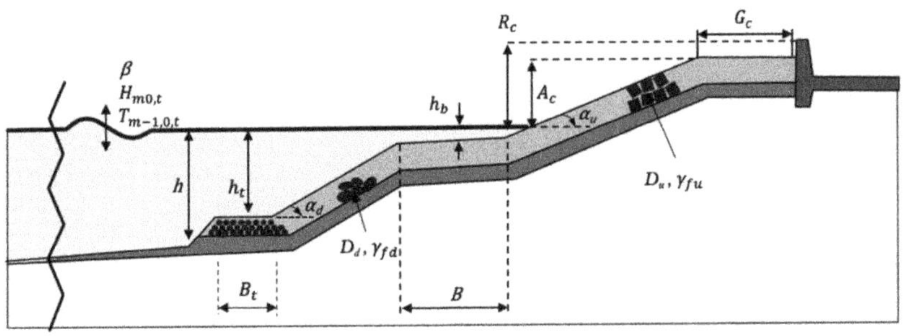

Figure 1. Schematic of structure based on CLASH, including geometric and hydraulic parameters [9]. $H_{m0,t}$: spectra significant wave height at the structure toe; $T_{m-1,0,t}$: spectral wave period at the structure toe; h: water depth at the structure toe; h_t: toe submergence; B_t: toe width; B: berm width; h_b: berm submergence; A_c: crest freeboard of armor structure; R_c: crest freeboard of wall; G_c: crest width of armor structure; α_u and α_d : angles of armor structure slope part above/below the berm; D_u and D_d: sizes of the armor elements along up/down slopes; γ_{fu} and γ_{fd}: roughness factors for up/down slopes; β: incident angle of wave direction.

Table 1. CNN input and output parameters.

#	Parameters		Definition of the Parameters
1	Hydraulic parameters of waves	$H_{m0,t}/L_{m-1,0,t}$	Wave steepness at the structure toe
2		$h/L_{m-1,0,t}$	Relative water depth at the structure toe
3		β	Wave obliquity
4	Structure parameters	$h_t/H_{m0,t}$	Relative submergence of the toe structure
5		$B_t/L_{m-1,0,t}$	Relative width of the toe structure
6		$cot\alpha_d$	Cotangent of the structure slope below the berm
7		$cot\alpha_u$	Cotangent of the structure slope above the berm
8		γ_{fd}	Roughness factor for $cot\alpha_d$
9		γ_{fu}	Roughness factor for $cot\alpha_u$
10		$D_d/H_{m0,t}$	Relative size of the armor elements along $cot\alpha_d$
11		$D_u/H_{m0,t}$	Relative size of the armor elements along $cot\alpha_u$
12		$A_c/H_{m0,t}$	Relative crest freeboard of armor structure
13		$R_c/H_{m0,t}$	Relative crest freeboard of wall
14		$G_c/L_{m-1,0,t}$	Relative crest width of armor structure
15		$B/L_{m-1,0,t}$	Relative berm width
16		$h_b/L_{m-1,0,t}$	Relative berm submergence
1	Predicted parameter	q_s	Normalized wave overtopping discharge per unit width

Following [11,16], the wave overtopping discharge q in the database is scaled by a normalized formula for training, which is defined as:

$$q_s = \frac{log_{10}(q_{ad}) - min[log_{10}(q_{ad})]}{|min[log_{10}(q_{ad})] - max[log_{10}(q_{ad})]|} \quad (1)$$

in which q_{ad} is defined as

$$q_{ad} = \frac{q}{\sqrt{gH_{m0,t}^3}} \quad (2)$$

where q is the original mean wave overtopping discharge value from the EurOtop database and g is the gravitational acceleration (9.8 m/s^2). Equation (2) represents a dimensionless formula known as the relative wave overtopping discharge. As elucidated in [13], the application of Equation (1) for normalization confines the fitted data within a range of 0 to 1, thereby mitigating the scale-dependent nature of the error. Subsequently, the predicted output value q_s generated by the trained CNN is reverted to obtain the corresponding value of q.

A weight factor WF = (4 − RF)·(4 − CF) defined in [4] was adopted in previous works for the selection of reliable data to minimize the influence of environmental uncertainties and irregular structures. Higher WFs for the overtopping database indicate greater reliability and simpler structural schemes. As the same in earlier prediction methods [11,16], excluding the data of very unreliable (RF = 4) or complex (CF = 4), $q = 0$, and missing values, a total of 8653 records labeled as core data in the overtopping database are used in the proposed CNN model. It is noted that the core data excluded the dataset of prototype overtopping.

3. Method Description

3.1. Model Description

In this section, we present how the proposed CNN model learns the inputs of the hydraulic and structural parameters corresponding to the output of wave overtopping discharge. The overtopping database contains many different structure types and related input parameters, but they may have the same distributions in the output parameters, as known as a covariate shift problem. The general neural network model may require specific feature engineering to select appropriate input parameters, preventing the model from possibly misclassifying the data. Alternatively, the CNN gains experience by observing a training set of input data and predicting the output without any feature assumptions. Accounting for this problem, the convolutional layer [13], layer normalization [17], and a dropout layer [18] are used for the proposed CNN, described as follows.

Convolutional Layer: This layer operates by traversing a compact filter across the input data, enabling the detection of patterns and distinctive features. The equation of the convolutional layers is expressed as follows:

$$\mathbf{x}^{(l)} = \mathbf{w}\cdot\mathbf{x}^{(l-1)} + \mathbf{b} = \sum_{\tau=0}^{T}\left[w[\tau]^{(l)}x[t-\tau+\phi]^{(l-1)}\right] + b[t] \tag{3}$$

where the superscript l represents the index of each neutral network layer and t denotes the index of the input parameters. In Equation (3), $x^{(l-1)}$ is a given input dataset, w is a set of one-dimensional convolutional filter coefficients, and ϕ denotes the shift number. Equation (3) can be described as the area under the function $w[\tau]$, which is weighted by the function $x[-\tau]$, which is shifted by an amount t. As t changes, the weighting function $x[t-\tau]$ emphasizes different components of the input weighting function $w[\tau]$ to extract the significant features determined by the filter matrices. $b[t]$ is the bias coefficient as the constant of a linear function.

Layer normalization layer: The main objective of layer normalization is to reduce the internal covariate shift that occurs during training, which helps stabilize the distribution of the inputs to a layer. Layer normalization works by normalizing the values within a layer, making their mean close to zero and their variance close to one. This normalization can speed up training and allow for the use of larger learning rates. Additionally, it helps with better gradient flow during backpropagation, which can lead to faster convergence and more stable training.

Dropout layer: It is used to prevent overfitting in neural networks. During training, a dropout layer randomly disables a fraction of the neurons or units in the layer by setting them to zero for that particular pass. By introducing noise and preventing neurons from relying too heavily on any particular input, dropout promotes learning more robust and generalized features. This regularization technique helps prevent overfitting and improves the generalization of neural networks, making them more capable of performing well on unseen data.

The proposed CNN consists of an input layer for inputting sequence data such as the parameters shown in Table 1, hidden layers, and an output layer as the fully connected layer to calculate a weighted sum of its inputs and is used to make predictions q_s. In any feed-forward neural network, the middle layers are referred to as hidden layers because their inputs and outputs are masked by the activation function and final convolution. To make a deeper CNN model, a bottleneck residual block [19] involving convolution layers, layer normalization, and a dropout layer are applied to mitigate degradation (accuracy saturation) and the overfitting problem [20] caused by combining many hidden layers. It may also accelerate learning for our proposed CNN. For the architecture of the overall connection model with the aforementioned network representations, the bottleneck residual block adds a route for simple addition into each hidden layer connection as shown in Figure 2.

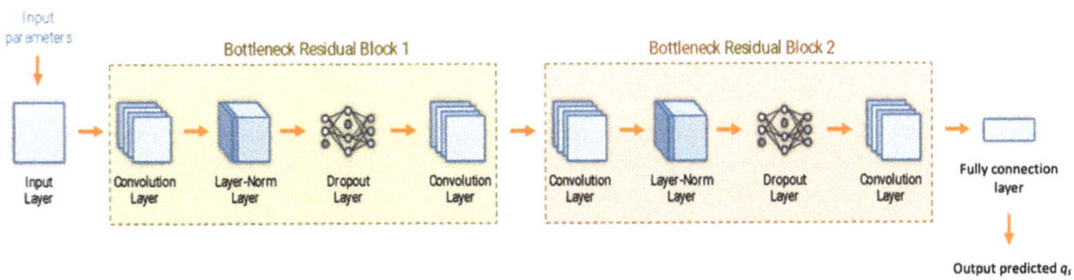

Figure 2. Architecture of proposed CNN with the combination of bottleneck residual blocks including convolutional, layer normalization, and dropout layers in serial for learning relevant feature values from inputs for output of wave overtopping discharge q_s.

3.2. CNN Training with Hyperparameter Optimization

To obtain an optimal model, it is common practice to limit the number of learning iterations or set a stopping threshold for the objective loss function in order to prevent overfitting. In this study, we employ the hyperparameter optimal tuning process (HOTP) [21,22] to minimize the objective loss function and mitigate overfitting [20]. The objective loss function combines the mean square loss function [23] and the Wasserstein distances W [24] between two measures $P_{\hat{q}}$ and P_q, which are the cumulative distribution functions of the model predictions $\hat{q}$ and the observed values q, respectively. The objective loss function L can be mathematically defined as follows:

$$L = \lambda_1 \frac{1}{I} \sum_{i=1}^{I} \sqrt{(\hat{q}_i - q_i)^2} + \lambda_2 W_s(P_{\hat{q}}, P_q) \tag{4}$$

In Equation (4), I is the subsample number to be randomly extracted from the training set. Each summand L is associated with the n-th observation in the subsample during training. Here, λ_1 and λ_2 are constraints on the sensitivity of the squared error loss and Wasserstein distances $W_s(P_{\hat{q}}, P_q)$. To train the CNN, the adaptive movement estimation algorithm, a backward propagation-based iterative method developed by Adam [25], is employed for optimizing the objective function with desirable smoothness properties. The learning rate is one of the key hyperparameters when training a deep learning neural network model. It is determined through HOTP to find the value that results in the best model performance.

Figure 3 provides an overview of the hyperparameter optimization (HOTP) process utilized to develop an optimal CNN model. The primary objective of HOTP is to identify the hyperparameters that yield the best performance of the CNN model, while avoiding overfitting. It is important to emphasize that the validation dataset used during HOTP is distinct from the testing dataset and is not part of the training process.

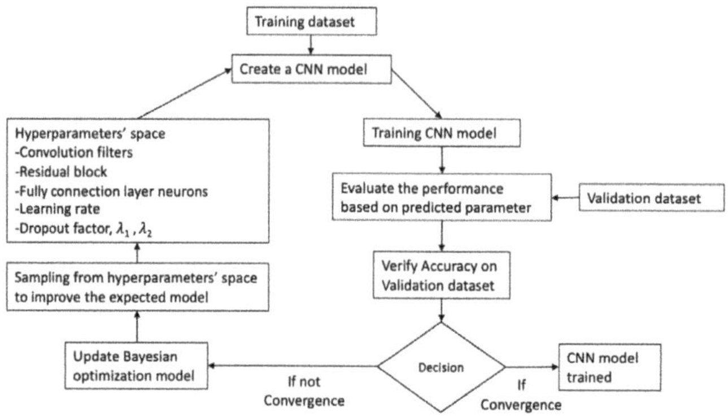

Figure 3. Schematic of hyperparameter optimal tuning process (HOTP) to create the CNN model.

In each iteration of HOTP, a new set of hyperparameters is evaluated by computing the associated loss using the validation dataset. The objective function aims to minimize this loss, enabling the identification of optimal hyperparameters that lead to a model with the lowest possible loss. Bayesian optimization [26] is employed in this study to conduct an exhaustive search of the hyperparameter space, focusing on a predetermined subset of hyperparameters. The optimization process involves exploring different combinations of hyperparameters based on previous observations, with the goal of refining the model's performance.

Once the CNN model is trained using HOTP and the optimal hyperparameters are determined, it can be utilized to make predictions for overtopping values. This comprehensive approach ensures that the CNN model is fine-tuned to achieve optimal performance and can provide accurate predictions for the given task.

To configure hyperparameters for a CNN, first we define the network's architectural aspects according to Figure 2 and initialize the corresponding hyperparameters randomly, such as the number of layers, filters, and size of neurons. Secondly, set general hyperparameters such as the neurons of convolution layers and fully connected layers, number of residual blocks, learning rate, batch size, dropout rate, and weighting coefficients of the loss function to fine-tune the training process. After establishing these foundational settings, fine-tune the hyperparameters with HOPT on a validation set. Table 2 displays a hypothetical interval of the hyperparameter space used during the HOTP process. The aim of HOTP is to identify the optimal values for these hyperparameters, which are crucial for the performance of the CNN model. Within Table 2, the optimal values of the hyperparameters are denoted by a star (*) symbol, indicating the configurations that yield the best performance for the CNN model.

Table 2. Setup of hypothetical interval of hyperparameters for HOTP.

Hyperparameter Name	Values	Definition of the Hyperparameter
Convolution_filters	[16, 24, 32, 64, 128, 256 *]	Number of filters w in each convolution layer
Residual_block	[1, 2 *]	Number of residual block
Fully_connection_layer_neurons	[32, 64, 128 *, 256, 512]	Neurons in a fully connected layer
Learning_rate	[0.0001 *, 0.0002, . . . , 0.1]	Given range of learning rate for CNN training
Dropout_rate	[0.1, 0.2, . . .0.7 *, 0.8, 0.9],	Determines the probability of a neuron being dropped out
λ_1	[0.5, 0.6, . . . , 1.0 *]	Given weights of RMSE term of L
λ_2	[0.1, 0.2,. . . , 0.5 *, . . . , 1.0]	Given weights for constraint term of L

4. Results and Discussion

4.1. Training Setup

For the CNN model in this study, the training process was carried out with specific configurations. In the context of neural network training, an epoch refers to a complete iteration over the entire training dataset. During an epoch, the model goes through multiple iterations, each involving a batch of training samples. The number of iterations per epoch is determined by the batch size. In this study, each batch size was limited to a maximum size of 256, while the remaining samples constituted the minimum quantity for the last batch size. The training was performed with a maximum of 5000 epochs, and the early stopping algorithm was employed to prevent overfitting. This algorithm continuously monitored the validation performance and stored the model with the best results when a predefined global training accuracy was achieved.

The training procedure took place on a computer equipped with a 3.4 GHz Intel Core i7-6700 CPU and two NVIDIA GeForce GTX 1080 Ti GPUs. The implementation of the CNN model was conducted using the Python programming language in conjunction with the TensorFlow library [27]. This setup ensured efficient training and allowed for the attainment of optimal outcomes.

In this study, the entire core dataset of the EurOtop overtopping database was utilized, following the approach outlined in [12]. The database was randomly divided into two portions: 80% for training the CNN model and 20% for validation. Similar to previous machine learning models [4,11,12,16], we employed the bootstrap resampling technique for the training of the CNN model. The bootstrap resampling technique involves creating N resamples from the original database. Each resample forms a distinct training and validation dataset. Subsequently, the CNN model is trained using each resample dataset, and the predicted overtopping discharge is obtained by calculating the ensemble mean of the predictions. This technique allows for the evaluation of the model's performance under various training conditions and provides a robust estimation of its predictive capabilities.

When dealing with a substantial volume of data, increasing the number of iterations in bootstrap resampling typically results in more precise estimates derived from these resamples [28]. However, in the context of the overtopping database, which contains relatively modest-sized samples compared to fields with datasets comprising hundreds of thousands of data points or more, excessively increasing the number of bootstrap resampling iterations may not yield substantial improvements in predicting future outcomes. Moreover, this practice can potentially create a deceptive sense of certainty if each model within the ensemble exhibits notable levels of uncertainty.

In cases where a model already displays minimal variance in its prediction outcomes, reducing uncertainty can be accomplished by examining only a limited number of bootstrap samples. This approach not only aids in reducing training time but has also proven to be effective. Therefore, it becomes crucial to determine the appropriate number of bootstrap resamples required to construct a CNN model ensemble. In this study, we evaluate the model's performance across various numbers of bootstrap resamples, ranging from $N = 1$ to $N = 500$. It is noteworthy that the scenario with $N = 500$ aligns with the same number of bootstrap resamples used in [4].

For the assessment of predicted quality, we employ the weighted root-mean-square error (RMSE) as the error criterion, as defined in [12]. The weighted RMSE is calculated using the following formula:

$$RMSE = \sqrt{\frac{1}{\sum_k^K WF_k} \sum_{k=1}^{K} WF_k \cdot [log_{10}(\hat{q}_k) - log_{10}(q_k)]^2} \quad (5)$$

where K is the observation number, and q_k and $\hat{q}_k$ are the observed and predicted values, respectively. This error criterion takes into account the weights assigned to different variables or data points, allowing for a more comprehensive evaluation of the prediction accuracy. By considering the weighted RMSE, we can assess the performance of the model in a more nuanced and representative manner.

In Figure 4, the variation of the weighted root-mean-square error (RMSE) is depicted from the ensemble mean with bootstrap resample numbers ranging from $N = 1$ to $N = 500$, along with the corresponding running time. It can be observed that as the bootstrap resample number increases, the RMSE decreases. The RMSE value converges at $N = 10$, as shown in Figure 4. Table 3 compares the RMSE values for $N = 1$, 10, and 500, indicating that satisfactory accuracy can be achieved using $N = 10$. It is worth noting that the computational time is significantly different between $N = 10$ (44 seconds) and $N = 500$ (2155 seconds). This finding suggests that employing $N = 10$ bootstrap resamples balances accuracy and efficiency for the predictions. Figure 5 displays the predictions made by the trained CNN model using $N = 10$ for both the training and validation datasets. The proposed CNN model incorporated a strategy to mitigate overfitting, which was validated by the strong performance observed in Figure 5, where the training and validation results exhibited good agreement.

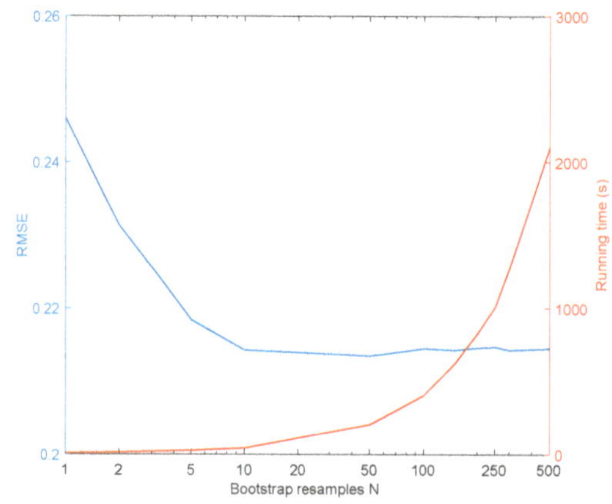

Figure 4. Comparisons of the RMSE and running time using different bootstrap resamples.

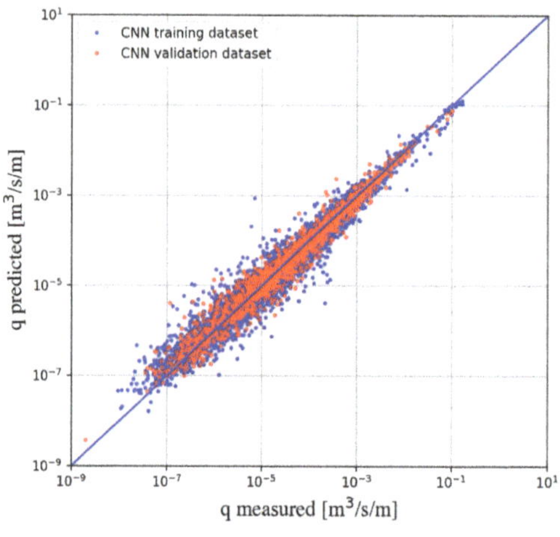

Figure 5. Predictions by the trained CNN for the training and the validation dataset with 10 bootstrap resamples.

Table 3. The comparison of predicted RMSEs for the training and validation datasets using different numbers of bootstrap resamples.

Dataset (Size)	Bootstrap Resamples		
	$N = 1$	$N = 10$	$N = 500$
Training data (6922)	0.238	0.230	0.225
Validation data (1731)	0.246	0.214	0.215

4.2. Verification on the Overtopping Database

Following the precedent of earlier ANN models [4,11,14], the trained CNN model is assessed for accuracy using the entire core dataset from the overtopping database. The trained CNN model can be accessed online at https://github.com/fcuyttsai/overtoppingEve.git/ (accessed on 1 September 2023). The predictions of existing ANNs can be directly estimated from their respective websites. These existing ANN models are the NN model [12] (online at https://www.deltares.nl/en/software/overtopping-neural-network/ (accessed on 1 May 2023)), the NNb model [11] (online at http://overtopping.ing.unibo.it/overtopping/ (accessed on 1 May 2023)), and the XGB model [12] (online at https://www.deltares.nl/en/software/overtopping-xgb/ (accessed on 1 May 2023)). In comparison to the deep learning CNN model, the NN, NNb, and XGB models fall within the realm of machine learning techniques. The NN model [4] employs the conventional back-propagation neural network as its learning method. Meanwhile, the NNb model [11], which utilizes an extended CLASH database as its training dataset, builds upon the fundamental NN model concept but integrates a quantitative classifier and three quantifiers. On the other hand, the XGB model [14], as an alternative to both the NN and NNb models, leverages an advanced machine learning technique known as XGBoost to predict overtopping discharge.

In Table 4, we present a comparison of the resulting weighted root-mean-square errors (RMSEs) for wave overtopping discharge predictions, along with the Pearson correlation coefficient (CC) and R^2 as measures of agreement. The Pearson correlation coefficient (CC) is given by:

$$CC = \frac{\sum_{i=1}^{n} (x_i - \overline{x})(y_i - \overline{y})}{(n-1)s_x s_y} \tag{6}$$

where x_i is the i-th value of observation, y_i is the i-th value of prediction, $\overline{x}$ is the mean value of observation, $\overline{y}$ is the mean value of prediction, s_x is the standard deviation of the observation, and s_y is the standard deviation of the prediction. R^2 is the square value of CC that is the coefficient of determination.

Table 4. The comparison of predicted RMSEs of different models.

Dataset (Size)	Model	RMSE	CC	R^2
Entire core data (8653)	NN	0.654	0.869	0.756
	NNb	0.639	0.858	0.736
	XGB	0.199	0.990	0.981
	CNN	0.112	0.996	0.991
A: rock permeable straight slopes (1131)	NN	0.702	0.768	0.590
	NNb	0.687	0.791	0.625
	XGB	0.247	0.977	0.955
	CNN	0.153	0.989	0.978
B: rock impermeable straight slopes (104)	NN	0.624	0.864	0.747
	NNb	0.823	0.710	0.504
	XGB	0.349	0.984	0.968
	CNN	0.136	0.994	0.989

Table 4. *Cont.*

Dataset (Size)	Model	RMSE	CC	R^2
C: armor units with straight slopes (934)	NN	0.506	0.872	0.760
	NNb	0.583	0.858	0.737
	XGB	0.219	0.983	0.966
	CNN	0.118	0.991	0.983
D: smooth and straight slopes (2069)	NN	0.480	0.972	0.946
	NNb	0.380	0.943	0.889
	XGB	0.120	0.996	0.991
	CNN	0.086	0.997	0.994
E: structures with combined slopes and berms (1631)	NN	0.784	0.838	0.703
	NNb	0.670	0.877	0.770
	XGB	0.204	0.987	0.975
	CNN	0.093	0.997	0.993
F: vertical walls (1677)	NN	0.685	0.863	0.745
	NNb	0.704	0.845	0.714
	XGB	0.206	0.988	0.977
	CNN	0.110	0.994	0.988
G: oblique wave attacks (1107)	NN	0.760	0.777	0.603
	NNb	0.872	0.772	0.595
	XGB	0.228	0.983	0.965
	CNN	0.128	0.989	0.978

As a result of the comparison, the RMSE value for the entire core dataset is 0.112 for the CNN model and 0.199 for the trained XGB. These values are significantly smaller than the RMSE values obtained from the trained NN (0.654) and NNb (0.639) models. Additionally, the CC and R^2 values further demonstrate that both the present deep learning model (CNN) and advanced machine learning model (XGB) outperform the conventional machine learning models (NN and NNb). The corresponding quantile–quantile plot of wave overtopping discharge predictions for each model is displayed in Figure 6.

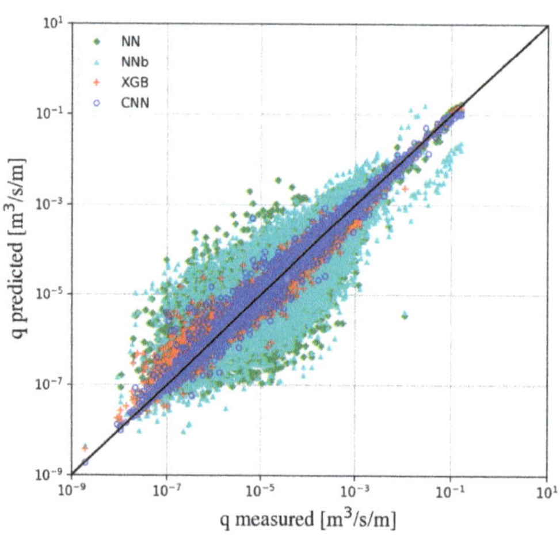

Figure 6. Comparison of predictions for the wave overtopping database predicted by the different models with the entire core dataset.

To demonstrate the adaptability of the current CNN model to different scenarios, we directly extract the results of Figure 6 for the groups A–G of the EurOtop database, based on structure type and oblique wave attack conditions. The predictions of each model are plotted in Figure 7a–g. The scatterplots in Figure 7a–g show a high correlation between the predictions and observations for each group using the CNN model. The results indicate that all models perform relatively well for group D, which represents the situation of coastal structures with smooth and straight slopes. The CNN model also performs better for group B data, despite it containing only a few observations compared to other categories. Regarding the oblique wave attack prediction (group G), it is noted by [12] that the low accuracy of the NN models may be due to the limited amount of training data, leading to an improvement in the XGB model by incorporating new oblique wave cases into the training data. However, as shown in Table 4, the group G dataset comprises 1107 cases, which is larger than one-eighth of the total dataset. It can be observed that the current CNN model achieves highly accurate oblique wave attack (group G) predictions without the need for additional datasets.

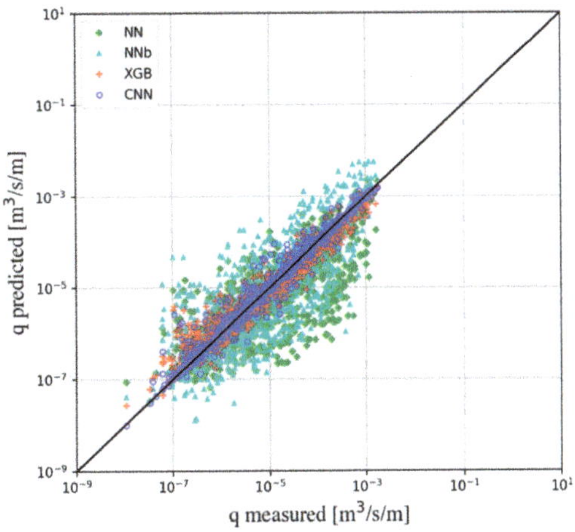

(**a**) Rock permeable straight slopes

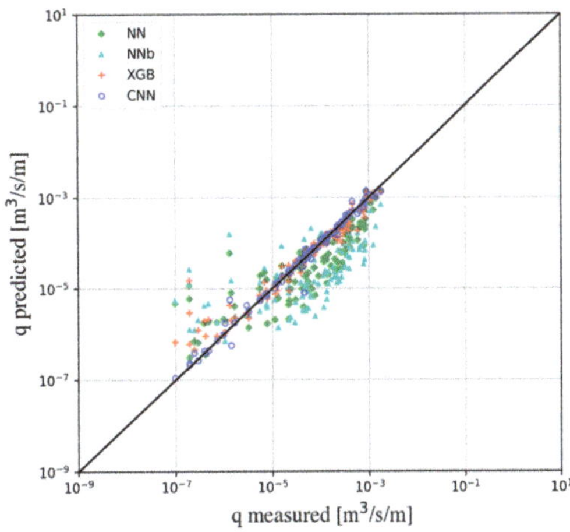

(**b**) Rock impermeable straight slopes

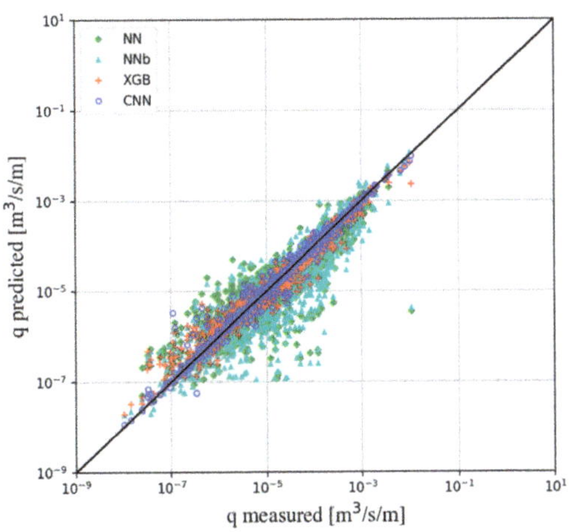

(**c**) Armor units with straight slopes

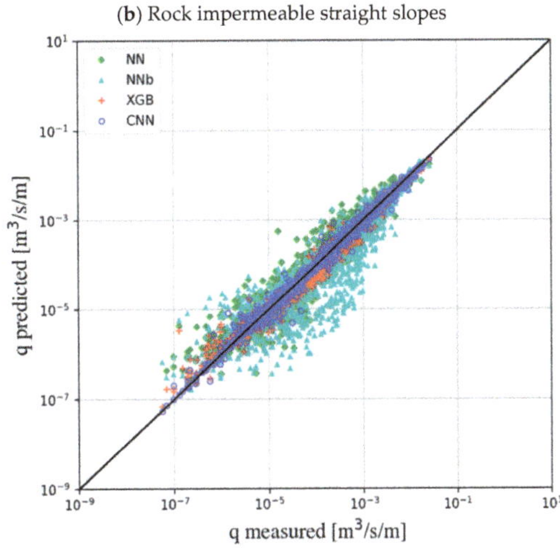

(**d**) Smooth and straight slopes

Figure 7. *Cont.*

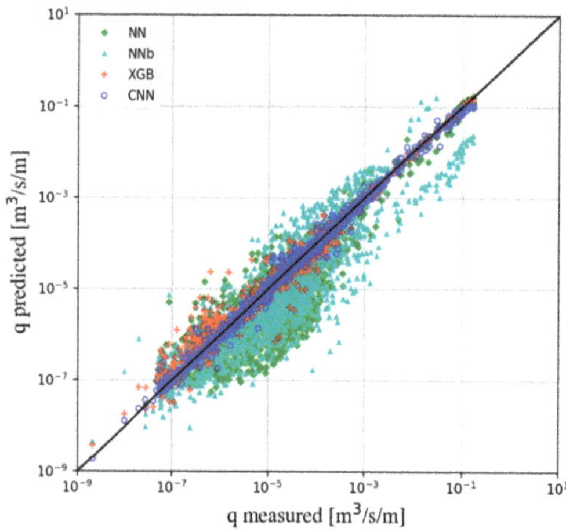

(**e**) Structures with combined slopes and berms

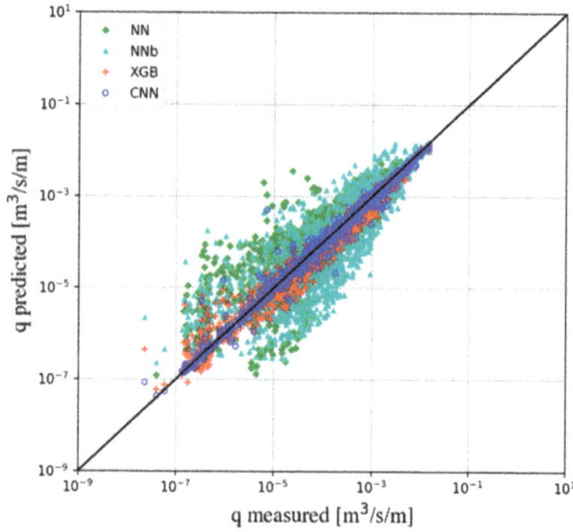

(**f**) Vertical walls

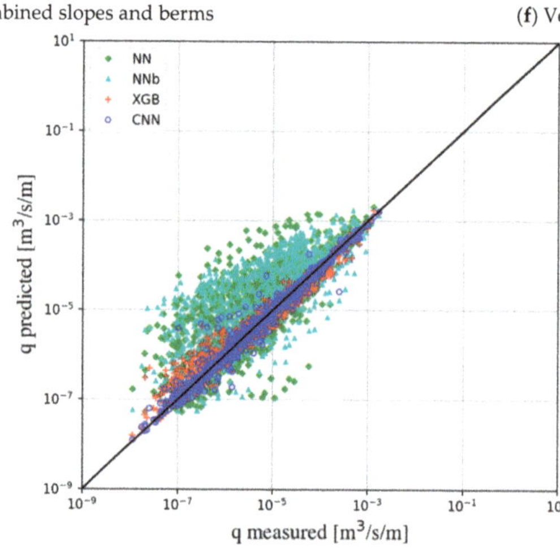

(**g**) Oblique wave attacks

Figure 7. Comparisons of predicted wave overtopping discharges of different models for various coastal structures and oblique wave attack conditions with the entire core data.

4.3. Testing/Prediction Using Prototype Overtopping Dataset

Designers expect prediction models to be applied to prototype scenarios. The testing performance of the trained CNN model is evaluated using the prototype overtopping dataset, which was not seen during the model's training and validation processes. The EurOtop database provides 150 wave overtopping measurement records from prototype situations. These records include 11 measurements labeled as C-59.1 to C-59.11 [28] from 9 storms between 1999 and 2004 at Zeebrugge breakwater, 77 measurements labeled as G-18.1 to G-18.77 [29] from 7 storms between 2003 and 2004 at Ostia breakwater, and 62 measurements labeled as F-18.1 to F-18.39 and G-12.1 to G-12.23 [30] from other sources.

The testing results of the trained CNN model are depicted in Figure 8, where the predictions of the NN, NNb, and XGB models are estimated directly from their respective websites using the same data inputs. It is important to note that all these models utilized the entire core dataset (i.e., laboratory measured data) during the training and validation processes, with the prototype dataset remaining unseen. As illustrated in Figure 8 and Table 5, the CNN model exhibits better performance in testing when applied to the prototype dataset. It is worth noting that all models achieve high correlation coefficient (CC) values. However, due to the larger scale of the prototype, the RMSE values reported in Table 5 are understandably larger than those observed in the laboratory-scale data presented in Table 4. Furthermore, real-world prototype scenarios can introduce unanticipated variables such as noise, surface irregularities, wind, spray, currents, and the scale effect itself. These sources of uncertainty can contribute to more pronounced prediction errors.

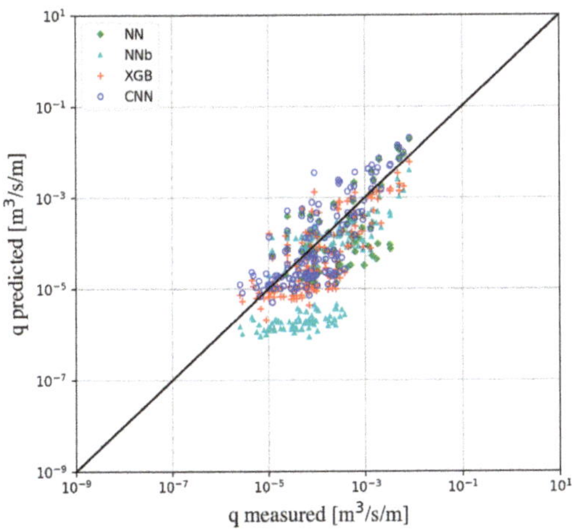

Figure 8. Comparison of predictions for the prototype wave overtopping discharge for different models.

Table 5. The comparison of predicted RMSEs of different models for the prototype dataset.

Dataset (Size)	Model	RMSE	CC	R^2
All testing data (150)	NN*	0.936	0.874	0.764
	NNb	1.218	0.895	0.802
	XGB	0.657	0.868	0.753
	CNN	0.555	0.932	0.868

* For the NN model, it is out of range for the estimation of G-18.1 to G-18.77. This result only calculates 23 cases of G-12.1 to G-12.23 for the NN model.

The prototype overtopping database can be divided into three categories: armor units with straight slopes (C-59.1 to C-59.11), vertical walls (F-18.1 to F-18.39), and oblique wave attacks (G-12.1 to G-12.23 and G-18.1 to G-18.77). Figure 9 presents the testing results for these three types of datasets, extracted directly from the results of Figure 9. As mentioned in [11], their model (i.e., NNb model) underestimated the predicted values of prototype overtopping for the Ostia breakwater (i.e., G-18.1 to G-18.77), while the predictions for Zeebrugge (i.e., C-59.1 to C-59.11) fell within the confidence bands. Figure 9a–c show similar tendencies observed in [11] by the NNb model.

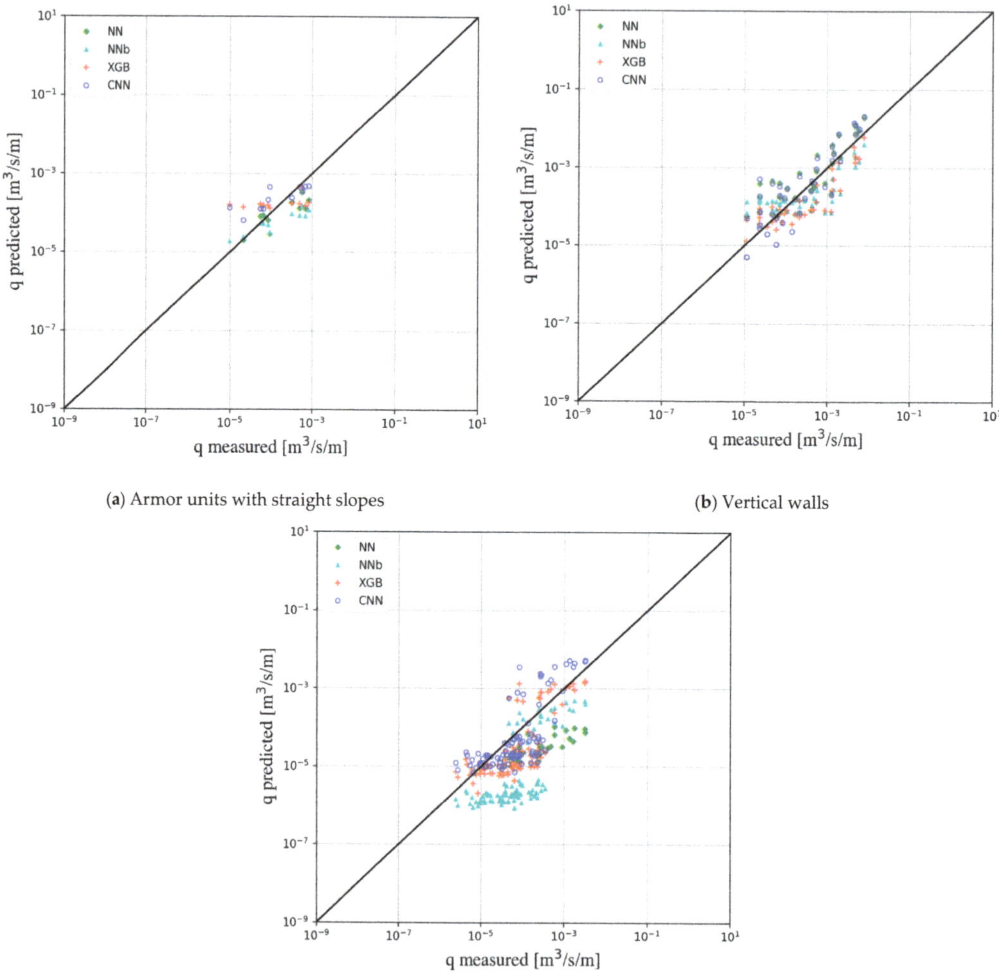

(a) Armor units with straight slopes

(b) Vertical walls

(c) Oblique wave attacks

Figure 9. Comparisons of predicted wave overtopping discharges of different models for various coastal structures.

4.4. Application of CNN for Prototype Various Crest Freeboards

In an example scenario, each model was applied to predict the prototype wave overtopping discharge, considering various crest freeboards (R_c) of coastal structures. The predictions were based on the observation labeled as G-12.22 in the prototype situations of the EurOtop database. The input conditions are presented in Table 6. The predictions started with R_c = 3.34 m and gradually increased to 7.50 m to observe the trend of the wave overtopping discharge, with a measured observation available at R_c = 7.36 m.

Table 6. Conditions for estimating overtopping discharge in example scenario.

Parameter	Value	Parameter	Value	Parameter	Value
$H_{mo,t}$ [m]	1.67	R_c [m]	3.34 to 7.50	D_d [m]	0.60
h_b [m]	1.03	A_c [m]	3.34	D_u [m]	0.00
h_t [m]	3.28	G_c [m]	1.00	γ_{fd}	0.55
h [m]	3.28	B [m]	8.00	γ_{fu}	1.00
$L_{m-1,0,t}$ [m]	59.39	$cot\alpha_d$	1.00	β [$^\circ$]	14
B_t [m]	0.00	$cot\alpha_u$	0.00		

In Figure 10, the depicted results illustrate the predicted outcomes of relative wave overtopping discharge concerning the relative crest freeboard for four different models: the current CNN, XGB, NN, and NNb models. These results reveal a consistent trend across all predictions, where a lower relative crest freeboard corresponds to a larger relative wave overtopping discharge. Along the projected curve, the predictions generated by the current CNN model closely align with the observed measurements at R_c = 7.36 m, specifically at $R_c/H_{m0,t}$ = 4.41. In contrast, the predictions produced by the NN, NNb, and XGB models notably underestimate the relative wave overtopping discharge for various relative $R_c/H_{m0,t}$ values.

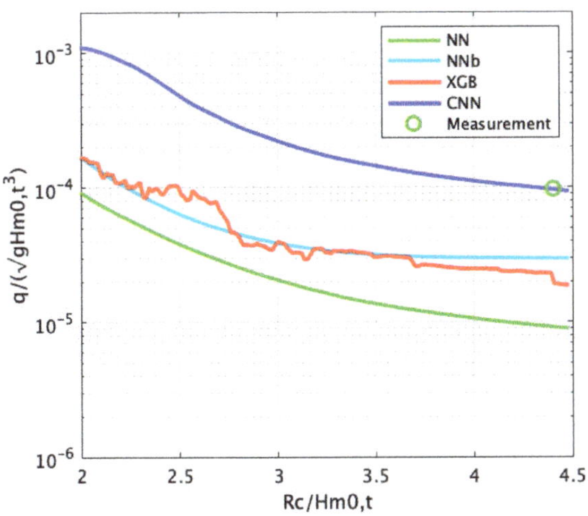

Figure 10. Comparisons of the predicted relative wave overtopping discharge versus relative crest freeboard using different models for prototype situations.

5. Conclusions

In this paper, a novel convolutional neural network (CNN) model with deep learning technologies is introduced for the accurate prediction of wave overtopping discharge at various types of coastal structures, including those with complex geometries and under a wide range of wave conditions. The proposed CNN model incorporates various hidden layers, such as bottleneck residual blocks, which effectively extract features and scale input sequences. For the enhancement of prediction accuracy, hyperparameters of the model are optimized through Bayesian optimization to mitigate overfitting, and the model is trained utilizing the bootstrap resampling technique. The results demonstrate that the proposed CNN model outperforms traditional machine learning models, as evident from comparisons of quantile–quantile plots and root-mean-square errors for estimations on the overtopping database. Notably, the CNN model exhibits remarkable agreement with observations across various structure types and oblique wave attack conditions. The model also demonstrates satisfactory performance for prototype overtopping prediction, highlighting its value as a valuable tool for coastal engineers involved in the design and maintenance of coastal structures.

Supplementary Materials: Our CNN model can be accessed online at https://github.com/fcuyttsai/overtoppingEve.git/ (accessed on 1 September 2023).

Author Contributions: Conceptualization, Y.-T.T. and C.-P.T.; Methodology, Y.-T.T. and C.-P.T.; Data curation, Y.-T.T. and C.-P.T.; Validation, Y.-T.T. and C.-P.T.; Formal analysis, Y.-T.T. and C.-P.T.; Visualization, Y.-T.T. and C.-P.T.; Software, Y.-T.T. and C.-P.T.; Writing—original draft, Y.-T.T. and C.-P.T.; Writing—review and editing, Y.-T.T. and C.-P.T. All authors have read and agreed to the published version of the manuscript.

Funding: This research was funded by the National Science and Technology Council of Taiwan for financial support under Grant Nos. MOST 109-2625-M-005-006-MY2, NSTC 111-2221-E-035-044, and MOST 109-2221-E-035-001-MY2.

Data Availability Statement: The authors do not have permission to share data.

Acknowledgments: The authors would like to thank the EurOtop team for providing the overtopping database.

Conflicts of Interest: The authors declare no conflict of interest.

References

1. Troch, P.; Geeraerts, J.; Van de Walle, B.; De Rouck, J.; Van Damme, L.; Allsop, W.; Franco, L. Full-scale wave-overtopping measurements on the Zeebrugge rubble mound breakwater. *Coast. Eng.* **2004**, *51*, 609–628. [CrossRef]
2. Briganti, R.; Bellotti, G.; Franco, L.; De Rouck, J.; Geeraerts, J. Field measurements of wave overtopping at the rubble mound breakwater of Rome–Ostia yacht harbour. *Coast. Eng.* **2005**, *52*, 1155–1174. [CrossRef]
3. Cáceres, I.; Sánchez-Arcilla, A.; Zanuttigh, B.; Lamberti, A.; Franco, L. Wave overtopping and induced currents at emergent low crested structures. *Coast. Eng.* **2005**, *52*, 931–947. [CrossRef]
4. van Gent, M.R.A.; van den Boogaard, H.F.P.; Pozueta, B.; Medina, J.R. Neural network modelling of wave overtopping at coastal structures. *Coast. Eng.* **2007**, *54*, 586–593. [CrossRef]
5. Losada, I.J.; Lara, J.L.; Guanche, R.; Gonzalez-Ondina, J.M. Numerical analysis of wave overtopping of rubble mound breakwaters. *Coast. Eng.* **2008**, *55*, 47–62. [CrossRef]
6. Reeve, D.E.; Soliman, A.; Lin, P.Z. Numerical study of combined overflow and wave overtopping over a smooth impermeable seawall. *Coast. Eng.* **2008**, *55*, 155–166. [CrossRef]
7. Verhaeghe, H.; De Rouck, J.; van der Meer, J. Combined classifier–quantifier model: A 2-phases neural model for prediction of wave overtopping at coastal structures. *Coast. Eng.* **2008**, *55*, 357–374. [CrossRef]
8. De Rouck, J.; Verhaeghe, H.; Geeraerts, J. Crest level assessment of coastal structures—General overview. *Coast. Eng.* **2009**, *56*, 99–107. [CrossRef]
9. EurOtop. *Manual on Wave Overtopping of Sea Defences and Related Structures. An Overtopping Manual Largely Based on European Research, but for Worldwide Application*; Van der Meer, J.W.; Allsop, N.W.H.; Bruce, T.; De Rouck, J.; Kortenhaus, A.; Pullen, T.; Schüttrumpf, H.; Troch, P.; Zanuttigh, B.: 2018. Available online: www.overtopping-manual.com (accessed on 1 November 2019).
10. van der Meer, J.W.; Verhaeghe, H.; Steendam, G.J. The new wave overtopping database for coastal structures. *Coast. Eng.* **2009**, *56*, 108–120. [CrossRef]
11. Zanuttigh, B.; Formentin, S.M.; van der Meer, J.W. Prediction of extreme and tolerable wave overtopping discharges through an advanced neural network. *Ocean Eng.* **2016**, *127*, 7–22. [CrossRef]
12. den Bieman, J.P.; van Gent, M.R.A.; van den Boogaard, H.F.P. Wave overtopping predictions using an advanced machine learning technique. *Coast. Eng.* **2021**, *166*, 103830. [CrossRef]
13. Heaton, J. Ian Goodfellow, Yoshua Bengio, Aaron Courville. Deep learning. *Genet. Program. Evolvable Mach.* **2018**, *19*, 305–307. [CrossRef]
14. Hochreiter, S.; Schmidhuber, J. Long short-term memory. *Neural. Comput.* **1997**, *9*, 1735–1780. [CrossRef]
15. Formentin, S.M.; Zanuttigh, B.; van der Meer, J.W. The New EurOtop Neural Network Tool for an Improved Prediction of Wave Overtopping. In *ICE Coasts, Marine Structures and Breakwaters*; ICE Publishing: Leeds, UK, 2017.
16. Formentin, S.M.; Zanuttigh, B.; van der Meer, J.W. A Neural Network Tool for Predicting Wave Reflection, Overtopping and Transmission. *Coast. Eng. J.* **2018**, *59*, 1750006-1–1750006-31. [CrossRef]
17. Ba, J.L.; Kiros, J.R.; Hinton, G.E. Layer Normalization. *arXiv* **2016**, arXiv:1909.11855.
18. Srivastava, N.; Hinton, G.; Krizhevsky, A.; Sutskever, I.; Salakhutdinov, R. Dropout: A Simple Way to Prevent Neural Networks from Overfitting. *J. Mach. Learn. Res.* **2014**, *15*, 1929–1958.

19. Wang, F.; Jiang, M.; Qian, C.; Yang, S.; Li, C.; Zhang, H.; Wang, X.; Tang, X. Residual attention network for image classification. In Proceedings of the IEEE Conference on Computer Vision and Pattern Recognition, Honolulu, HI, USA, 21–26 July 2017; pp. 3156–3164.
20. Ying, X. An Overview of Overfitting and its Solutions. *J. Phys. Conf. Ser.* **2019**, *1168*, 022022. [CrossRef]
21. Isabona, J.; Imoize, A.L.; Kim, Y. Machine Learning-Based Boosted Regression Ensemble Combined with Hyperparameter Tuning for Optimal Adaptive Learning. *Sensors* **2022**, *22*, 3776. [CrossRef]
22. Sóbester, A.; Leary, S.J.; Keane, A.J. On the Design of Optimization Strategies Based on Global Response Surface Approximation Models. *J. Glob. Optim.* **2005**, *33*, 31–59. [CrossRef]
23. Friedman, J.H. Greedy function approximation: A gradient boosting machine. *Ann. Stat.* **2001**, *29*, 1189–1232. [CrossRef]
24. Mueller, J.; Jaakkola, T. Principal Differences Analysis: Interpretable Characterization of Differences between Distributions. *arXiv* **2015**, arXiv:1510.08956.
25. Diederik, P.; Kingma, J.B. Adam: A Method for Stochastic Optimization. In Proceedings of the 3rd International Conference for Learning Representations, San Diego, CA, USA, 7–9 May 2015.
26. Scikit-Optimize. Sequential Model-Based Optimization in Python. Available online: https://scikit-optimize.github.io/ (accessed on 1 May 2023).
27. Abadi, M.; Barham, P.; Chen, J.; Chen, Z.; Davis, A.; Dean, J.; Devin, M.; Ghemawat, S.; Irving, G.; Isard, M. Tensorflow: A system for large-scale machine learning. In Proceedings of the 12th USENIX Symposium on Operating Systems Design and Implementation, Savannah, GA, USA, 2–4 November 2016; pp. 265–283.
28. Geeraerts, J.; Boone, C. *Report on Full Scale Measurements Zeebrugge, 2nd Full Winter Season*; Ghent University: Gent, Belgium, 2004.
29. Franco, L.; Briganti, R.; Bellotti, G. *Report on Full Scale Measurements, Ostia, 2nd Full Winter Season*; Modimar: Rome, Italy, 2004.
30. Pullen, T. *Final Report on Laboratory Measurements, Samphire Hoe*; HR Wallingford: Wallingford, UK, 2004.

Journal of
Marine Science and Engineering

MDPI

Article

Hydraulic Response and Overtopping Performance of Single-Layer Double Cube Unit Armored Mound Breakwater

Iman Safari [1,*], Dominique Mouazé [2], Soroush Aliasgary [3], Guillaume Carpentier [1] and François Ropert [4]

1 Builders Ecole D'Ingenieurs, 1 Rue Pierre et Marie Curie, 14610 Epron, France; guillaume.carpentier@builders-ingenieurs.fr
2 Normandie University, UNICAEN, UNIROUEN, CNRS, M2C, 14000 Caen, France; dominique.mouaze@unicaen.fr
3 School of Civil Engineering, College of Engineering, University of Tehran, 16th Azar St., Enghelab Sq., Tehran 1417466191, Iran; soroush.asgary@ut.ac.ir
4 Cerema Eau Mer Fleuves, 60280 Margny-Lès-Compiègne, France; francois.ropert@cerema.fr
* Correspondence: iman.safari@builders-ingenieurs.fr

Abstract: A newly developed friction-interlocking armor unit called the 'Double cube' (DC) was designed to improve the performance of a concrete armor unit used in river/marine structures against currents/waves actions. The DC unit is an octagonal-shaped block made up of two parts: an upper cube set on a lower base that is either square or octagonal in shape. The innovative design aims to provide good performance in terms of stability, to allow for high tolerance placement with various contact points, and to allow for ease of placement. The DC's shape and placement enhance the unit's stability by bringing the center of gravity closer to the underlayer and by providing a large contact surface with the surrounding blocks that reduces the chance of extraction and limits movements (rocking, lifting) via the "keystone" effect. The characteristics of this new unit provide a relatively high hydraulic stability number for the armor layer (N_S = 2.9), a favorable hydraulic performance due to energy dissipation from turbulence and aeration, as well as a high roughness coefficient ($\gamma_f \approx 0.46$), helping to reduce overtopping.

Keywords: armor unit; hydraulic stability; hydraulic performance; turbulence; roughness; marine structures

1. Introduction

Among all the various parts composing a rubble mound breakwater, the armor layer represents the key component of the structure's stability. It is composed either of quarry stones or of artificial concrete armor units. The latter can be efficient alternatives to replace the rock armor materials when one cannot provide the appropriate natural rocks in sufficient quantity or size to ensure stability (CIRIA [1]).

In the context of climate change and significant sea level rise, coastal engineers and scientists are still conducting extensive research on developing new types of artificial armor units to improve performance, reducing the amount of concrete used, and limit the visual and environmental impacts.

Key considerations for an artificial unit are:

- Hydraulic stability.
- Good performance in terms of run-up and overtopping.
- Structural robustness.
- Ease and speed of placement with sufficient tolerances, ease of handling and simplicity of lifting.
- Easy prefabrication with a simple mold composed of limited parts.
- Ease of storage.

In recent decades, there has been a resurgence of interest in structurally robust forms: Accropode®I, Coreloc®, Xbloc® (CIRIA [1]), and Crablock (Salauddin et al. [2,3]), and particularly for simpler shapes, Cubipod® (Gómez-Martín and Medina [4,5]) and C-ROC (Perrin et al. [6]). It is also worth mentioning that during recent years, many new regular-placed armor blocks have been introduced in the literature, such as Xbloc® Plus (Reedijk et al. [7]), Chi (Park et al. [8]), Starblock (Safari et al. [9–12]), and TB-CUBE (Peng et al. [13]).

The stability of concrete armor units against waves is achieved through the units' own weight, friction, and interlocking (CERC [14], CIRIA [1]). Setting aside the hollowed units, which are placed in a tight regular pattern side by side and rather confined to revetments, the best performing units in terms of hydraulic stability are those that mobilize strong interlocking, for example, the Dolos. However, since the failure of the Sines breakwater, engineers have been hesitant to use slender units, which are thought to be structurally fragile (Bakker et al. [15,16]). Bulky units such as Accropode®I appear to be more preferred by coastal structure designers. For the same reason—the significance of the structural integrity of armor blocks—some of the most basic armor blocks, such as cubes, are frequently employed in the design of breakwaters. Concrete armor blocks may be placed in one or two layers according to a random or a regular pattern (Van Gent and Luis [17]).

Nowadays, the most commonly used armor blocks are double-layer cube-shaped blocks and interlocking-type armor blocks. Interlocking type armor blocks can be placed in single-layer or double-layer placements. Single-layer interlocking armor blocks are particularly interesting for economic reasons, due to their cost efficiency (CIRIA [1]).

The objective of this group was primarily to propose optimized shapes leading to high strength and high stability, and secondarily to reduce overtopping. These characteristics led to the development of high-interlocking units with more empty spaces between units, leading to an increased porosity of the armor layer. The major drawback regarding the placement of these types of units is that the placement can strongly affect the hydraulic stability of the armor layer (Jensen [18]). The random placement method, used for most single-layer blocks, is constrained by several rules governing the blocks' orientation, as well as their positioning or number of contacts. The latter affects the placement process and thus, the cost of the entire project (CIRIA [1]).

This is one of reasons that authors such as Bhageloe [19], d'Angremond et al. [20], Van Gent et al. [21–23], Van Buchem [24], Van Gent and Luis [17], Van der Lem et al. [25], Van Gent and van der Werf [26], and Vieira et al. [27–29] proposed that single-layer-placed cubes could be an economically interesting alternative solution for the construction of the armor layer.

However, simple single-layer cubes suffer from a huge disadvantage, as they tend to fit face-to-face during storms. (Medina et al. [30]). In such cases, the placement tends to form a heterogeneous porous armor layer with areas of high cube concentration to the detriment of other parts with less protective cover (Figure 1).

Low porosity cubes arranged closely together are likely to experience excessive pore pressure inside the breakwater, which is unfavorable for stability. Consequently, a damaged armor layer cannot not provide sufficient cover for exposed underlayers; therefore, such an arrangement would increase the risk of material extraction. A single armor layer of cubes, if not designed with a very low porosity, would likely experience significant settlement. Moreover, the edges and sharp corners of the cubes would be subjected to localized breakages, not only during placement but also throughout the life of the structure (Safari [9]).

Therefore, even if the hydraulic stability performance is acceptable, the outer surface of the armor would be rather smooth, as the units would be parallel to the breakwater's slope, thus enhancing the overtopping rates.

Figure 1. Disarrangement of cube placement (Safari [9]).

In general, cubes placed in a single layer exhibit a stability number (N_s) between 2 and 3, depending on the packing density of the armor layer (Van Gent et al. [21–23]). Regarding overtopping, the roughness coefficient is close to 0.5 for the single-layer cube armor layer, whereas it decreases to 0.47 for cubes in two layers, resulting in an unexpectedly small gap (CIRIA [1]).

The Cubipod® recently developed by the PCUPV laboratory (Ports and Coasts of the University of Valencia Polytecnic) also illustrates the strong interest in the simple cube shape. This unit is based on a cube shape with a protrusion (half pyramid) located on each face. In order to avoid face-to-face contact, the cubes are moved away from one another; protrusions are used. This design is intended to obtain a greater, more uniform porosity. (Gómez-Martín and Medina [4,5]). Nevertheless, this unit does not provide higher stability in a single layer compared to other existing irregularly placed armor units. (Salauddin et al. [2]).

This demonstrates that a more effective concrete armor can be proposed with a simple shape, while still assuring an easy placement method and construction. From a general point of view, stability is closely related to interlocking. In contrast, a strong interlocking results in a low surface porosity and permeability, leading to unsatisfactory hydraulic responses (Safari et al. [11]). This contradiction between stability and overtopping provides an incentive to look for innovative unit forms able to reach an optimal compromise.

The main purpose of this research is to develop a new artificial unit, namely the 'Double Cube' or DC, for breakwater protection against wave action. DC is comprised of a cubic base with chamfered corners and an upper half with a cubic shape that has a smaller cross section than the base part. The basic design idea consists in developing a bulky armor unit (Dupray and Roberts [31]) that fulfills the following main criteria:

- Able to be placed in a single layer on a 3V: 4H slope.
- A homogeneous porosity.
- Easy to place; it might be an advantage that the unit could be placed on a filter made of rather small size elements to avoid surface irregularities.
- High hydraulic stability (expected N_S (K_D) $\approx$ 2.9 (18)).
- High rough surface armoring ($\gamma_f \approx 0.46$).

All physical modeling tests were performed in 2D wave flume, the units being placed on a uniform slope and subjected to irregular waves. Section 2 presents the DC's shape and the logic of the geometry unit. Section 3 focuses on two proposed placements for DC units,

'Direct placement' and 'Random placement'. Section 4 presents physical model tests and explains the armor damage analysis and overtopping measurement used in this research. Sections 5 and 6 describe the 2D hydraulic stability tests of DC models and provide a comparison with the other single armor units. Finally, the most relevant conclusions of this research are provided in Section 7.

2. Logic of the New Designed Geometry

It is well known that the stability of an armor layer increases with increasing packing density and decreases with decreasing porosity (Medina and Gómez-Martín [32]). It has also been proven that a decrease in porosity leads to an unfavorable increase in overtopping (Safari et al. [10]). The DC's goal is to take advantage of these functions by combining two parts into one unit.

The DC consists of a cubic base with chamfered corners (hexagonal form), topped by a cubic form with a reduced cross-sectional area compared to the base part (Figure 2). The upper part and lower part are connected with a short transition part to reduce the structural tensions.

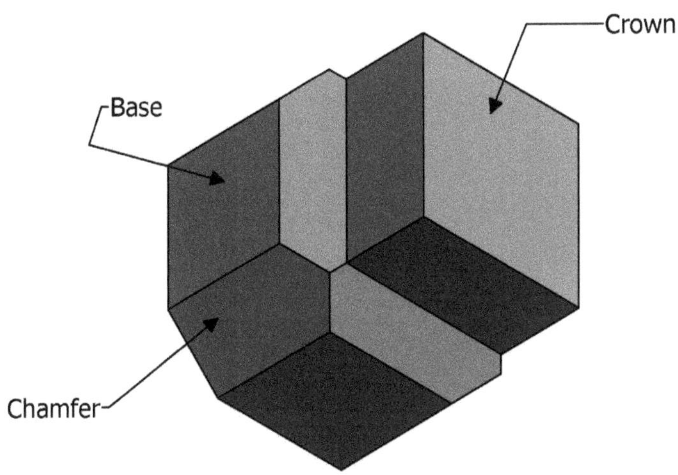

Figure 2. Plan view of the Double Cube unit.

The rough pattern created by the high porosity of upper part of the block increases the dissipation of wave energy. The upper part, which can be of different shapes such as square or octagonal, creates a rough outer layer that is able to dissipate the energy of the flows close to the armor layer. The upper part also offers a grip for clamps and cranes due to the large gaps around the upper part of units.

The lower part of the new unit contributes the most to the block's weight as well as frictional forces, therefore ensuring the stability of the block. The unit's base has been chamfered to ensure a minimum porosity, even with a tight placement or different orientations of the units. The transition zone eliminates the sharp corners that are most exposed to contact. Chamfers also provide better quality supports and contact surfaces for the blocks.

The chamfers also offer the possibility for a number of different regular placements, with a homogeneous porosity over the entire surface and without the risk of uneven settlement of the armor layer (surface porosity).

The dimensions of chamfered DCs were standardized on the following ratios based on the characteristic dimension "D", the side length of the unit in which the element fits (Figure 3):

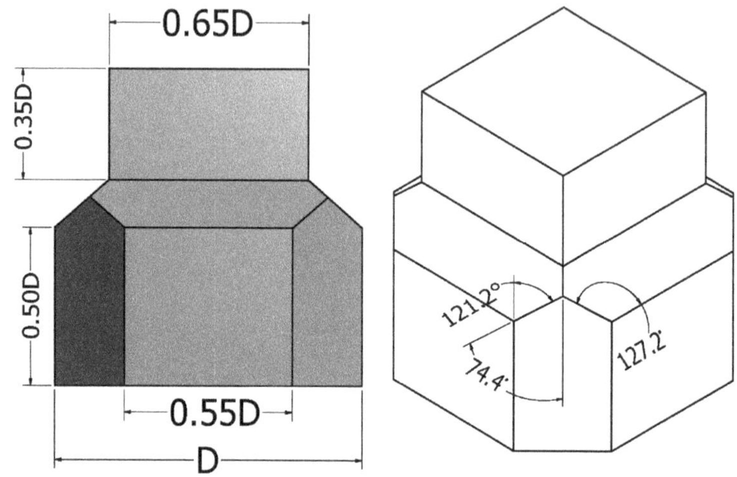

Figure 3. DC unit geometry.

The volume of an individual unit, as a function of characteristic dimension, can be given by:

$$V = 0.698\ D^3 = D_n^3 \tag{1}$$

where D_n is the nominal diameter of the unit and D is the 'primary or characteristic' length of the unit.

3. Placement Method

One of the most important features of each block is the ease of placement, as a difficult or complicated placement method can be time consuming and therefore increase the final cost of the project. A series of 3D virtual models were used to assess the possible placement configurations and their properties.

Different placements were modelled to find the best performing configuration regarding the following criteria:

- Interlocking of the units, taking into account the blocks' geometry to avoid loose connections.
- Optimal porosity of the armor layer, to increase wave dissipation and minimize the run-up as well as the uplift pressure.

In this study, the packing density coefficient (ϕ) and the armor layer porosity (n_v) are defined by the following equations (CIRIA [1]):

$$\phi = \frac{N_a}{A} {D_n}^2 = t_a(1 - n_v)/D_n \tag{2}$$

$$n_v = 100\,[1 - \frac{N_a D_n^3}{A t_a}] \tag{3}$$

where N_a is the total number of armor units in the studied area, A is the surface area of the armor layer parallel to the slope, D_n is the nominal diameter of armor unit, t_a is the armor layer thickness.

In the following sections, two proposed placements for the DC units, 'Direct placement' and 'Random placement', will be discussed.

3.1. Direct Placement (DP)

This placement method can be described as a regular placement. All units are placed uniformly in one direction (Figure 4). In the first line, all units are placed with the bases inclined at 0° to the line of the greatest slope of the embankment.

Figure 4. Schematic of Direct placement method (DP).

In the second row, the units are secured between two units on the row below, so the chamfers of the unit are in contact with the two chamfered parts of the lower units. The placement of the following lines will be carried out similarly to the two previous lines, up to the crest of the structure. In this pattern, the theoretical horizontal (D_x) and upslope (D_y) distances are 1.550 D and 0.775 D, respectively. Therefore, a theoretical packing density coefficient of $(1.550\ D \times 0.775\ D)^{-1} = 0.832\ D^{-2}$ ($0.655\ D_n{}^{-2}$) would be the result.

Similar to other blocks with regular placement, the main disadvantage of such placements is the precision needed and the low tolerance for errors. It was observed during the small-scale modelling of the DC that the correct placement of the first line is very important for the accurate placement of the next lines. Since the units' positions must be very precise, any inaccuracy will be transmitted and exacerbated in higher lines.

3.2. Random Placement (RP)

In the Random Placement (RP) method, the units are placed randomly line-by-line. The pattern does not follow any strict rules or specific orientations (Figure 5).

Figure 5. Schematic of Random placement method (RP).

For the first line, the units are placed with various orientations. There is no predetermined orientation for the units and the units are placed in various positions. For the second line, the units are placed between the lower units and their orientations can vary, without any specific rule.

In this configuration, the horizontal distance of the blocks is the only factor that should be controlled. The less horizontal distance, the more upslope distance (the upslope distance decreases as the horizontal distance increases), although the total packing density remains constant.

The horizontal placing distance varies between 1.50 $D < D_x < 1.70$ D and the upslope spacing varies between 0.625 $D < D_y < 0.750$ D. As previously stated, the theoretical total packing density coefficient remains constant at around 0.893 D^{-2} (0.702 $D_n{}^{-2}$).

For both placements mentioned above, the blocks have shown a strong interlocking with the surrendering units, and it is difficult to withdraw a unit without dislodging the surrounding units. It is important to note that the interlocking effect for a DC unit is mainly the result of the friction mechanism between adjacent units. Scaravaglione et al. [33] described that the interlocking for some massive units does not come from an entanglement or hooking effect.

4. Hydraulic Model Tests

4.1. Experimental Set-Up

The physical mode tests were carried out in the wave flume of the Coastal and Continental Morphodynamics laboratory of the University of Caen.

This wave flume has a length of 22 m, a width of 0.8 m, and a depth of 1.0 m, as shown in Figure 6. The flume has an Edinburgh Designs piston wave generator that enables the generation of regular and irregular waves with an active wave reflection compensation system. The wave-maker can produce a significant wave height of about 0.1 m with a period of 1 to 3 s (Edesign.co.uk. [34]). All tests were conducted on a flat bottom.

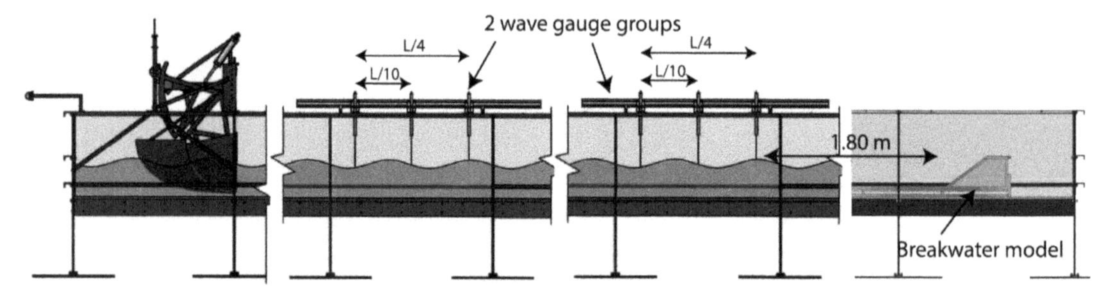

Figure 6. Schematic diagram of wave flume and instrumentation.

The sidewalls of the flume are made of glass, allowing for visual observations and optical measurement of wave–structure interactions (Figure 6).

The sketch of the breakwater cross-section, as well as the material characteristics used in this model, are presented in Figure 7. The 2D model was built with a Froude scaling (model to prototype) of 1:57. As gravity forces are predominant in the physical processes, the Froude number is the recommended scaling factor. Viscous effects are still important, but it is impossible to take them into account in such a small flume.

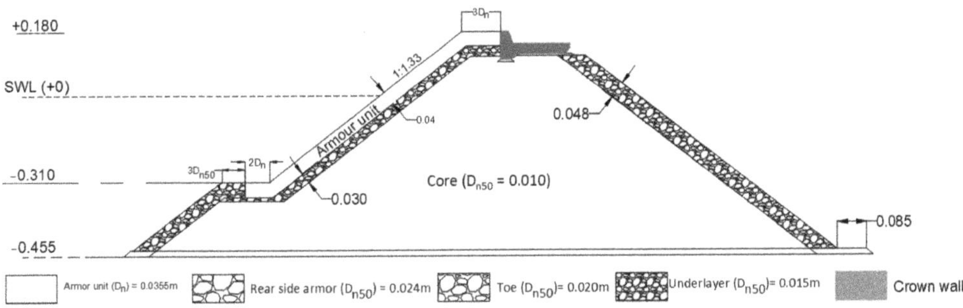

Figure 7. Cross section of the rubble mound breakwater (all dimensions in m).

The armor layer is built using DCs with an average mass of 0.072 kg, and a nominal diameter of 0.0355 m. The model units were cast using a concrete made with cement, perlite concrete, and with a water/cement ratio of 0.5. It must be noted that the mean mass density of each unit is 1620 kgm^{-3}, which is lower than the normal concrete elements (2400 kgm^{-3}). The reason for this was that the dimensions of the model were constrained by the capabilities of the wave generator (significant wave height and period). Nevertheless, the major dominant forces are reproduced in correct proportion (Hughes [35]) at the initiation of damages. This is relevant to estimate the armor stability. The same techniques (using light units) have been validated and used successfully by Gómez-Martín and Medina [4] to study a highly stable unit (Cubipod) and Safari et al. [11] on the Starbloc unit.

There are methods used to estimate scale effects in core permeability such as those introduced by Burcharth et al. [36], Vanneste and Troch [37], and Wolters et al. [38]. In this study, the dimension of core materials has been determined according to the method proposed by Burcharth et al. [36]. In this method, an empirical model is based on pore pressure calculations, leading to the hydraulic gradient of pore velocity, which is used for Reynolds comparisons. Burcharth et al. [36] proposed using time and space averaged interstitial velocity for the calculation of the Reynolds number. For the evaluation of the interstitial velocity in the core, the extended Forchheimer formulation was used. Finally, the core nominal diameter (D_{n50}) is equal to 0.010 m.

To calculate the dimension of the filter material, the methods recommended by CERC [14] are used. The filter support consists of a narrow grading of natural rocks with a median nominal diameter and a mass density of 0.015 m and 0.009 kg, respectively. The thickness of the underlayer was about $2D_{n50}$ = 0.030 m.

The properties of the materials used in the breakwater are listed in Table 1. The tests were carried out for varying wave parameters such as the wave height and wave period, as well as properties of the armor layer such as placement, packing density, and freeboard position. The test matrix is summarized in Table 2. The water depth at the toe of the slope was 0.455 m, the crown height R_c was 0.18 m (stability and overtopping tests) or 0.08 m (for additional overtopping tests). The crown width is equal to the length of three rows of units.

Table 1. Properties of unit and model parameters.

Elements	D_n–D_{n50} (m)	ρ_s (kgm^{-3})	M_{50} [kg]
Armor layer	0.0355	1620	0.0725
Underlayer	0.0150	2650	0.0090
Core	0.0100	2650	0.0026
Rear-side armor	0.0240	2650	0.0360
Toe	0.0200	2650	0.0210

Table 2. Summary of hydraulic stability wave conditions.

Armour Layer	Placement Pattern	Slope Angle	Under Layer D_{n50} (m)	T_p (s)	No. of Tests	R_c (m)	Packing Density
DC	Random	3V:4H	0.0100	1.2	16	0.08	0.68
				1.5	8	0.08	
					6	0.11	
	Direct	3V:4H	0.0125	1.2	8	0.08	0.64
				1.5	4	0.08	
					3	0.11	
	Random			1	8	0.08	0.68
				1.2	40	0.08	
					16	0.11	
				1.5	20	0.08	
					27	0.11	
	Random	3V:4H	0.0150	1	6	0.07	0.68
					3	0.105	
				1.2	11	0.07	
					18	0.105	
				1.5	4	0.09	
					12	0.105	
	Direct			1	1	0.105	0.64
				1.2	5	0.105	
				1.5	4	0.105	
	Random	2V:3H		1	10	0.08	0.68
				1.2	42	0.08	
				1.5	6	0.08	
	Direct			1	4	0.07	0.64
				1.2	22	0.07	
				1.5	2	0.07	
Cube	Simple	3V:4H	0.0100	1.2	10	0.08	0.69
				1.5	8	0.08	
			0.0150	1	6	0.07	
				1.2	11	0.07	
					6	0.105	
				1.5	3	0.09	
		2V:3H		1	7	0.07	
				1.2	19	0.07	
					6	0.105	
				1.5	3	0.07	
					4	0.105	

4.2. Wave Measurements

Two groups of three resistance-type wave gauges, with a precision of ±2%, are used to measure the water surface elevations in the flume (Figure 6).

The first group was positioned 1.5 m seaward of the structure toe and the second group was placed 10 m away from the wave-maker. Incident and reflected waves were analyzed using the least-squares method proposed by Mansard and Funke [39].

Each test was performed with a target mean peak period and an Iribarren number, ξ, which varied from one test to another:

$$\xi = tan\alpha / \left(H_s \Big/ L_0 \right)^{1/2} \tag{4}$$

where H_s is the significant wave height at the toe of the structure, $L_0 = gT_p{}^2/2\pi$, T_p is the peak wave period, and α is the armor slope angle. Here, the significant wave height H_{m0} (=H_s) and T_p (peak wave period) are obtained using frequency domain analysis.

Each test starts with a lower wave height than the target wave to induce the initial settlement. Subsequently, the wave height is increased (with a constant wave period) incrementally up to a wave height resulting in damage (failure). Therefore, each test series consisted of five to six steps of increasing wave height. With T_p being fixed and H_s variable during each series, the Iribarren number changes within the same series. This approach is therefore different from the one adopted by Medina et al. [40], where the Iribarren number was kept constant. All tests were conducted in a non-breaking waves condition (Table 2).

All tests were conducted using irregular waves with a JONSWAP spectrum (γ = 3.3). The stability test for each wave height was performed for fixed acquisition duration, from 1024 s to 2048 s. This corresponds to 1000 to 1700 waves, depending on the tested wave period. Wave statistic significance was already achieved for 1000 waves. The measurement data were obtained using a sample frequency of 32 Hz. In order to obtain the accurate generation of waves, all wave conditions were calibrated through a transfer function with the model in place. Before starting each test, the wave gauges were calibrated in still water through three fixed positions.

4.3. Damage Analysis

Armor damage measurement in this study was performed by visual observations of the displacements of the units of the armor layer. Photographs of the armor layer were taken after each series of tests to measure the evolution of the damage (before, during, and after the test).

To improve the visualization of displacement and the orientation change of the units during damage, each row was colored differently.

In this study, three different levels of damages were considered (CIRIA [1]):

- Start of damage.
- Intermediate damage.
- Failure.

Start of damage corresponds to no damage or no movement of the units. Intermediate damage corresponds to the first significant movements detected on an armor layer, and finally, with increasing wave height, the failure of the armor layer is reached when the removal of a number of units leads to the exposure of the filter layer or the core. Damages were not repaired during successive test series. In this way, the cumulative damage during the test series was determined. The armor layer was reconstructed, if necessary, only after the completion of each test series.

The damage level corresponds to the value of N_{od}, which is the number of displaced armor units to a width (along the longitudinal axis of the breakwater) of one nominal diameter D_n. The value of (accepted) N_{od} is affected by the unit type (massive, interlocking) as well as by the number of layers. For example, the allowed N_{od} for a single-layer cube armor layer is N_{od} = 0.0 for the start of damage and N_{od} = 0.2 for failure. The N_{od} for the failure level of a Double-layer cube armor layer increases to about N_{od} = 2 (Van Gent and Luis [17]).

Moreover, the stability of the armor units can be represented using the dimensionless parameters, such as the stability number N_s (Van de Meer [41]), against the Iribarren number (ξ_p):

$$N_s = \frac{H_s}{\Delta D_n} = \left(K_D cot\alpha\right)^{1/3} \quad (5)$$

where N_s = stability number.

K_D = stability coefficient (introduced by Hudson [42]).
H_{m0} = H_s = significant wave height in front of the structure.
Δ = ($\rho_a/\rho_w - 1$).
ρ_a = mass density of the armor unit.
ρ_w = mass density of the water.
D_n = nominal diameter of the unit = $(m/\rho_a)^{1/3}$.
m = mass of the armor unit.
α = slope angle.

4.4. Overtopping Measurement

The breakwater's overtopping is affected by different parameters such as the slope geometry, crest level, presence of a crown wall, etc. Overtopping has been investigated in various studies (Bradbury et al. [43], Owen [44], Van der Meer and Stam [45], Aminti and Franco [46], Van Gent et al. [22], Bruce et al. [47,48], Molines and Medina [49], EurOtop [50,51]).

In this study, the mean overtopping rate is measured for all tests using the same standard method described by researchers such as Möller et al. [52]. The overtopping discharge ($m^3s^{-1}m^{-1}$) is measured using a collection container placed behind the breakwater model, as shown in Figure 6. This container is made of 10 mm-thick PVC plates, with dimensions of 0.795 m × 0.785 m × 0.360 m (length × width × height).

The post analysis allows us to calculate the average overtopping rate, i.e., the quantity of collected water in the container during a sequence of N incident waves (a storm or period considered), per unit length of the breakwater's width.

For this purpose, the discharge, q, is calculated according to the following formula:

$$q = \frac{V}{tB} \quad (6)$$

where q: mean overtopping discharge.

V: accumulated wave overtopping volume.
t: test duration.
B: width of wave flume.

For tests with high overtopping rates, water is pumped into the leeward part of the wave flume during the test run to maintain a constant level of water in the front of the structure.

The accuracy of overtopping measurement is:

- A container with an uncertainty of 1.3% (calibration with given input water volumes).
- A chronometer with an operational accuracy precision of 1 s.
- A digital scale balance with an accuracy of 5 g (test weights).
- Wave gauges with a precision of 2% (calibration in still water).

5. Hydraulic Stability Results

Among the great variety of factors affecting the design of a breakwater, hydraulic stability is one of the key design criteria that should be carefully investigated, particularly in a one-layer system. The goal was to find the hydraulic stability of the single armor layer for the two placement methods mentioned earlier.

The 'Direct placement' method has a theoretical packing density of about 0.83 D^{-2} (0.65 $D_n{}^{-2}$). However, due to the irregularities of the under-layer, the final packing density was decreased to 0.81 D^{-2} (0.64 $D_n{}^{-2}$). Similarly, for the 'Random placement' method, the theoretical packing density of 0.89 D^{-2} (0.70 $D_n{}^{-2}$) decreased to 0.86 D^{-2} (0.68 $D_n{}^{-2}$).

Figure 8 presents the value of stability number N_s versus the Iribarren number (ξ_p) for the two studied placements.

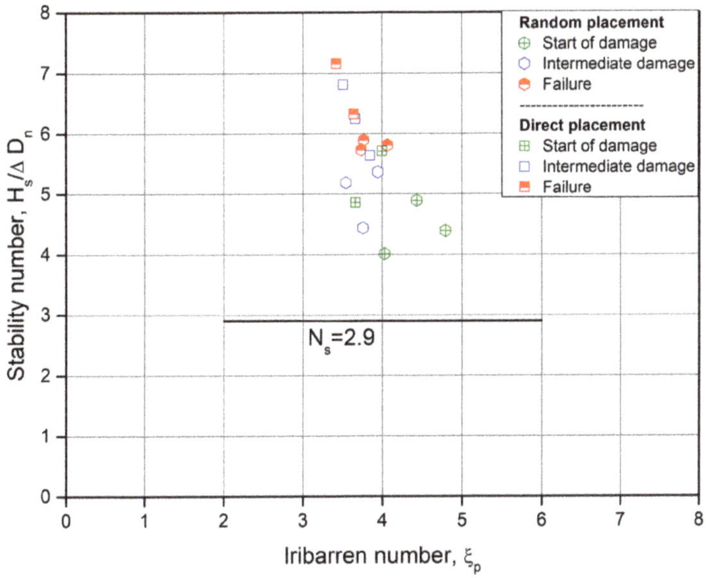

Figure 8. Stability test results (Filter D_{n50} = 0.015 m).

The resultant packing densities in the experiments were 0.68 for random placement and 0.64 for direct placement. These tests were repeated at least two times in order to assess the reliability of the results.

The square sign represents the stability numbers related to the 'Direct placement' method (DP); the diamond sign corresponds to the 'Random placement' method (RP). Center-crossed signs correspond to the start of the damage, and full white and half black symbols correspond to the intermediate damage and failure of the armor layer, respectively. As previously mentioned, the start of damage is accordance with the standard practice of 'no damage', as defined for randomly placed armor units in a single layer (CIRIA [1]).

In the 'Random placement' tests, the start of damage was observed at the N_s close to 4.0 and failure was reached at N_s values between 5.7–5.9.

During the 'Direct placement' tests, the start of damage occurred at the N_s close to 5.6–6.2 and failure happened at a stability number in the range of 6.4–7.2.

As can be seen in Figure 8, higher values of N_s were obtained for 'Direct placement' than for the random placement in all tests for the three levels of damages.

5.1. Discussion

Figure 9 demonstrates the damage level as a function of the stability number for the two studied placements. It is noticeable that the behavior of this unit is comparable to other single armor units. In general, for a single-layer armor, one can expect to observe failure of the structure after the occurrence of the first damage in the armor layer. In light of this, it is concluded that the criterion of failure and the start of damage are very close (Van der Meer [41]).

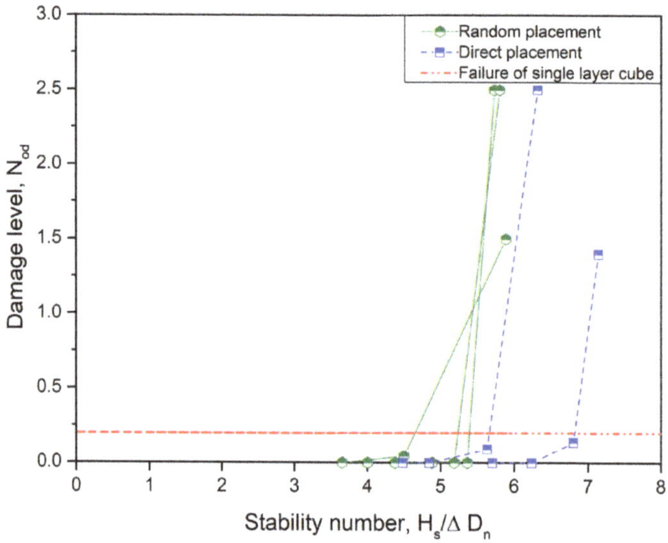

Figure 9. Relation between N_{od} and $H_s/\Delta D_n$.

According to observations made from both placements, only minor changes in the damage progression process were noted. A more gradual damage progress of the armor layer and a wider gap between 'no damage' and 'failure' criteria were observed (intermediate damage). Indeed, the gradual extraction of more units from the initial damaged zone leads to failure. Moreover, it was noticed that one unit can be extracted without affecting the stability of the adjacent units. In fact, the armor layer can remain stable after the occurrence of the first damage, and the failure of the armor layer may originate from a different region in the following test. In this case, it can be considered that such a block was initially poorly positioned with its neighbors during placement, and that it suffered, as a result, from increased vulnerability. It can also be assumed that this lack of support can lead to a redistribution of specific forces for the neighboring units by mechanisms similar to the 'arching effect'. In fact, the placement and specific shape of this unit (base hexagonal) lead to a 'Keystone' feature which allows it to keep the stability of the units adjacent to a unit extracted from the armor layer. There is no 'unravelling' effect observed on any single-layer interlocking units.

Reedjik et al. [7] also stated that the irregularities in the surface of the underlayer might cause mispositions of some units that leads to lower stability.

Additionally, vertical settlement of the units parallel to the slope, particularly with a direct placement, was observed. Vertical settlement of units can be extended toward the crest. This phenomenon tends to increase the packing density on the lower rows of the armor layer and reduce the packing density in the higher areas (Figure 10).

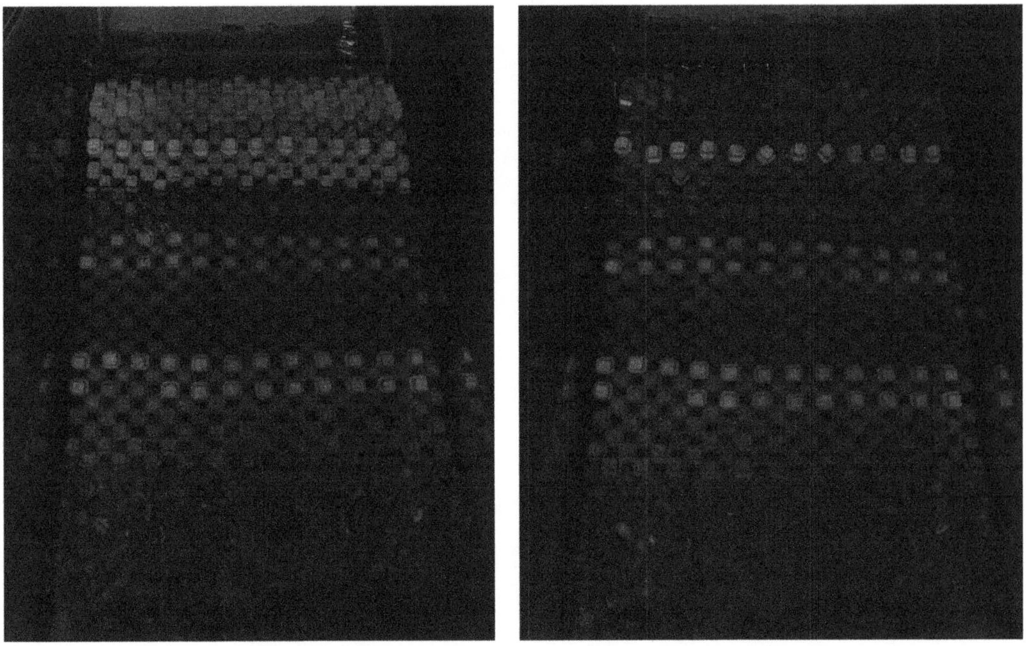

(**a**) (**b**)

Figure 10. Vertical settlement of upper armor layer: (**a**) before test; (**b**) after some tests.

By analyses of the test results, the following formulas were derived for the start of the damage and failure (CIRIA et al. [1]):

$$\frac{H_s}{\Delta D_n} = 4.0 \qquad \text{Start of damage} \qquad \phi = 0.68 \tag{7}$$

$$\frac{H_s}{\Delta D_n} = 5.7 \qquad \text{Failure} \qquad \phi = 0.68 \tag{8}$$

These formulas are related to the 'random placement' with a packing density of 0.68 and $3.4 < \xi_p < 4.5$.

Medina and Gómez-Martín [32] and Jacobs et al. [53] stated that a safety factor in the case of single armor layer should be considered.

As the above values are close to other single armor units such as Accropode (N_s = 3.7) or Coreloc (N_s = 4.2), it is recommended that a safety factor of 1.5 be used on the $(H_s/\Delta D_n)$ (CIRIA [1]). As the design stability criteria refers to the start of damage, the following equation could be proposed as the design rule for the DC unit:

$$\frac{H_s}{\Delta D_n} = 2.7\ (K_D = 15) \qquad \text{For design} \tag{9}$$

5.2. Influence of Slope Angle on Stability

A steeper slope provides a more economical solution for breakwater construction (reduced number of armor units and less material). Figure 11 shows the test results for an armor layer with random placement with two different slopes cotα = 3/4 and cotα = 2/3. For the two slopes, the packing densities of the armor layer were ϕ = 0.68.

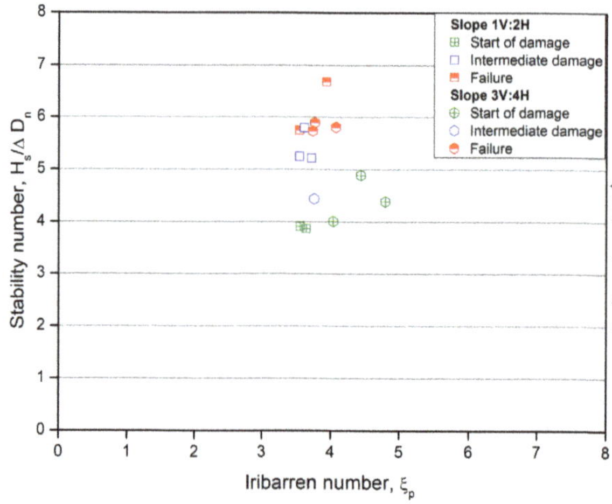

Figure 11. Influence of the slope angle on stability on a 'Random placement' with a packing density of 0.68.

It is evident that the stability number N_s for the start of damage for the two slopes was nearly the same. However, the failure level for the 2V:3H slope was slightly higher than the 3V:4H slope. This result is in agreement with those of many others researchers who investigated the effect of slope steepness (Reedijk et al. [7]).

5.3. Comparison Tests with Cubes

In order to investigate the reliability of the previous tests and the influence of the scale effects, similar experimental tests with cubes as a single-layer armor were conducted. The characteristics of the tests were exactly the same as previous tests (number of waves, specific mass of units, etc.).

The cubes had a nominal diameter of 0.0355 m and were placed in an irregular pattern on a 4H:3V slope, as described by Van Gent and Luis [17]. The cubes were placed on an underlayer of quarry stone with a D_{n50} = 0.015 m, with a theoretical packing density of 0.70.

Figure 12 shows the results for the stability numbers obtained from tests with DCs in a random placement and tests with cubes in a single top layer placement.

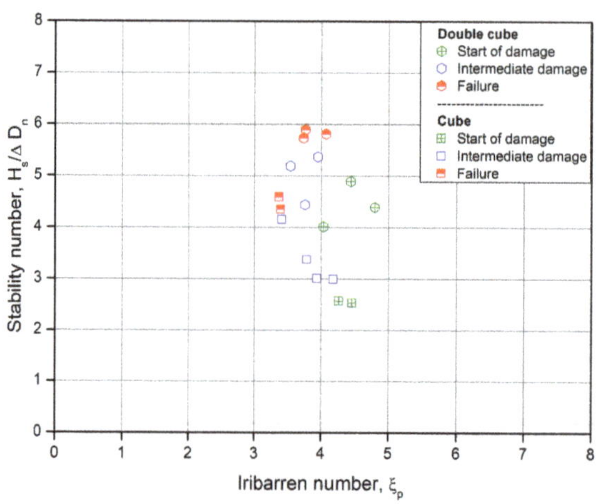

Figure 12. Comparison of stability test results for DC unit with a single-layer cube.

For these single top layers of cubes, the stability results show an N_s 30% higher than the values given in the CIRIA [1], and stability values close to those of Van Gent and Luis [17] and Vieira et al. [28]. The stability numbers N_s obtained from tests with DCs, demonstrate a considerable improvement of 30% for all criteria of damage in comparison to the cubes.

It is well known that porosity plays an important role in armor stability. Considering that packing density and volume porosity were almost the same for both cases, the differences in stability number can be regarded as the effect of unit shape and surface porosity of the armor layer described by Safari et al. [10]. The ratio of porous area in any section is critically compared to the more common global main armor porosity. It is then demonstrated that the minimum section porosity should be the critical retained parameter. The DC shape provides a nonhomogeneous porosity within the armor's section. This feature can enhance the energy dissipation characteristics of an armor block through pressure differentials (Safari et al. [10]).

5.4. Comparison with Other Types of Armor Units

To compare the Double cube with other single-layer units, the following table demonstrates the data related to the stability, packing density, and concrete consumption (Table 3).

Table 3. Comparative table for hydraulic stability and consumption of concrete for different units.

	Accropode® I	X-bloc®	Cubipod®	Cube 1 Layer	DC	
Placement pattern	Random	Random	Random	Random	Random	Direct
Breakwater slope	3V:4H	3V:4H	2V:3H	2V:3H	3V:4H	3V:4H
N_s no damage	3.7 [1]	3.3–5.5 [15]	33	2.9–3.0 [1]	4.0–5.4	4.9–5.7
N_s failure	4.1 [1]	3.7–6.0 [15]	3.7 [32]	3.5–3.75 [1]	5.7–5.9	6.3–7.2
N_s project	2.7 [1]	2.8 [1]	2.6 [32]	2.2 [1]	2.9	3.1
Min N_s/N_s project	1.37	1.18	1.15	1.32	1.38	1.58
Packing density	0.62	0.58	0.65	0.7	0.67	0.63
Consumption of concrete unit related to a cubic shape	72.2	65.1	78.6	100	72.6	64.9

The packing density is optimized through the new design, allowing an effective use of concrete for the breakwater armor unit. The Double cube's performance is comparable to the units available on the market, and the tests indicate that in terms of concrete consumption, there will be no significant increases in the total concrete volume. Regarding concrete consumption, Xbloc®, Accropode®, and DC (random placement) use less concrete than cube and Cubipod®. It should be noted that the proposed design N_s is as a preliminary result and that it should be confirmed by complementary tests.

To compare and summarize the performance aspects of some the most common single-layer armor blocks with DC, the following table is presented (Table 4).

This assessment has been conducted considering the following points:

- Easy placement of new unit without strict rules or specific positioning.
- In comparison to simple units, complex-shaped units typically require a more complicated handling, building, and storage processes.
- Regarding structural integrity, simple-shaped units (massive units) are more robust than complex-shaped units (slender units).

Table 4. Comparative table for hydraulic stability and consumption of concrete for different units.

	Cube [1]	Cubipod®	Accropode® I	DC	Xbloc®	Core-loc®
Number of layers	++	++	++	++	++	++
Hydraulic stability	+− [2]	+	++	++	++	++
Overtopping	+−	+	+	+	+	+
Structural Integrity	+	+	+−	+	+−	+−
Porosity	−	+	+	+	+	+
Ease of placement	+−	+−	−−	+	+−	−−
Ease of build	++	+	−−	+	−−	−−
Storage	++	+	−	+	−	+−
Safety	+	+−	+−	+	+−	+−
Economy	−	+	+−	+	+−	+−

[1] Cubes are considered as the basis of the comparison. [2] The +− sign means that we did not make a determination either because of too much variation in the criterion within the group, or because of insufficient available data (economy), or because of the strong dependency on other criteria (safety).

6. Analysis of Wave Overtopping

Overtopping tests were conducted following the conventional method described by TAW [54] to estimate the unit's performance in terms of roughness. In the basic proposed equations, it is assumed that the overtopping rate can be estimated based on the relative crest freeboard R_c/H_s, R_c being the crest height of the structure above the still water level.

Van de Meer and Janssen [55] provided an empirical overtopping formula, in case of non-breaking waves ($\xi_{m-1,0} > \approx 2$):

$$Q = q \Big/ \sqrt{gH_{m0}^3} = 0.2exp\left(-2.6\frac{R_c}{H_{m0}}\frac{1}{\gamma_f}\right) \tag{10}$$

where q is the average specific overtopping discharge, R_c is the elevation of crest above SWL (m), H_{m0} is the spectral wave height at the toe of the structure, $\xi_{m-1,0}$ is the local surf similarity parameter, and γ_f is the reduction factor for the effect of slope roughness.

In total, more than 100 tests were conducted with a structure armored with DC. The ranges of all of the parameters used in the tests are provided in Table 5.

The water level for all tests was adjusted prior to each test. Depending upon the wave period, the tests had a duration of 1024 s to 2048 s, corresponding to 850–1365 waves, respectively.

Figure 13 shows the results for the dimensionless mean overtopping discharge as a function of crest relative height. The square symbol corresponds to direct placement and the polygon symbol corresponds to random placement. All the tests were conducted using non-breaking wave conditions, which means that the surf-similarity parameter is greater than 2 ($\xi_{m-1,0} > \approx 2$). The graph shows an almost identical trend for both placement patterns, a decrease in overtopping with increasing R_c/H_{m0}.

Furthermore, for the test with Direct placement, the results show a little wider dispersion than the tests with Random placement.

6.1. Effect of Filter Layer on Overtopping

As discussed in a previous section, we studied the effect of different filter layers on the stability of the armor layer. In order to complete these series of tests, we also investigated the effect of filter materials on overtopping.

In Figure 14, all overtopping results from Random placement are illustrated as a relative mean overtopping rate against the relative freeboard R_c/H_{m0}. In this figure, the results for a smooth slope are also illustrated.

Table 5. Summary of overtopping test conditions.

Armour Layer	Under Layer D_{n50} (m)	Slope Angle	Placement Pattern	T_p (s)	No. of Tests	R_c (m)	Packing Density
DC	0.0100	3V:4H	Random	1.2	16	0.08	0.68
				1.5	8	0.08	
					6	0.11	
	0.0125	3V:4H	Direct	1.2	8	0.08	0.64
				1.5	4	0.08	
					3	0.11	
			Random	1	8	0.08	0.68
				1.2	40	0.08	
					16	0.11	
				1.5	20	0.08	
					27	0.11	
	0.0150	3V:4H	Random	1	6	0.07	0.68
					3	0.105	
				1.2	11	0.07	
					18	0.105	
				1.5	4	0.09	
					12	0.105	
			Direct	1	1	0.105	0.64
				1.2	5	0.105	
				1.5	4	0.105	
		2V:3H	Random	1	10	0.08	0.68
				1.2	42	0.08	
				1.5	6	0.08	
			Direct	1	4	0.07	0.64
				1.2	22	0.07	
				1.5	2	0.07	
Cube	0.0100	3V:4H	Simple	1.2	10	0.08	0.69
				1.5	8	0.08	
	0.0150			1	6	0.07	
				1.2	11	0.07	
					6	0.105	
				1.5	3	0.09	
		2V:3H		1	7	0.07	
				1.2	19	0.07	
					6	0.105	
				1.5	3	0.07	
					4	0.105	

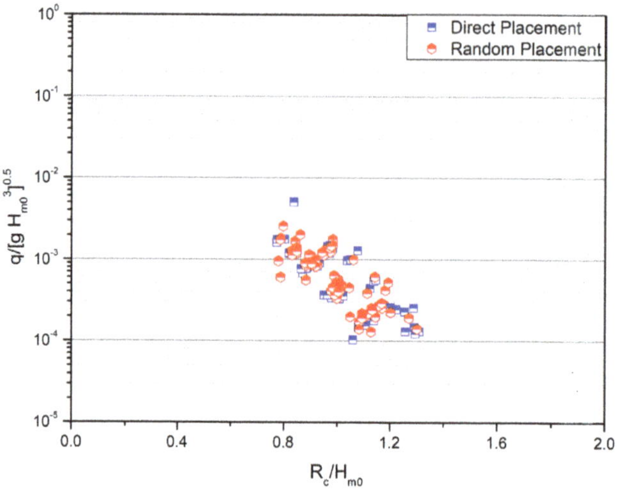

Figure 13. Dimensionless plot of overtopping tests for two studied placements.

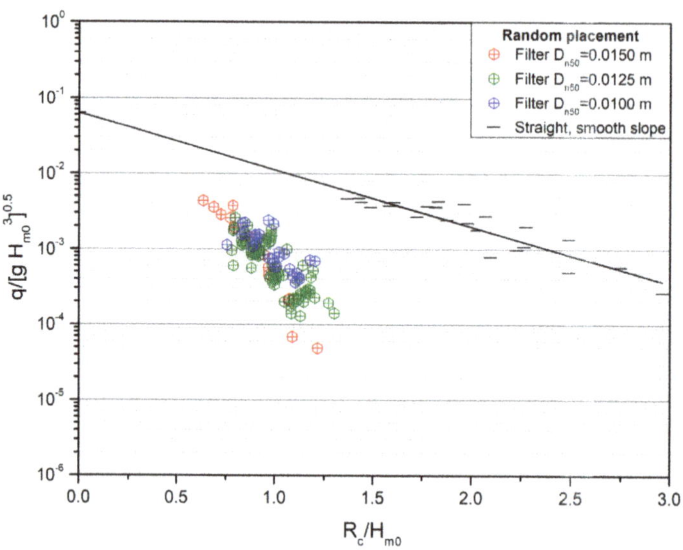

Figure 14. Comparison of tests results for three under layers.

From this figure, the following points are evident:

- Among the different studied filter layers, the results obtained from the tests with D_{n50} = 0.0150 m have a larger scattering compared to those of other sizes.
- The responses for the two underlayers D_{n50} = 0.0150 m and 0.0100 m are similar, but less overtopping was collected for the underlayer D_{n50} = 0.0125 m.

6.2. Comparison with Other Units

The model profile used in this study, particularly the crest of the structure (small freeboard), is similar to the designs that had been used in the CLASH program. The CLASH program intended to provide roughness coefficients for natural stones and various types of artificial units (Bruce et al. [47,48]).

Table 6 provides some recent roughness coefficients for single armor units, extracted from the CLASH program (Bruce et al. [48] data), Reedijk et al. [7] data and Perrin et al. [6] data.

Table 6. Roughness coefficient for single placed armor layer, from synthesis of new data and other comparable tests (Bruce et al. [48] data; Reedijk et al. [7] data; Perrin et al. [6] data; Safari et al. [11] data).

Type of Armour	No. of Layers	Slope Angle	γ_f	γ_f	γ_f
			Mean	95% CI, Low	95% CI, High
Smooth	-	1.5	1		
Rock (permeable core)	1	1.5	0.45		
Cube	1	1.5	0.49	0.46	0.52
Accropode I	1	1.5	0.46	0.43	0.48
Core-Loc	1	1.5	0.44	0.41	0.47
Xbloc	1	1.5	0.44	0.41	0.46
Xbloc^{PLUS}	1	1.5	0.45		
Starbloc	1	1.5	0.45	0.43	0.47
C-ROC	1	1.5/1.33	0.67		
DC	1	1.33	0.46	0.43	0.48
		1.5	0.43	0.40	0.45

As stated by Bakker et al. [15], it is difficult to compare different results obtained from different laboratories, due to the effect of different parameters such as crest width, packing density of the armor layer, geometric characteristics of the underlayer and the core, and the scale effects (Safari et al. [11]).

Therefore, in order to avoid confronting the same problems as those encountered during the tests by Delft Hydraulics for the Xbloc® (Bakker et al. [15]), it was necessary for our model to perform a new series of tests on a reference unit that had been used during the CLASH program studies. This was the simple cube placed in a monolayer, and it was chosen due to the similarity in its configuration to that of a new unit and also due to the easy casting of such a cube in our laboratory.

The overtopping volume was measured for a single-layer regular cube placed on a slope of 3V:4H, followed by a packing density of 70% (Figure 15). The test resulted in a roughness coefficient of $\gamma_f = 0.51$ instead of the expected 0.50. In this study, the roughness coefficient for the DC unit was determined to be 0.46, a 10% decrease when compared to the single layer regular placed cubes.

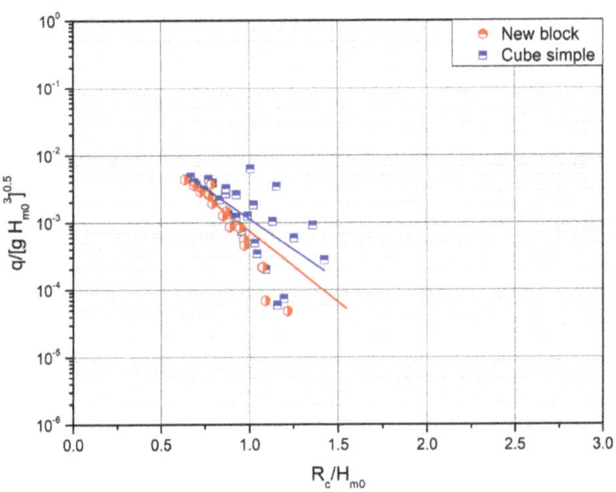

Figure 15. Comparison of measured wave overtopping for two units (3V:4H).

Additional tests were performed on a 2V:3H slope for the single-layer regular cube and for the DC (Figure 16). In the tests with a cube armor layer, the results revealed a roughness coefficient of approximately 0.48, a 6% decrease when compared to the slope of 3V:4H.

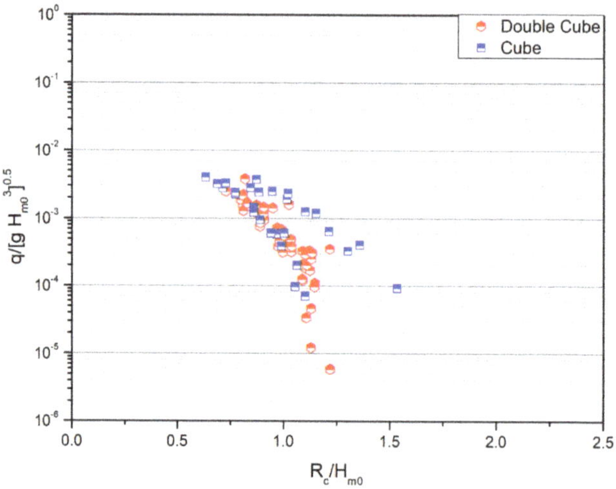

Figure 16. Comparison of measured wave overtopping for cubes and DC (2V:3H).

The roughness coefficient measured for the DC was approximately 0.43, meaning a decrease of about 7% compared to the slope of 3V:4H (Table 4).

The tests conducted with a 2V:3H slope also confirm that the DC provides an approximately 10% lower roughness coefficient compared to a single layer regular placed cube.

7. Conclusions

The Double cube is a new friction-type bulky armor unit designed to protect breakwaters, shorelines, and riverbanks. The DC is composed of two cubic parts joined by a transition section. It is inspired by the simplicity and effectiveness of the cube; however, the new design is aimed to address some of the cube's performance flaws. The basic design idea consists of placing a roughness element onto a sturdier lower part, enabling an assembly of such elements to constitute a stable armor layer.

The innovative design seeks an optimized shape leading to high structural integrity, high hydraulic stability, and a low concrete consumption, while providing an easy placement method. Here, two placement methods were studied, "Direct" and "Random" placements.

Another goal was to increase the roughness coefficient of the armor layer and subsequently, to reduce the overtopping.

The following conclusions can be drawn from the 2D hydraulic laboratory investigations:

- The Double cube allows for simple random placement with no special requirements for the toe or the filter layer.
- The hydraulic stability of the DC is comparable to that of a single armor layer such as Accropode®I or Xbloc®.
- In terms of hydraulic performance, the DC has a roughness parameter that is roughly 10% lower than the single-layer regular placed cube. The new block's roughness parameter is comparable to that of Xbloc® or Accropode®I.
- Regarding concrete consumption, it is comparable to that of the most efficient units, such as the Xbloc® or the Accropode®I.
- Because of its new innovative form, fewer units can be used in a single layer. This will reduce the time and cost of manufacturing, storing, and placing units.

- Greater hydraulic stability allows for the use of smaller cranes.

Promising results have been observed from testing of the newly developed block. This can be regarded as a start to further studies and optimizations on the proposed block. Furthermore, these 2D experiments were focused only on one size of armor unit. In the laboratory, it is easy to place units into their assigned positions. However, this may not be the case for placing the blocks in prototype conditions, which can especially be the case for blocks under the water level while undergoing wave agitation. It is therefore necessary to consider placement tolerances and anticipate the influence of the displacement of some blocks on the stability of the armor layer. Further structural integrity tests must be realized at a prototype scale (drop test) or using a numerical model.

Author Contributions: Conceptualization, I.S. and D.M.; methodology, I.S. and D.M.; experiment, I.S.; formal analysis and investigation, I.S. and D.M.; writing—original draft preparation, I.S.; writing—review and editing, D.M. and S.A.; supervision, D.M.; funding acquisition, G.C. and F.R. All authors have read and agreed to the published version of the manuscript.

Funding: This research received no external funding.

Institutional Review Board Statement: Not applicable.

Informed Consent Statement: Not applicable.

Data Availability Statement: Not applicable.

Acknowledgments: The study was carried out as part of the research project of Iman Safari in the framework of a research fellowship program. The authors gratefully acknowledge the laboratory of Continental and Coastal Morphodynamics of the University of Caen, for the use of research facilities. Financial support for this study was granted to the INOUCO Company.

Conflicts of Interest: The authors declare no conflict of interest.

References

1. CIRIA/CUR/CETMEF. *The Rock Manual. The Use of Rock in Hydraulic Engineering*, 2nd ed.; CIRIA: London, UK, 2007; 1267p.
2. Salauddin, M.; Broere, A.; Van Der Meer, J.W.; Verhagen, H.J.; Bijl, E. First tests on the symmetrical breakwater armor unit crablock. *J. Coast. Eng.* **2017**, *59*, 1–33. [CrossRef]
3. Salauddin, M.; Broere, A.; Van Der Meer, J.W.; Verhagen, H.J.; Bijl, E. A New Symmetrical Unit for Breakwater Armour: First Tests. In Proceedings of the Coastal Structures and Solutions to Coastal Disasters, Boston, MA, USA, 9–11 September 2015; pp. 891–901. Available online: https://ascelibrary.org/doi/abs/10.1061/9780784480304.091 (accessed on 28 June 2023).
4. Gómez-Martín, M.E.; Medina, J.R. Cubipod Concrete Armour Block and Heterogeneous Packing. In Proceedings of the 5th Coastal Structures International Conference, Venice, Italy, 2–4 July 2007; World Scientific: Venice, Italy; pp. 140–151.
5. Gómez-Martín, M.E.; Medina, J.R. Erosion of cubes and Cubipods armour layers under wave attack. In Proceedings of the 31th International Conference on Coastal Engineering, ASCE, Hambourg, Germany, 31 August–5 September 2008; pp. 3461–3473.
6. Perrin, S.; Giraudel, C.; Collinsworth, S.; Melby, J. Hydraulic response & placement methods for a new single-layer concrete armour unit called C-ROC™. In Proceedings of the Coasts, Marine Structures and Breakwaters Conference, ICE, Liverpool, UK, 5–7 September 2017; pp. 321–330.
7. Reedijk, B.; Eggeling, T.; Bakker, P.; Jacobs, R.; Muttray, M. Hydraulic stability and overtopping performance of a new type of regular placed armor unit. In Proceedings of the 36th International Conference on Coastal Engineering, Baltimore, MD, USA, 30 July–3 August 2018; Volume 1.
8. Park, Y.H.; Oh, Y.-M.; Ahn, S.M.; Han, T.H.; Kim, Y.-T.; Suh, K.-D.; Won, D. Development of a new concrete armor unit for high waves. *J. Coast. Res.* **2019**, *35*, 719–728. [CrossRef]
9. Safari, I. Analyse de la Performance Hydraulique d'un Nouveau Type de Bloc Artificiel Utilisé Pour la Protection Côtière. Ph.D. Thesis, University of Caen, Caen, France, 2011.
10. Safari, I.; Mouazé, D.; Ropert, F.; Haquin, S.; Ezersky, A. Influence du plan de pose sur les distributions de porosité au sein d'une carapace de digue à talus. *XIIèmes Journée Natl. Génie Côtière-Génie Civ.* **2012**, 791–798.
11. Safari, I.; Mouazé, D.; Ropert, F.; Haquin, S.; Ezersky, A. Hydraulic stability and wave overtopping of Starbloc® armored mound breakwaters. *Ocean. Eng.* **2018**, *151*, 268–275. [CrossRef]
12. Safari, I.; Mouazé, D.; Ropert, F.; Haquin, S.; Ezersky, A. Experimental study to determine forces acting on starbloc armor units and velocities occurring in a single-layer rubble mound breakwater. *J. Waterw. Port Coast. Ocean. Eng.* **2022**, *148*, 04022007. [CrossRef]
13. Peng, C.; Wang, H.; Zhang, H.; Chen, H. Parametric design and numerical investigation of hydrodynamic characteristics of a new type of armour block TB-CUBE based on SPH method. *J. Mar. Sci. Eng.* **2022**, *10*, 1116. [CrossRef]

14. CERC. *Shore Protection Manual [SPM]*, 4th ed.; Coastal Engineering Research Center, US Army Corps of Engineers: Vicksburg, MS, USA, 1984.
15. Bakker, P.; Klabbers, M.; Muttray, M.; Van den Berge, A. Hydraulic performance of Xbloc Armour units. In Proceedings of the 1st International Conferenec on Coastal Zone Management and Engeering in the Middle East, Dubai, United Arab Emirates, 27–29 November 2005.
16. Bakker, P.; Van den Berge, A.; Hakenberg, R.; Klabbers, M.; Muttray, M.; Reedijk, B.; Rovers, I. Development of concrete breakwater armour blocks. In Proceedings of the 1st Coastal, Estuary and Offshore Engineering Specialty Conference of the Canadian Society for Civil Engineering, Moncton, NB, Canada, 4–7 June 2003.
17. Van Gent, M.R.A.; Luis, L. Application of cubes in a single layer. In Proceedings of the 6th Conference on Applied Coastal Research (SCACR), Lisbon, Portugal, 4–7 June 2013.
18. Jensen, O.J. Safety of Breakwater Armour Layers with Special Focus on Monolayer Armour Units: (Discussion Based Upon Four Decades Experience of Breakwater Damage). In *From Sea to Shore–Meeting the Challenges of the Sea: (Coasts, Marine Structures and Breakwaters 2013)*; ICE Publishing: London, UK, 2014; pp. 33–44.
19. Bhageloe, G.S. Breakwaters with a Single Toplayer (In Dutch: Golfbrekers Met een Enkele Toplaag). Master's Thesis, Delft University of Technology, Delft, The Netherlands, 1998.
20. D'Angremond, K.; Berendsen, E.; Bhageloe, G.S.; Van Gent, M.R.A.; Van der Meer, J.W. Breakwaters with a Single Armour Layer. In Proceedings of the Conference on Coastal and Port Engineering in Developing Countries, Cape Town, South Africa, 19–23 April 1999; pp. 1441–1449.
21. Van Gent, M.R.A.; d'Angremond, K.; Triemstra, R. Rubble mound breakwaters: Single armour layers and high-density blocks. In *Coastlines, Structures and Breakwaters*; ICE: London, UK, 2001.
22. Van Gent, M.R.A.; Van den Boogaard, H.F.P.; Pozueta, B.; Medina, J.R. Neural network modelling of wave overtopping at coastal structures. *Coast. Eng.* **2007**, *54*, 586–593. [CrossRef]
23. Van Gent, M.R.A.; Spaan, G.B.H.; Plate, S.E.; Berendson, E.; Van der Meer, J.W.; d'Angremond, K. Single-Layer Rubble Mound Breakwaters. In Proceedings of the 3th International Conference on Coastal Structures 1999, Satander, Spain, 7–10 June 1999; Volume 1, pp. 231–239.
24. Van Buchem, R.V. Stability of a Single Top Layer of Cubes. Master's Thesis, Delft University of Technology, Delft, The Netherlands, 2009.
25. Van der Lem, C.; Stive, R.; Van Gent, M.R.A. Sal Rei breakwaters with single layer cubes. In Proceedings of the PIANC-Copedec, Rio de Janeiro, Brazil, 16–21 October 2016.
26. Van Gent, M.R.A.; Van der Werf, I.M. Single Layer Cubes in a Berm. In Proceedings of the Applied Coastal Research, Santander, Spain, 3 October–6 October 2017.
27. Vieira, F.; Taveira-Pinto, F.; Rosa-Santos, P. Single-layer cube armoured breakwaters: Critical review and technical challenges. *Ocean Eng.* **2020**, *216*, 108042. [CrossRef]
28. Vieira, F.; Taveira-Pinto, F.; Rosa-Santos, P. Damage evolution in single-layer cube armoured breakwaters with a regular placement pattern. *Coast. Eng.* **2021**, *169*, 103943. [CrossRef]
29. Vieira, F.; Taveira-Pinto, F.; Rosa-Santos, P. New developments in assessment of wave overtopping on single-layer cube armoured breakwaters based on laboratory experiments. *Coast. Eng.* **2021**, *166*, 103883. [CrossRef]
30. Medina, J.R.; Gómez-Martín, M.E.; Corredor, A. Influence of armour unit placement on armour porosity and hydraulic stability. In Proceedings of the 32nd International Conference on Coastal Engineering, ASCE, Shanghai, China, 30 June–5 July 2010.
31. Dupray, S.; Roberts, J. Review of the Use of Concrete in the Manufacture of Concrete Armour Units. In Proceedings of the International Conference on Coastal, Marine Structures and Breakwaters, Sochi, Russia, 10–14 November 2009; Volume 1, pp. 245–259.
32. Medina, J.R.; Gómez-Martín, M.E. KD and safety factors of concrete armour blocks. In Proceedings of the 33nd International Conference on Coastal Engineering, ASCE, Santander, Spain, 1–6 July 2012; Volume 1, p. 29.
33. Scaravaglione, G.; Latham, J.-P.; Xiang, J.; Francone, A.; Tomasicchio, G.R. Historical overview of the structural integrity of Concrete Armour Units. *Coast. Offshore Sci. Eng.* **2022**, *1*, 68–98.
34. Edesign.co.uk. Piston Coastal Wave Generators | Edinburgh Designs. 2012. Available online: http://www.edesign.co.uk/product/piston-wave-generators (accessed on 16 March 2012).
35. Hughes, S.A. Physical models and laboratory techniques in coastal engineering. In *Advanced Series on Ocean Engineering*; Coastal Engineering Research Center: Vicksburg, MS, USA, 1993; Volume 7. [CrossRef]
36. Burcharth, H.F.; Liu, Z.; Troch, P. Scaling of core material in rubble mound breakwater model tests. In Proceedings of the 5th International Conference on Coastal and Port Engineering in Developing Countries (COPEDEC V), Cape Town, South Africa, 19–23 April 1999.
37. Vanneste, D.; Troch, P. An improved calculation model for the wave induced pore pressure distribution in a rubble-mound breakwater core. *Coast. Eng.* **2012**, *66*, 8–23. [CrossRef]
38. Wolters, G.; Van Gent, M.R.A.; Hofland, B.; Wellens, P. Wave damping and permeability scaling in rubble mound breakwaters. In Proceedings of the 5th International Conference on the Application of Physical Modelling to Port and Coastal Protection, Varna, Bulgaria, 29 September–2 October 2014.
39. Mansard, E.P.D.; Funke, E.R. The measurement of incident and reflected spectra using a least squares method. In Proceedings of the 17th International Conference on Coastal Engineering, ASCE, Sydney, Australia, 23–28 March 1980; Volume 1, pp. 154–172.

40. Medina, J.R.; Molines, J.; Gómez-Martín, M.E. Influence of armour porosity on the hydraulic stability of cube armour layers. *J. Ocean Eng.* **2014**, *88*, 289–297. [CrossRef]
41. Van der Meer, J.W. Design of concrete armour Layers. In Proceedings of the Coastal Structures 99, Santander, Spain, 7–10 June 1999; pp. 213–221.
42. Hudson, R.Y. Laboratory investigations of rubble mound breakwaters. *J. Waterw. Harb. Div.* **1959**, *85*, 93–121. [CrossRef]
43. Bradbury, A.P.; Allsop, N.W.H.; Stephens, R.V. *Hydraulic Performance of Breakwater Crown Walls*; Report SR146; H.R. Wallingford: Oxfordshire, UK, 1988.
44. Owen, M.W. *Design of Seawalls Allowing for Wave Overtopping*; Report 924; Hydraulic Research Wallingford: Oxfordshire, UK, 1980.
45. Van der Meer, J.W.; Stam, C.J.M. Wave run-up on smooth and rock slopes of coastal structures. *J. Waterw. Port Coast. Ocean. Eng.* **1992**, *118*, 534–550. [CrossRef]
46. Aminti, P.; Franco, L. Wave overtopping on rubble mound breakwaters. *Coast. Eng. Proc.* **1988**, *1*, 770–781. [CrossRef]
47. Bruce, T.; Van der Meer, J.W.; Franco, L.; Pearson, J.M. A comparison of overtopping performance of different rubble mound breakwater armours. In Proceedings of the 30th Conference on Coastal Engineering, San Diego, CA, USA, 19 November 2006; Volume 5, pp. 4567–4579.
48. Bruce, T.; Van der Meer, J.W.; Franco, L.; Pearson, J.M. Overtopping performance of different armour blocks for rubble mound breakwaters. *Coast. Eng.* **2009**, *56*, 166–179. [CrossRef]
49. Molines, J.; Medina, J.R. Calibration of overtopping roughness factors for concrete armor units in non-breaking conditions using the CLASH database. *Coast. Eng.* **2015**, *96*, 62–70. [CrossRef]
50. EurOtop. *Overtopping Manual, Wave Overtopping of Sea Defences and Related Structures: Assessment Manual*; Pullen, T., Allsop, N.W.H., Bruce, T., Kortenhaus, A., Schüttrumpf, H., Van der Meer, J.W., Eds.; EurOtop: Frickenhausen, Germany, 2007. Available online: www.overtopping-manual.com (accessed on 1 August 2007).
51. EurOtop. *Manual on Wave Overtopping of Sea Defences and Related Structures. An Overtopping Manual Largely Based on European Research, but for Worldwide Application*; Van der Meer, J.W., Allsop, N.W.H., Bruce, T., De Rouck, J., Kortenhaus, A., Pullen, T., Schüttrumpf, H., Troch, P., Zanuttigh, B., Eds.; EurOtop: Frickenhausen, Germany, 2018. Available online: www.overtopping-manual.com (accessed on 15 March 2019).
52. Möller, J.; Kortenhaus, A.; Oumeraci, H.; de Rouck, J.; Medina, J.R. Wave run-up and wave overtopping on a rubble mound breakwater-Comparison of prototype and laboratory investigations. *Coast. Struct.* **2003**, *2003*, 456–468. [CrossRef]
53. Jacobs, R.; Bakker, P.; Vos-Rovers, I.; Reedijk, B. Xbloc-plus development of a regular placed interlocking armour unit. *Coast. Eng. Proc.* **2018**, *36*, 45. [CrossRef]
54. TAW. *Technical Report Wave Run-Up and Wave Overtopping at Dikes*; Technical Report; Technical Advisory Committee on Flood Defence TAW: Delft, The Netherlands, 2002. Available online: http://resolver.tudelft.nl/uuid:d3cb82f1-8e0b-4d85-ae06-542651472f49 (accessed on 1 May 2002).
55. Van der Meer, J.W.; Janssen, J.P.F.M. Wave Run-Up and Wave Overtopping at Dikes. In *Wave Forces on Inclined and Vertical Wall Structures*; Kobayashi, N., Demirbilek, Z., Eds.; ASCE: New York, NY, USA, 1995; pp. 1–27.

Journal of
Marine Science and Engineering

MDPI

Article

Wave Transmission over Rubble-Mound Submerged Breakwaters

Elpidoforos Repousis *, Ioannis Roupas and Constantine Memos

Laboratory of Harbour Works, School of Civil Engineering, National Technical University of Athens, 15780 Zografos, Greece

* Correspondence: elpirep@hotmail.com

Abstract: This study focuses on the prediction of technical efficiency of narrow-crested submerged permeable rubble-mound breakwaters, in terms of wave attenuation. A number of existing formulae for estimating wave transmission coefficient for submerged breakwaters can be found in the literature, whereas in this work further improvement for that estimation has been achieved mainly through physical modelling. A series of 2D experiments under scale were conducted for regular and random waves providing data on wave transmission coefficient and respective wave breaking characteristics. A Boussinesq-type wave model capable of simulating wave propagation for regular waves over porous submerged breakwaters was also used in order to provide additional wave transmission information. Data analysis showed that wave breaking mechanism significantly affects wave energy dissipation and, therefore, wave breaking occurrence and type can be directly linked to wave transmission coefficient for a given structure's geometry and sea state. The result of this work is the proposal of a set of simple semi-empirical equations for predicting wave transmission coefficient over small profile porous submerged breakwaters in relevance to the parameterization of the expected dominant wave breaking mechanism.

Keywords: submerged breakwaters; wave transmission; wave breaking

1. Introduction

With a view to minimizing environmental impact of coastal protection measures, submerged breakwaters have been adopted as a shore protection alternative to their emerged counterparts. This is because they can enhance water renewal of inshore waters, avoid degradation of the aesthetic value of the landscape, occupy relatively smaller seabed areas, etc. Beyond their basic role of coastal stabilization, submerged breakwaters also gain attention as it has been deduced that they may function as artificial marine habitats like natural reefs [1]. This is applicable especially to rubble-mound submerged permeable breakwaters. In this work, technical efficiency of those coastal protection structures of minimum sea bed occupation is studied, specifically the trapezoidal, narrow-crested, homogenous rubble-mound, steep-sloped submerged breakwater layout (SPB). When addressing wave energy abstraction, such structures are expected to be effective for up to average surface wave exposure due to their submergence. Also, they are intended for micro-tidal environments as in the Mediterranean, the Caribbean, and the Baltic Sea due to the critical impact the freeboard has on the efficiency of a submerged structure in reducing wave heights. Nevertheless, due to environmental considerations such as the ones mentioned above, they do gather research interest. Aiming at furthering the acceptance of small profile SPBs as coastal defense structures, this study focuses on parameterizing wave energy dissipation, expressed in terms of wave transmission coefficient (K_t), in relation to certain physical processes giving satisfactory prediction accuracy, acceptable for technical design.

Several attempts on predicting K_t for single submerged porous breakwaters based on various data sets have been carried out. Existing formulae' reviews summarizing them and assessing their accuracy can be found in the literature [2–4]. Also, a few studies (e.g., [5,6]) addressed multiple trapezoidal structures. In contrast to most approaches [7–19], the method considered in this study is based on physical wave breaking characteristics. It has been indicated that wave breaking is a significant factor affecting wave transmission over submerged breakwaters [2]. This is the core novelty advanced in this study by highlighting the strong dependence of wave transmission level to the corresponding type of wave breaking induced by the submerged obstacle. Surely technical efficiency of SPBs also depends on other wave energy dissipating phenomena like the turbulence generated by water movement through the pores between the stones. This work suggests correlating wave breaking macro-features over SPBs with K_t for predicting the latter in a rather reasonable way by taking into account such crucial physical processes. The method applied to acquire the information needed for this task was mainly based on carrying out a series of 2D experiments combining K_t with relevant wave breaking characteristics. The models tested were under scale and overall dimensions considered real life applications of SPBs. It is typical in such research to evaluate K_t by assuming 2D conditions. However, a few works also treated 3D configurations, thus capturing diffraction effects by submerged [20] or low-crested [21] breakwater heads.

Beyond semi-empirical approaches on estimating K_t, several wave models of different types have been developed that can predict the wave transformation due to the presence of submerged obstacles, by treating the surface elevation time-series. The authors of [22] presented a one-dimensional Boussinesq model with improved linear dispersion characteristics capable of predicting the non-linear wave transformations over impermeable barred topographies. The Boussineq-type model was extended to include the effect of structural porosity [23–25]. Another Boussinesq-type model in conjunction with a macroscopic drag formula was introduced by [26]. A fully dispersive and highly non-linear Boussinesq-type model was presented [27] and extended by [28], to include the effects of structural porosity, steep slopes, and relevant wave breaking prediction. In another approach, ref. [29] developed a 1DH model based on the non-linear shallow water equations and two vertical dimensional (2DV) model based on Reynolds-averaged Navier–Stokes equations (RANS) that are capable of simulating wave interaction with permeable and impermeable coastal structures. Ref. [30] simulated wave interaction with porous structures by introducing an improved smoothed particle hydrodynamic (SPH) model based on RANS equations and a large eddy simulation (LES) model. The authors of [31] formulated a non-hydrostatic model based on a non-linear shallow water equation (NSWE) to investigate wave evolution on a water channel with a submerged trapezoidal breakwater.

Data analysis showed that if the main wave breaking mechanism is predicted for a given structure's geometry and sea state, it can then lead to K_t by simple linear semi-empirical equations suitable mainly for small profile porous submerged breakwaters, the type of structure addressed herein. The method proposed was also compared to a widely accepted Formula [9] based on a large number of experimental results, showing satisfactory performance on estimating K_t.

2. Materials and Methods

2.1. Experimental Setup

Experiments associated with the present study were conducted in the three-dimensional basin D1 of the Laboratory of Harbour Works at the National Technical University of Athens, Greece. Part of the basin was transformed into four parallel tangent canals with 0.80 m width and 22.00 m length suitable for carrying out 2D experiments. More details for this facility can be found in previous experimental work [32]. Experiments involving regular and irregular waves under breaking and non-breaking conditions over various SPB physical models were conducted, and layouts according to three different stone mean diameters (d_{n50}) that were tested were constructed in three of the total existing four canals.

Five physical models of submerged breakwaters made of uniform natural stones of porosity close to 0.5 were made with crest widths (B) varying from $3d_{n50}$ to $5d_{n50}$, namely

SPB1 with d_{n50} around 0.10 m and B equal to 0.30 m; SPB2a and SPB2b with d_{n50} around 0.08 m and respective B equal to 0.28 m and 0.40 m; SPB3a and SPB3b with d_{n50} around 0.06 m and respective B equal to 0.25 m and 0.35 m. All structures displayed slope (m) of 1:1.5 at both sides and height (h_s) equal to 0.33 m. The SPB1 was tested for a sea depth (d) of 0.43 m giving a free board (F) of 0.10 m, while the other four models were tested for two sea depths of 0.38 m and 0.43 m giving F of 0.05 m and 0.1 m, respectively.

In regard to SPB geometry and water depth, in total nine different sections of SPBs were formed and tested. The wave maker's paddle was located 7.5 m from the upstream toe of the SPB models and a dissipative layer, made of gravel beach topped with wire mess cylinders, was formed starting around 4.0 m from the downstream toe. The plane of the sea bed was horizontal along the flumes. Testing scale was around 1:10 following Froude scaling similarity, and the submerged bars were designed as statically stable. A view from the facility during experiments is given in Figure 1.

Figure 1. View of the experimental layout from the downstream area during wave tests.

In order to calculate K_t as the ratio of the incoming to the transmitted wave height, two resistant-type wave gauges were used to measure water surface elevation, placed at two specific locations along the canals. Each structure was tested separately from the others. During wave runs, in order to minimize effects between the models, wire mess cylinders functioning as wave energy absorbers were used. The first gauge, used to measure the characteristics of the incident wave, was placed 1.0 m up-wave from the offshore slope toe while the other was positioned 1.6 m inshore from the downslope toe of the SPBs in order to take measurements of the transmitted wave characteristics. The gauge signals were recorded by a personal computer loaded with the WaveData acquisition and analysis software of HR Wallingford. The software controlling the wave paddle did not cope with re-reflection absorption. The issue was addressed by a suitable treatment of the numerical output, which gave higher weight to the measurements in the time window prior to the re-reflection arriving at the gauges. In general, contamination of the used signal was insignificant especially for the runs with irregular waves. In each run, data were recorded simultaneously from the two stations at a sampling frequency of 100 Hz (dt = 0.01 s). In total, 16,400 data points per gauge for each regular wave run and 92,200 data points for the irregular ones were taken. In Figure 2, the overall geometry of the physical models and location of wave gauges placed for measuring K_t is displayed. The structures were tested for regular waves with incoming wave height (H_i) ranging from 0.045 m to 0.173 m and wave period (T) ranging from 1.0 s to 2.2 s. As for irregular waves, the respective range covered from 0.031 m to 0.132 m for the incoming significant height (H_s) and from 1.0 s to 2.2 s for the wave peak period (T_p). Incident wave characteristics correspond to statistical analysis of free surface elevation measurements for the upstream section. Summary of B, D_{n50}, and sea state for each SPB model is given in Table 1. Incoming wave data for all scenarios are given in Table A1 for regular and Table A2 for irregular waves.

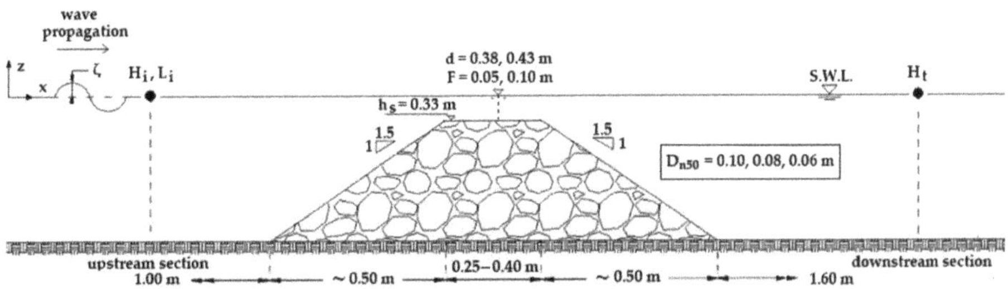

Figure 2. Layout of the physical rubble-mound models tested and location of measurements.

Table 1. Sea state and basic structural features for SPB models tested.

Model	SPB1	SPB2a	SPB2b	SPB3a	SPB3b
B (m)	0.30	0.28	0.40	0.25	0.35
D_{n50} (m)	0.10	0.08	0.08	0.06	0.06
F (m)	0.10	0.05, 0.10	0.05, 0.10	0.05, 0.10	0.05, 0.10 [1]
			Regular Waves		
No of tests	5	21	22	20	12
T (s)	1.0–1.85	1.0–2.2	1.0–2.2	1.0–2.2	1.0–2.2
$\mathbf{H_i}$ (m)	0.078–0.145	0.045–0.152	0.045–0.173	0.052–0.15	0.053–0.145
			Irregular Waves		
No of tests	9	30	21	29	27
$\mathbf{T_p}$ (s)	1.0–2.0	1.0–2.2	1.0–2.2	1.0–2.2	1.0–2.2
$\mathbf{H_s}$ (m)	0.066–0.102	0.04–0.124	0.039–0.124	0.043–0.122	0.031–0.132

[1] For SPB3b, no data were obtained for regular waves for F = 0.10 m.

Wave breaking occurrence was visually observed, and its main characteristics were noted down. For regular waves, a certain wave breaking mechanism was deduced for each wave test, while for irregular wave conditions, following a Jonswap spectrum, various wave breaking types were observed for each test as expected. In general, with the interest mostly concentrated on the visual observations for regular waves, wave breaking classification considered that wave breaking evolution over submerged structures is a complex phenomenon and may vary: from mixed spilling to plunging with or without bore formation, bore, collapsing to surging, and two steps spilling with or without plunging wave breakers. Details for this wave breaking type classification can be found in the literature [33]. In order to simplify the wave breaking types, transmitted waves were sorted in four groups according to the dominant breaker type evolution: the non-breaking, the spilling, the plunging, and the "other" (such as bore and collapsing) with a view to parameterization of each type, in order to implement it subsequently for irregular waves. Data results of K_t for regular waves in relation to the above simplified wave breaking classification are given in Table A1. The K_t values for irregular wave tests are also presented in Table A2.

2.2. Numerical Model

Considering wave models as described in the introduction, a Boussinesq-type numerical wave model capable of simulating wave propagation of incoming regular waves over porous submerged breakwaters was also used in order to provide additional wave transmission coefficient information. Wave propagation over the SPBs tested experimentally was simulated utilizing the fully dispersive and highly nonlinear Boussinesq-type model [28], referred to as a modified version (mCM14) of a previous one [27] that was based on Boussinesq equations by [34]. Validation with the experimental data showed that the numerical model is capable of reproducing K_t for short waves. Thus, it was applied for wave tests with incoming wave height (H_i) ranging from 0.06 m to 0.12 m and wave period (T) ranging from 1.0 s to 1.25 s. In this way, certain gaps on K_t values for wave tests not conducted experimentally were filled. Results are given in Table A3.

2.3. Breaker Type Classification

2.3.1. Regular Waves

Parameterization of breaker type was performed following a previous approach for submerged breakwaters, based on two non-dimensional parameters [33] given as R^*_d and ξ_B, their relevance to certain limit values defining wave breaking features for regular waves, with:

$$R^*_d = \frac{m}{\sqrt{\frac{H_i}{L_o}}}\frac{H_i}{F}\exp(-1.7P) \quad (1)$$

and

$$\xi_B = \frac{h_s}{\frac{h_s}{m} + B}\sqrt{\frac{H_i}{L_o}} \quad (2)$$

where L_o is the deep water wave length, H_i the incident wave height, F the free board in the absence of waves, m the sloping, B the crest width, and h_s the structure's height.

In Equation (1), P denotes a notional permeability [35], equal to 0.6 for permeable rubble-mound homogenous structures. This value was applied to the physical models tested in this study. The P value decreases for lower porosity rubble-mounds and may take a value down to 0.4 for structures made of smaller or graded stones or in structures with a core.

Certain modifications were made on these parameters in order to correlate them closer with the main type of wave breaking also including the non-breaking occurrence not considered in previous studies. Initially, in R^*_d and ξ_B, L_o was replaced by the incoming wave length L_i computed at the depth close to the front toe of the breakwater but not that close to be affected by shoaling due to the structure, where the incident wave characteristics were measured.

Then, R^*_d was multiplied by two additional non-dimensional factors c_1 and c_2. The first one takes into account a width (B_{ch}) that loosely can characterize wave-structure interaction in terms of momentum flux. Its calculation is based on the estimation of a characteristic water depth (d_{ch}), where the mean of the product of $u_x \cdot C$, integrated along the height of the water column at the structure location, matches the local $u_x \cdot C$ value. Parameter u_x is the orbital horizontal velocity and C is the wave celerity, both calculated through linear theory. After applying this process for a wide range of wave scenarios and for depths up to 1.0 m, considering testing scale around 1:10, B_{ch} was defined as follows:

$$B_{ch} = \begin{cases} B + 2\frac{1}{m}[h_s - (0.22 - 0.022L_i)], \text{for } h_s < 0.4 \text{ m} \\ B + 2\frac{1}{m}\left[h_s - \left(0.22 - 0.022L_i + (h_s - 0.4)^{1.35}\right)\right], \text{for } 0.95 > h_s > 0.4 \text{ m} \end{cases} \quad (3)$$

Now factor c_1, denoting somehow the wave-structure geometrical interaction, can be expressed by:

$$c_1 = \frac{L_i}{2B_{ch}} \quad (4)$$

Factor c_2 takes into account B_{ch} as well, but also d_{n50}, and an equivalent freeboard F_{eq} defined as [33]:

$$F_{eq} = d - \frac{A_s}{L_m} \tag{5}$$

where A_s is the cross-sectional area of the structure and L_m its width along the base with c_2 defined by:

$$c_2 = \left(\frac{H_i}{F_{eq}} - \frac{d_{n50}}{B_{ch} - d_{n50}}\right)^{-1} \tag{6}$$

where d_{n50} is the median nominal diameter of the breakwater structure. It should be added here that the freeboard in the absence of waves is actually modified by waves through their setup resulting from eventual wave breaking forced by the structure; see e.g., [21]. This contribution was indirectly taken into account in this study by the identification of the breaker type associated to the corresponding measurements that finally supported the proposed formulae, as it will be seen in the following.

In accordance with the above modifications made mainly on R^*_d, the set of surf similarity type non-dimensionless parameters named R and ξ_m herein becomes:

$$R = \frac{L_i}{2B_{ch}} \frac{1}{\frac{H_i}{F_{eq}} - \frac{d_{n50}}{B_{ch}-d_{n50}}} \frac{m}{\sqrt{\frac{H_i}{L_i}}} \frac{H_i}{F} \exp(-1.7P) \tag{7}$$

and

$$\xi_m = \frac{h_s}{\frac{h_s}{m} + B}\sqrt{\frac{H_i}{L_i}} \tag{8}$$

The R and ξ_m values were calculated for each regular wave test carried out on physical models and was correlated to the main wave breaking type visually observed, namely the plunging, spilling, "other", and non-breaking as previously mentioned. Plots of ξ_m-R showed gathering of scenarios with the same wave breaking type to certain areas of the diagram as presented in Figure 3. These sub areas where the majority of tests had a certain breaking type were delineated by T_1 and T_2 curves described by:

$$T_1 = \exp(1.2\xi_m) - 2.5, \quad \xi_m < 2.35 \tag{9}$$

and

$$T_2 = \begin{cases} \exp(1.02\xi_m) - 2, & \xi_m < 2.35 \\ 11.5\xi_m - 18.5, & \xi_m \geq 2.35 \end{cases} \tag{10}$$

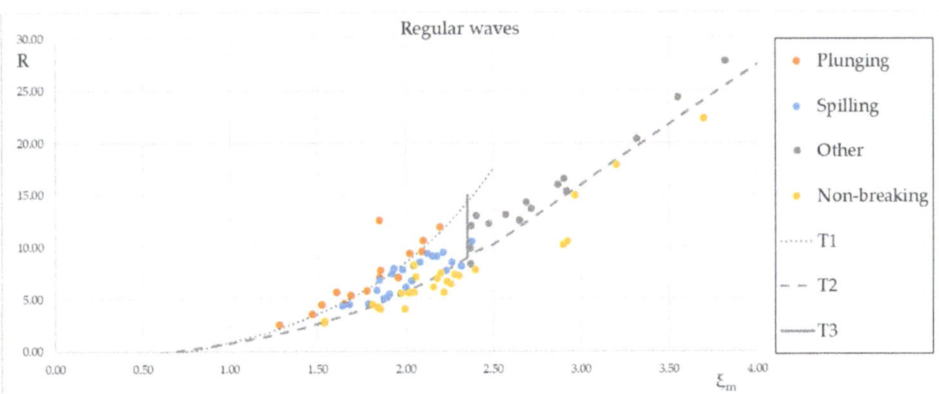

Figure 3. Breaker types parameterization for regular waves.

Curve T_1 separates plunging from spilling breaking events, while T_2 functions as a boundary for spilling and "other" breakers from cases for non-breaking wave occurrence. Moreover, there is a certain ξ_m value that separates spilling from "other" types of breaking equal to 2.35 (line T_3 in Figure 2). In general, spilling and plunging breakers appear for ξ_m lower than 2.35. Sub areas for each wave breaking type are distinguished in Table 2.

Table 2. Breaker type in relevance to R, ξ_m parameters for regular waves.

Sub Area	Breaker Type
if $\xi_m < 2.35$ and $R \geq T_1$	Plunging [1]
if $\xi_m < 2.35$ and $T_1 > R > T_2$	Spilling [1]
if $\xi_m \geq 2.35$ and $R > T_2$	Other
if $R \leq T_2$	Non-breaking

[1] For wave tests with plunging and spilling breakers, data gave values for R lower than 14 (Figure 3).

As it can be seen in Figure 3, determination of the boundary line T_1 (Equation (9)) for ξ_m greater than 2.35 was not possible as testing conditions did not cover this area of the diagram. Thus, the relevant limit where plunging breaking type occurs was not defined there.

The next step was to investigate the relevance of K_t to the breaker type. To do so, additional K_t data derived from the numerical model described in Section 2.2 were also used. The breaker type identification for the numerical simulations followed Table 2. The relation of R to K_t values showed that for each breaker type there is a different range of K_t values, quite distinct for plunging and non-breaking waves with spilling and "other" types covering more or less a common area on the K_t–R plane and showing a trend with a mean value of K_t between the relevant mean values for plunging and non-breaking wave tests (Figure 4). In the following, types spilling and "other" were unified in one type for simplicity. The unified type was named spilling+ breaking type. The K_t values that were numerically calculated contributed in examining further the consistence of the wave breaking type's sub-areas defined by the experimental data, to the expected wave energy abstraction level. Results showed that the derived boundaries distinguished reasonably well the K_t values between sub-areas in the ξ_m–R plane of the numerical simulations' wave tests.

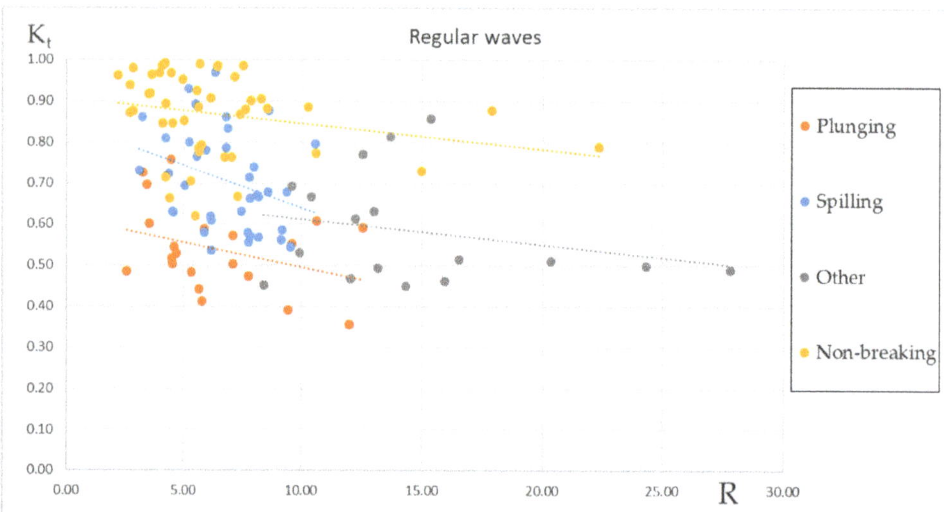

Figure 4. Correlation of K_t values to R parameter and linear approximation from regression analysis for each wave breaking type (dotted lines) for regular waves.

2.3.2. Irregular Waves

Data analysis for regular waves showed that the K_t value depends significantly on wave breaking occurrence and type. This observation was extended to the irregular wave tests. As a single breaker type cannot be identified visually for such runs, calculation of ξ_m and R values for each irregular wave test was performed considering the peak wave period T_p instead of T in regular waves, the relevant peak wave length L_p instead of L_i at the sea depth associated to the incoming wave train, and the significant wave height H_s instead of H_i. Table 2 was then applied in order to estimate the dominant breaking form in irregular waves. Results are presented in Figure 5.

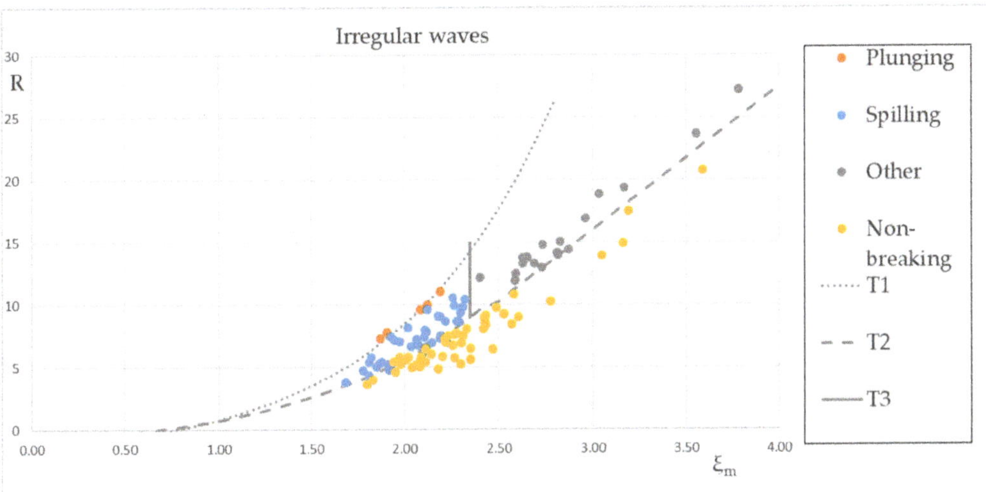

Figure 5. Breaker types parameterization for irregular waves.

In order to estimate the T_1 boundary line for ξ_m greater than 2.35 not defined for regular waves, K_t data for irregular waves for submerged structures of geometries close to the tested models but of lower porosity [36] were used. For this data series, P was taken around 0.4 in Equation (7) and the extension of the curve T_1 (see Figure 5) distinguished reasonably the wave breaking type according to the K_t values. This process is further discussed in Section 4.

Investigation of the relevance of K_t to R for any breaker type showed that for each breaker type likewise for regular waves, there is a certain range of K_t values close to a straight line (Figure 6).

As noted for regular waves, spilling and "other" breakers present a similar range for K_t values more distinct for the irregular wave tests and for simplicity can be gathered in one general group denoted as spilling+. This is further supported in irregular waves, where no clear boundary between typical spilling and bore/surging breaking types can be accurately defined.

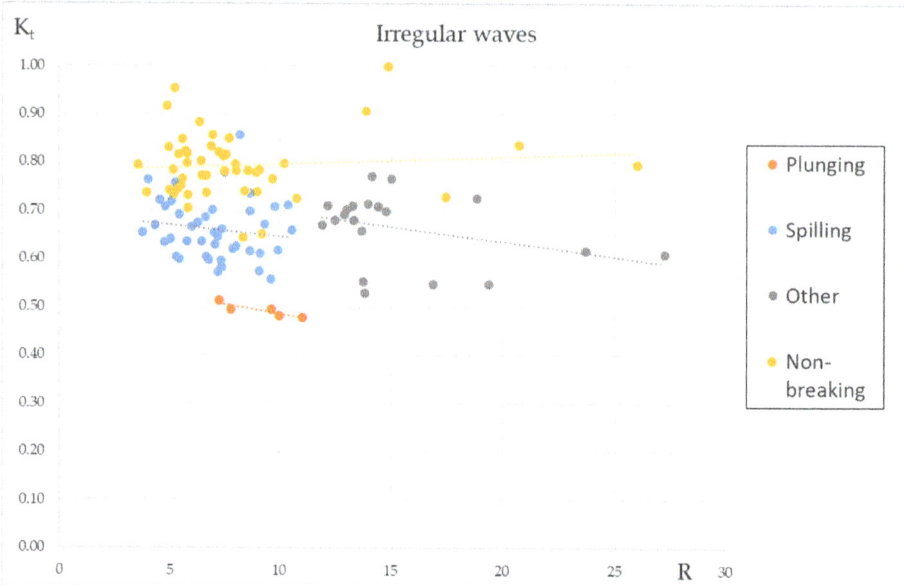

Figure 6. Correlation of K_t values to R parameter and linear approximation from regression analysis for each wave breaking type (dotted lines) for irregular waves.

3. Results

The K_t parameterization for irregular waves was based on the predicted dominant breaker characteristic. Three main types are considered: plunging, spilling+, and non-breaking. Similarly to the methodology applied for regular wave conditions, the set of R and ξ_m values is calculated as:

$$R = \frac{L_p}{2B_{ch}} \frac{1}{\frac{H_P}{F_{eq}} - \frac{d_{n50}}{B_{ch} - d_{n50}}} \frac{m}{\sqrt{\frac{H_s}{L_p}}} \frac{H_s}{F} \exp(-1.7P) \tag{11}$$

where

$$B_{ch} = \begin{cases} B + 2\frac{1}{m}\left[h_s - \left(0.22 - 0.022L_p\right)\right], \text{for } h_s < 0.4 \text{ m} \\ B + 2\frac{1}{m}\left[h_s - \left(0.22 - 0.022L_p + (h_s - 0.4)^{1.35}\right)\right], \text{for } 0.95 > h_s > 0.4 \text{ m} \end{cases} \tag{12}$$

F_{eq} from Equation (5), and

$$\xi_m = \frac{h_s}{\frac{h_s}{m} + B}\sqrt{\frac{H_s}{L_p}} \tag{13}$$

Consecutively, for every given ξ_m, the relation of the R value to T_1 and T_2 boundary lines (Equations (9) and (10)) is checked according to Table 3, and the breaker type is defined.

Table 3. Breaker type in relevance to R parameter for irregular waves.

Sub Area	Breaker Type
$R \geq T_1$	Plunging
$T_1 > R > T_2$	Spilling+
$R \leq T_2$	Non-breaking

After the prediction of the wave breaking feature, the K_t value is calculated by simple linear equations, one for each breaker type, extracted through linear regression of the experimental K_t data in conjunction to the non-dimensional parameter described as:

$$p = \frac{L_m}{L_p}\frac{F}{H_s} \tag{14}$$

The result of the semi-empirical K_t prediction model proposed herein is the following:

$$K_t = \begin{cases} 0.42p + 0.41, & \text{for plunging} \\ 0.11p + 0.62, & \text{for spilling+} \\ 0.1p + 0.73, & \text{for non-breaking} \end{cases} \tag{15}$$

The above formulation is aligned with what one would expect by considering the physical process of wave breaking that controls the phenomenon. Indeed, the three main breaking classes extract a progressively lower amount of wave energy in the order noted in Equation (15), giving thus a correspondingly higher transmission coefficient.

4. Discussion

The applicability range of the model, derived solely by the experiments framing this work is for H_s/F values from 0.40 to 1.50 and for L_m/L_p from 0.3 to 1.0. Additionally, the relative depth F/d was approximately between 0.10 and 0.25. It should be pointed out that only a few wave scenarios were categorized as plunging for the irregular wave tests, thus information from regular waves giving an average K_t value of 0.5 was also considered in defining the relevant K_t equation. Root mean squared error (RMSE) was employed as a reliability indicator to assess the proposed formula as compared to the experimental data, where:

$$RMSE = \sqrt{\frac{\Sigma_N \left(\frac{K_{tm} - K_{te}}{K_{te}}\right)_i^2}{N}} \tag{16}$$

N is the number of data, and K_{tm} and K_{te} are the predicted by the model and experimentally measured values of K_t, respectively. Their relation is shown in Figure 7. For the total number of all 115 data measurements, RMSE calculated was equal to a quite satisfactory value of 0.08.

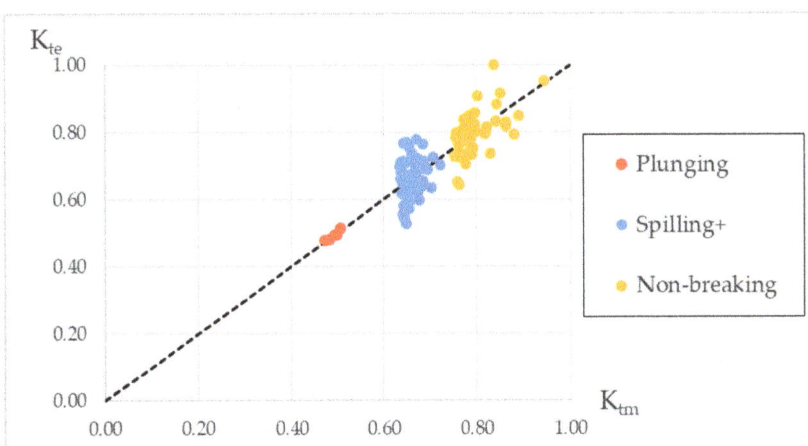

Figure 7. Comparison between K_t calculated with the proposed model (K_{tm}) and K_t measured experimentally (K_{te}).

Despite the fact that various data sets of K_t values for submerged structures are mentioned in [3], for the small profile SPBs investigated in this study (narrow crested, homogenous, statically stable rubble-mound and steep sloped trapezoidal bars of a relative depth F/d approximately between 0.10 and 0.25 for up to average wave exposure) and especially for irregular waves, such data were limited. Thus, the need for the experiments framing this work was apparent. Additionally, information of combined K_t values and the physical process of wave breaking has not been previously conducted systematically as herein, despite the fact that wave breaking type obviously dominates the resulting transmission coefficient. Moreover, comparison to a widely accepted existing formula for irregular waves [12] was performed, initially introduced by [9] and incorporated in commonly used technical manuals [37,38] by implementing it on the experimental set-ups and K_t data of this study. The approach according to [25] is described by the equation:

$$K_t = 0.4\frac{F}{H_{si}} + 0.64\left(\frac{B}{H_{si}}\right)^{-0.31} \cdot \left[1 - \exp(-0.5\xi_{op})\right] \begin{cases} 0.075 \le K_t \le 0.8 \\ \frac{B}{H_{si}} < 10 \end{cases} \quad (17)$$

where H_{si} is the significant wave height of the incident wave and ξ_{op} a breaker parameter based on wave steepness given by:

$$\xi_{op} = m\left(\frac{H_{si}}{L_{op}}\right)^{-0.5} \quad (18)$$

where L_{op} is the deep water peak wave length and m is the leeward sloping of the structure. It is noted that Formula (17) is based on a number of experimental results including earlier published data.

Despite the fact that this formula is considered to present a high level of accuracy, it is limited for K_t values lower than 0.8. This means that it cannot be used for a range of cases with non-breaking waves thus not giving information for cases under which wave energy dissipation is low. A check for 88 of the total 115 irregular wave tests for which K_t values from Equation (17) were calculated lower than 0.8 gave an RMSE score close to 0.125, greater than the relevant value for the proposed method covering all scenarios.

The proposed method was compared to some K_t values from a previous experiment on a narrow-crested SPB in order to examine its reliability. Data on K_t measurements, not used in the formulae production of this study, on a narrow-crested SPB were identified [35], and according to [36], 11 irregular wave tests were assorted. The geometry of the structure was h_s = 0.40 m, B = 0.30 m, d_{n50} = 0.035 m, m = 1:2, and F around 0.09 m, while wave characteristics ranged for T_p from 1.94 s to 2.6 s and H_s from 0.11 m to 0.254 m. The H_s/F values ranged from 1.20 to 2.70 and L_m/L_p from 0.25 to 0.80. The proposed method resulted in RMSE equal to 0.065 while Formula (17) gave an RMSE equal to 0.12.

Moreover, a comparison with experimental data for submerged rubble-mound breakwaters with core of small d_{n50} was executed in order to also investigate the extension of boundary line T_1 (Equation (9)) for ξ_m values greater than 2.35, and it was assumed that the algebraic expression of curve T_1 would not change. To this end, experimental data from the literature for narrow-crested submerged breakwaters with core were used [36]. The geometry of two given structures with armor layer of d_{n50} = 0.04 m and 0.06 m, respectively, was for both h_s = 0.46 m, B = 0.34 m, m = 1:1:5, and F ranging from 0.04 to 0.09 m, T_p from 0.99 s to 1.91 s and H_s from 0.067 m to 0.14 m. In total, K_t data for 14 irregular wave tests was distinguished. By applying the hypothesis that the T_1 boundary line between plunging and spilling breaking waves is suitable for every ξ_m value, the proposed method resulted in RMSE equal to 0.14 while Formula (17) had the same RMSE. For further investigation, another set of experimental data for a submerged breakwater with a core was used for comparison [39]. The relevant structure with armor layer of d_{n50} = 0.16 m had h_s = 0.66 m, B = 0.3 m, m = 1:1:5 and F = 0.09 m, T_p from 1.33 to 2.0 s, and H_s from 0.12 to 0.187 m. In total, six irregular wave tests were used. The proposed method resulted in RMSE equal to 0.27 while Formula (17) gave an RMSE close to 0.38. Considerable differentiation on K_t

values was observed for these wave tests defined as plunging breakers according to the proposed method. Based on the above experiments for submerged breakwaters with core, the plunging leg of the equation was re-examined, and for 11 predicted plunging breaking wave tests from [36,39], linear regression analysis of K_t gave the following equation:

$$K_t = -0.2p + 0.52, \qquad \text{for plunging} - \text{With Core} \tag{19}$$

When this expression was applied, RMSE slightly improved to 0.13 for the first set of experiments [36], while it dropped significantly to 0.07 for the other one [39]. An overall view of the final results is presented in Figure 8. It should be noted that wave height for the scenarios from literature was expressed in H_{mo}. Thus H_{mo} values were converted to H_s prior to comparison with the present model.

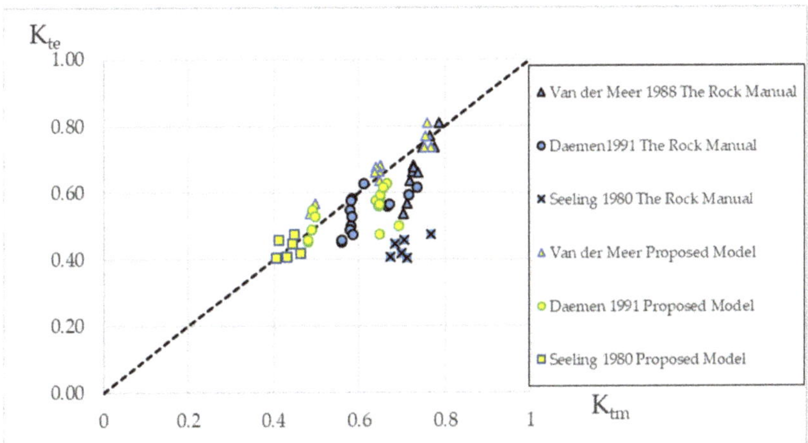

Figure 8. Comparison for K_t calculated with the proposed model, Formula (19) (K_{tm}-axis), and K_t measured experimentally from literature (K_{te}).

Extension of the derived model to layered submerged rubble-mound breakwaters is foreseen as it is common practice to build layouts with a core, when a higher level of technical efficiency, i.e., lower K_t, is sought after. Examination of both the existing formula applied and the proposed method's performance for such cases showed that further research would be welcome. However, based on the analysis on some previous experiments for submerged breakwaters with core, an approximation was derived for cases with plunging wave breaking prediction (Equation (19)), but requires further investigation. More experimental data on correlating K_t with breaker type are needed in order to better determine the subareas between spilling and plunging breaking waves. Specifically, lowering of the permeability due to a core is more probable to induce intense wave breakers meaning more energy loss, leading probably to modification of the corresponding breaker type areas in the ξ_m–R plane especially for plunging waves in comparison to homogenous counterparts. Thus, further evaluation of the proposed equation predicting K_t for plunging waves should be undertaken for low permeability (or multi-layered) breakwaters.

5. Conclusions

In this article, a simple but reliable method is presented to calculate the transmission coefficient for irregular waves propagating over submerged breakwaters. It is based mainly on process-based considerations, such as type of wave breaking and nominal interaction between wave and structure. The method has been developed based on numerous experimental measurements for such breakwaters of high permeability that display enhanced environmental merits, enriched by results from a suitable numerical model. Its applicability range covers primarily the said breakwater types, but a number of further checks showed that it can be extended to also include submerged breakwaters of reduced permeability. In general, the proposed model shows a good performance, superior to that from a well-established formula, reproduced also in EurOtop. Further research is suggested to confirm the previously mentioned extension of the method to cover virtually all types of practically applicable submerged breakwaters for coastal protection.

Author Contributions: Conceptualization, E.R. and C.M.; methodology, E.R. and C.M.; software, I.R.; validation, I.R.; formal analysis, E.R.; writing—original draft preparation, E.R.; writing—review and editing, E.R., I.R., and C.M.; visualization, E.R. and C.M.; supervision, C.M. All authors have read and agreed to the published version of the manuscript.

Funding: This research received no external funding.

Institutional Review Board Statement: Not applicable.

Informed Consent Statement: Not applicable.

Data Availability Statement: Experimental data framing this work are available in Appendix A.

Acknowledgments: The authors acknowledge the head and staff of the laboratory of Harbour Works, National Technical University of Athens, for providing the installations for the experiments framing this work.

Conflicts of Interest: The authors declare no conflict of interest.

Appendix A

Table A1. List of regular incident wave tests characteristics, K_t values, and breaker type where (P) is for plunging, (S) is for spilling, (O) is for "other", and (x) is for non-breaking waves.

	#	T (s)	H_i (m)	K_t	Breaker Type
SPB1 F = 0.10 m	1	1	0.078	0.85	x
	2	1.15	0.066	0.98	x
	3	1.3	0.08	0.76	x
	4	1.45	0.122	0.89	S
	5	1.85	0.145	0.79	S
SPB2a F = 0.05 m	6	1	0.098	0.72	S
	7	1.15	0.141	0.50	P
	8	1.25	0.051	0.45	O
	9	1.25	0.066	0.53	O
	10	1.25	0.096	0.57	P
	11	1.45	0.079	0.80	S
	12	1.6	0.083	0.61	O
	13	1.7	0.125	0.55	P
	14	1.85	0.107	0.47	O
	15	2	0.045	0.49	O
	16	2	0.052	0.50	O
	17	2	0.08	0.46	O
	18	2.2	0.152	0.36	P
SPB2a F = 0.10 m	19	1	0.078	0.85	x
	20	1	0.113	0.88	x
	21	1.15	0.066	0.98	x
	22	1.3	0.07	0.90	x
	23	1.3	0.08	0.76	S
	24	1.45	0.12	0.93	x
	25	1.45	0.13	0.80	S
	26	2	0.14	0.57	S

Table A1. *Cont.*

	#	T (s)	H_i (m)	K_t	Breaker Type
SPB2b F = 0.05 m	27	1	0.093	0.60	P
	28	1.25	0.087	0.41	P
	29	1.25	0.097	0.48	S
	30	1.5	0.081	0.68	S
	31	1.5	0.089	0.66	S
	32	1.5	0.095	0.63	S
	33	1.6	0.08	0.56	S
	34	1.6	0.082	0.59	S
	35	1.7	0.118	0.47	P
	36	1.85	0.11	0.39	P
	37	2	0.045	0.51	O
	38	2	0.075	0.49	O
	39	2.1	0.062	0.52	O
SPB2b F = 0.10 m	40	1	0.085	0.87	x
	41	1.25	0.173	0.49	P
	42	1.3	0.065	0.91	x
	43	1.5	0.094	0.77	S
	44	1.6	0.075	0.67	O
	45	1.6	0.141	0.63	S
	46	1.8	0.115	0.61	S
	47	2	0.093	0.45	O
	48	2.2	0.108	0.67	S
SPB3a F = 0.05 m	49	1	0.07	0.79	x
	50	1.15	0.105	0.58	S
	51	1.25	0.094	0.96	x
	52	1.3	0.082	0.88	S
	53	1.45	0.141	0.50	P
	54	1.6	0.112	0.54	S
	55	1.7	0.131	0.68	S
	56	1.85	0.074	0.73	x
	57	2	0.052	0.79	x
	58	2	0.084	0.86	x
	59	2.2	0.138	0.63	O
SPB3a F = 0.10 m	60	1	0.073	0.99	x
	61	1.25	0.122	0.89	S
	62	1.3	0.089	0.99	x
	63	1.45	0.159	0.63	S
	64	1.5	0.126	0.78	x
	65	1.5	0.15	0.69	S
	66	1.8	0.079	0.89	x
	67	1.8	0.105	0.90	x
	68	2	0.088	0.77	x
SPB3b F = 0.05 m	69	1	0.066	0.83	S
	70	1.15	0.053	0.87	x
	71	1.15	0.101	0.54	P
	72	1.3	0.079	0.86	S
	73	1.45	0.145	0.44	P
	74	1.5	0.115	0.59	P
	75	1.6	0.102	0.57	S
	76	1.7	0.123	0.74	S
	78	1.8	0.07	0.77	x
	79	2	0.075	0.81	O
	80	2.1	0.057	0.88	x

Table A2. List of irregular incident wave tests characteristics and K_t values.

	#	T (s)	H_s (m)	K_t
SPB1 F = 0.10 m	1	1	0.066	0.82
	2	1.15	0.079	0.8
	3	1.3	0.086	0.8
	4	1.3	0.102	0.75
	5	1.45	0.087	0.82
	6	1.5	0.095	0.82
	7	1.6	0.102	0.81
	8	1.85	0.102	0.78
	9	2	0.107	0.77
SPB2a F = 0.05 m	10	1	0.064	0.69
	11	1.15	0.063	0.70
	12	1.15	0.074	0.66
	13	1.25	0.04	0.73
	14	1.25	0.052	0.69
	15	1.25	0.07	0.67
	16	1.25	0.083	0.63
	17	1.3	0.049	0.71
	18	1.3	0.087	0.62
	19	1.45	0.062	0.71
	20	1.45	0.095	0.61
	21	1.6	0.073	0.66
	22	1.6	0.113	0.56
	23	1.7	0.092	0.54
	24	1.85	0.087	0.55
	25	2	0.046	0.61
	26	2	0.075	0.55
	27	2.2	0.073	0.55
SPB2a F = 0.10 m	28	1	0.065	0.83
	29	1.15	0.076	0.82
	30	1.3	0.093	0.75
	31	1.3	0.11	0.71
	32	1.3	0.124	0.67
	33	1.45	0.092	0.77
	34	1.45	0.108	0.73
	35	1.6	0.12	0.68
	36	1.7	0.116	0.66
	37	1.7	0.129	0.64
	38	1.85	0.107	0.64
	39	2	0.109	0.65
SPB2b F = 0.05 m	40	1	0.063	0.63
	41	1.15	0.075	0.60
	42	1.25	0.037	0.70
	43	1.25	0.061	0.64
	44	1.45	0.089	0.57
	45	1.6	0.109	0.51
	46	1.7	0.113	0.50
	47	1.85	0.099	0.49
	48	2	0.039	0.62
	49	2	0.07	0.53
	50	2	0.11	0.48
	51	2.2	0.115	0.48

Table A2. *Cont.*

	#	T (s)	H_s (m)	K_t
SPB2b F = 0.10 m	52	1	0.063	0.79
	53	1.15	0.076	0.74
	54	1.3	0.107	0.65
	55	1.45	0.09	0.73
	56	1.6	0.117	0.64
	57	1.7	0.124	0.60
	58	1.85	0.106	0.60
	59	2	0.109	0.60
	60	2	0.122	0.60
SPB3a F = 0.05 m	61	1	0.049	0.80
	62	1.15	0.06	0.78
	63	1.15	0.077	0.70
	64	1.25	0.043	0.91
	65	1.25	0.068	0.78
	66	1.3	0.099	0.65
	67	1.45	0.073	0.73
	68	1.45	0.099	0.62
	69	1.6	0.115	0.57
	70	1.7	0.088	0.67
	71	1.7	0.115	0.62
	72	1.85	0.079	0.71
	73	1.85	0.097	0.68
	74	2	0.044	0.80
	75	2	0.07	0.73
	76	2	0.103	0.68
	77	2.2	0.106	0.70
SPB3a F = 0.10 m	78	1	0.053	0.85
	79	1.15	0.06	0.88
	80	1.3	0.085	0.82
	81	1.3	0.101	0.74
	82	1.3	0.115	0.72
	83	1.45	0.092	0.77
	84	1.45	0.105	0.70
	85	1.6	0.126	0.67
	86	1.7	0.122	0.74
	87	1.85	0.104	0.74
	88	2	0.111	0.74
	89	2.2	0.109	0.80
SPB3b F = 0.05 m	90	1	0.042	0.83
	91	1.15	0.056	0.86
	92	1.25	0.031	1.00
	93	1.25	0.06	0.85
	94	1.3	0.068	0.78
	95	1.3	0.099	0.64
	96	1.45	0.072	0.74
	97	1.45	0.097	0.63
	98	1.6	0.115	0.58
	99	1.7	0.086	0.71
	100	1.7	0.113	0.65
	101	1.85	0.094	0.71
	102	2	0.043	0.84
	103	2	0.07	0.77
	104	2	0.109	0.66
	105	2.2	0.107	0.71

Table A2. *Cont.*

	#	T (s)	H_s (m)	K_t
SPB3b F = 0.10 m	106	1	0.043	0.95
	107	1.15	0.06	0.92
	108	1.3	0.102	0.76
	109	1.45	0.093	0.78
	110	1.6	0.121	0.72
	111	1.7	0.12	0.77
	112	1.7	0.132	0.76
	113	2	0.11	0.78
	114	2.2	0.12	0.78

Table A3. List of regular incident wave tests characteristics and K_t values, simulated by the mCM14 Boussinesq-type numerical model [28].

	#	T (s)	H_i (m)	K_t
SPB2a F = 0.05 m	10	1	0.067	0.97
	11	1	0.099	0.76
	12	1.25	0.063	0.67
	13	1.25	0.082	0.67
	14	1.25	0.119	0.59
SPB2a F = 0.10 m	28	1	0.063	0.95
	29	1	0.082	0.96
	30	1.25	0.059	0.88
	31	1.25	0.082	0.79
	32	1.25	0.117	0.99
SPB2b F = 0.05 m	40	1	0.058	0.93
	41	1	0.102	0.73
	42	1.25	0.061	0.56
	43	1.25	0.08	0.54
	44	1.25	0.122	0.52
SPB2b F = 0.10 m	52	1	0.061	0.92
	53	1	0.08	0.94
	54	1.25	0.063	0.62
	55	1.25	0.08	0.66
	56	1.25	0.123	0.73
SPB3a F = 0.05 m	61	1	0.062	0.98
	62	1	0.106	0.81
	63	1.25	0.065	0.69
	64	1.25	0.084	0.71
	65	1.25	0.119	0.78
SPB3a F = 0.10 m	78	1	0.062	0.97
	79	1	0.102	0.98
	80	1.25	0.058	0.88
	81	1.25	0.079	0.89
	82	1.25	0.124	0.97
SPB3b F = 0.05 m	90	1	0.061	0.85
	91	1	0.102	0.70
	92	1.25	0.062	0.58
	93	1.25	0.084	0.62
	94	1.25	0.123	0.53
SPB3b F = 0.10 m	106	1	0.06	0.92
	107	1	0.104	0.96
	108	1.25	0.064	0.70
	109	1.25	0.084	0.71
	110	1.25	0.122	0.86

References

1. Kontaxi, C.; Memos, C.D. Submerged Breakwaters as Artificial Habitats. In Proceedings of the 31st IAHR Congress, Seoul, Republic of Korea, 11–16 September 2005.
2. Makris, V.C.; Memos, C.D. Wave Transmission over Submerged Breakwaters: Performance of Formulae and Models. In Proceedings of the 16th International Offshore and Polar Engineering Conference, Lisbon, Portugal, 1–6 July 2007.
3. Brancasi, A.; Leone, E.; Francone, A.; Scaravaglione, G.; Tomasicchio, G.R. On Formulae for Wave Transmission at Submerged and Low-Crested Breakwaters. *J. Mar. Sci. Eng.* **2022**, *10*, 1986. [CrossRef]
4. Hassanpour, N.; Vicinanza, D.; Contestabile, P. Determining Wave Transmission over Rubble-Mound Breakwaters: Assessment of Existing Formulae through Benchmark Testing. *Water* **2023**, *15*, 1111. [CrossRef]
5. Behera, H.; Khan, M.B.M. Numerical Modeling for Wave Attenuation in Double Trapezoidal Porous Structures. *J. Ocean Eng.* **2019**, *184*, 91–106. [CrossRef]
6. Khan, M.B.M.; Behera, H. Analysis of Wave Action through multiple Submerged Porous Structures. *J. Offshore Mech. Arctic Eng.* **2019**, *142*, 011101. [CrossRef]
7. Van der Meer, J.W. *Data on Wave Transmission Due to Overtopping*; Delft Hydraulics: Delft, The Netherlands, 1990.
8. Van der Meer, J.W.; Daemen, I.F. Stability and wave transmission at low-crested rubble-mound structures. *J. Waterw. Port Coast. Ocean. Eng.* **1994**, *120*, 1–19. [CrossRef]
9. d'Angremond, K.; van der Meer, J.W.; de Jong, R.J. Wave Transmission at Low-Crested Structures. In *Coastal Engineering 1996*; American Society of Civil Engineers: New York, NY, USA, 1997; pp. 2418–2427.
10. Seabrook, S.R.; Hall, K.R. Wave Transmission at Submerged Rubblemound Breakwaters. In *Coastal Engineering 1998*; American Society of Civil Engineers: Reston, VA, USA, 1999; pp. 2000–2013.
11. Calabrese, M.; Vicinanza, D.; Buccino, M. Large-Scale Experiments On The Behaviour Of Low Crested And Submerged Breakwaters In Presence Of Broken Waves. In *Coastal Engineering 2002*; World Scientific Publishing Company: Singapore, 2003; pp. 1900–1912.
12. Briganti, R.; van der Meer, J.; Buccino, M.; Calabrese, M. Wave Transmission Behind Low-Crested Structures. In *Coastal Structures 2003*; American Society of Civil Engineers: Reston, VA, USA, 2004; pp. 580–592.
13. van der Meer, J.W.; Briganti, R.; Zanuttigh, B.; Wang, B. Wave Transmission and Reflection at Low-Crested Structures: Design Formulae, Oblique Wave Attack and Spectral Change. *Coast. Eng.* **2005**, *52*, 915–929. [CrossRef]
14. Buccino, M.; Calabrese, M. Conceptual Approach for Prediction of Wave Transmission at Low-Crested Breakwaters. *J. Waterw. Port Coast. Ocean. Eng.* **2007**, *3*, 213–224. [CrossRef]
15. Goda, Y.; Ahrens, J.P. New Formulation of Wave Transmission Over And Through Low-Crested Structures. In *Coastal Engineering 2008*; World Scientific Publishing Company: Singapore, 2009; pp. 3530–3541.
16. Tomasicchio, G.R.; D'Alessandro, F. Wave Energy Transmission through and over Low Crested Breakwaters. *J. Coast. Res.* **2013**, *65*, 398–403. [CrossRef]
17. Zhang, S.X.; Li, X. Design Formulas of Transmission Coefficients for Permeable Breakwaters. *Water Sci. Eng.* **2014**, *7*, 457–467.
18. Sindhu, S.; Shirlal, K.G.; Manu. Prediction of Wave Transmission Characteristics at Submerged Reef Breakwater. *Procedia Eng.* **2015**, *116*, 262–268. [CrossRef]
19. Kurdistani, S.M.; Tomasicchio, G.R.; D'Alessandro, F.; Francone, A. Formula for Wave Transmission at Submerged Homogeneous Porous Breakwaters. *Ocean. Eng.* **2022**, *266*, 113053. [CrossRef]
20. Metallinos, S.A.; Klonaris, T.G.; Memos, C.; Dimas, A.A. Hydrodynamic Conditions in a Submerged Porous Breakwater. *J. Ocean Eng.* **2019**, *172*, 712–725. [CrossRef]
21. Vicinanza, D.; Cáceres, I.; Buccino, M.; Gironella, X.; Calabrese, M. Wave Disturbance behind Low-Crested Structures: Diffraction and Overtopping Effects. *Coast. Eng.* **2009**, *56*, 1173–1185. [CrossRef]
22. Madsen, P.A.; Sørensen, O.R. A New Form of the Boussinesq Equations with Improved Linear Dispersion Characteristics 2: A Slowly Varying Bathymetry. *Coast. Eng.* **1992**, *18*, 183–204. [CrossRef]
23. Avgeris, I.; Karambas, V.T.; Prinos, P. Boussinesq Modeling of Wave Interaction with Porous Submerged Breakwaters. In *Coastal Engineering 2004: (In 4 Volumes)*; World Scientific Publishing Company: Singapore, 2004; pp. 604–616.
24. Metallinos, S.A.; Memos, C. Wave-induced Kinematics inside Submerged Porous Structures. *J. Hydraul. Res.* **2012**, *50*, 388–394. [CrossRef]
25. Metallinos, S.A.; Emmanouilidou, A.M.; Memos, C. Wave-induced Pore Pressures in Submerged Rubble-mound Breakwaters Simulated by a Compound Boussinesq Model. *J. Hydraul. Res.* **2014**, *52*, 24–35. [CrossRef]
26. Hsiao, S.C.; Hu, K.C.; Hwung, H.H. Extended Boussinesq Equations for Water-Wave Propagation in Porous Media. *J. Eng. Mech.* **2010**, *136*, 625–640. [CrossRef]
27. Chondros, M.K.; Memos, C.D. A 2DH non Linear Boussinesq-type Wave Model of Improved Dispersion, Shoaling and Wave Generation Characteristics. *Coast. Eng.* **2014**, *91*, 99–122. [CrossRef]
28. Metallinos, A.; Repousis, E.; Memos, C. Wave Propagation over a Submerged Porous Breakwater with steep slopes. *Ocean Eng.* **2016**, *111*, 424–438. [CrossRef]
29. Van Gent, M.R.A. The Modelling of Wave Action on and in Coastal Structures. *Coast. Eng.* **1994**, *22*, 311–339. [CrossRef]
30. Bing, R.; Wen, H.; Dong, P.; Wang, Y. Numerical Simulation of Wave Interaction with Porous Structures using an Improved Smoothed Particle Hydrodynamic Method. *Coat. Eng.* **2014**, *88*, 88–100.

31. Magdalena, I.; Rif'atin, H.Q.; Kusuma, M.; Reeve, D.E. A non-Hydrostatic Model for Wave Evolution on a Submerged Trapezoidal Breakwater. *Results Appl. Math.* **2023**, *18*, 100374. [CrossRef]
32. Memos, C.; Malliouri, D.; Kazakidou, E.; Tsoukala, V. Impact of Wavelength on the Stability of Rubble Mound Breakwaters. In Proceedings of the 7th International Conference on the Application of Physical Modelling in Coastal and Port Engineering and Science, Santander, Spain, 22–26 May 2018.
33. Calabrese, M.; Buccino, M.; Pasanisi, F. Wave Breaking Macrofeatures on a Submerged Rubble Mound Breakwater. *J. Hydro-Environ. Res.* **2008**, *1*, 216–225. [CrossRef]
34. Madsen, P.A.; Schäffer, H.A. Higher-order Boussinesq-type Equations for Surface Gravity Waves: Derivation and Analysis. *Philos. Trans. R. Soc. Lond.* **1998**, *356*, 3123–3184. [CrossRef]
35. Van der Meer, J.W. Rock Slopes and Gravel Beaches under Wave Attack. Ph.D. Thesis, Delft University of Technology, Delft, The Netherlands, 1988. no. 396.
36. Daemen, I.F.R. *Wave Transmission at Low Crested Structures*; Delft Hydraulics Report H 462; Delft University of Technology: Delft, The Netherlands, 1991.
37. CIRIA; CUR; CETMEF. *The Rock Manual. The Use of Rock in Hydraulic Engineering*; C683; CIRIA: London, UK, 2007.
38. EurOtop. Manual on Wave Overtopping of Sea Defences and Related Structures. An Overtopping Manual Largely Based on European Research, but for Worldwide Application. 2018. Available online: www.overtopping-manual.com (accessed on 15 May 2023).
39. Seeling, W.N. *Two-Dimensional Tests of Wave Transmission and Reflection Characteristics of Laboratory Breakwaters*; Technical Report No. 80-1; US army, Corps of Engineers, CERC: Fort Belvoir, VA, USA, 1980.

Journal of
*Marine Science
and Engineering*

MDPI

Article

Experimental Investigation of Wave Loads on U-OWC Breakwater

Alessandra Romolo *, Bruna Timpano, Valentina Laface, Vincenzo Fiamma and Felice Arena

Natural Ocean Engineering Laboratory NOEL, University "Mediterranea" of Reggio Calabria, Loc. Feo di Vito, 89100 Reggio Calabria, Italy
* Correspondence: aromolo@unirc.it

Abstract: A small-scale field experiment was conducted on a U-OWC incorporated into a caisson breakwater at the NOEL laboratory of Reggio Calabria (Italy). The U-Oscillating Water Column (U-OWC) or REWEC (REsonant Wave Energy Converter) is a device belonging to the family of OWCs. Such a device is very innovative, being able to absorb a very high percentage of incoming sea waves energy and to produce electrical power via proper PTO. The focus of the paper has been the analysis of the impact wave loads acting on the modified U-OWC structure during extreme wave events. A total of 250 records of pure wind waves were analyzed to verify the behaviors of wave loads acting on a U-OWC breakwater during operating conditions. The occurrence of both "quasi-standing wave" loads due to non-breaking waves and "impulsive wave loads", exerted by a wave breaking against the U-OWC model, were observed. Then, Goda's model was applied to predict the wave pressure distribution on the external wall of the U-OWC pneumatic chamber, and the theoretical results were compared to those obtained via small-scale field experiment.

Keywords: wave load; U-OWC breakwater; pressure distribution; small-scale field experiment; random waves

1. Introduction

Breakwaters and upright marine structures are designed and built with the main purpose to define protected basins, and thus they must be realized to resist under the action of extreme sea wave impacts during their lifetime.

Depending on the behavior of the extreme wave pressures, pressure distributions, and total wave forces acting on the vertical structures, two main configurations have been identified: (i) the "quasi-static" loads, associated to irrotational wave motion and "quasi-standing" wave patterns, and (ii) "impulsive" loads, related to the occurrence of breaking waves, for which a dynamic response of the structure is determined.

Many theoretical and numerical approaches have been developed for "quasi-standing" sea waves, based on inviscid, irrotational, and incompressible flow assumptions, which can accurately evaluate wave pressure distributions and related wave forces on an upright coastal structure, in the absence of breaking waves, with both linear and non-linear theories [1–6].

Instead, the estimation of impulsive loads due to breaking waves have been the subject of numerous investigations and estimates, but they are still affected by considerable uncertainties. First of all, experimental investigations on impulsive wave loads have been widely pursued [7–12]. A summary of all the experimental investigations on vertical coastal structures are given in the following main projects; Monlithic Coastal Structures (MCS) and PRObabilistic design tools for VERtical Breakwaters (PROVERBS), funded by the European Union's Marine Science and Technology (MAST) program [13].

These experimental investigations have showed as impulsive wave loads can be much more intense than those produced by quasi-standing irrotational sea waves, and their impacts can produce very huge damage and destruction on coastal structures. Starting from evidence, some analytical and numerical models have been developed by Cooker &

Peregrine [14], and Wood et al. [15], while some other approaches are based on probabilistic methods [16].

In the last decades, vertical coastal structures had a common evolutionary trend, they have been developed to incorporate wave energy converters. This approach turns out to be consistent, because the same structure, that is traditionally an effective port infrastructure, can become 'active' and produce green energy from ocean waves with only a limited increase in costs. In this perspective, the most reliable wave energy converter has been demonstrated to be an Oscillating Water Column (OWC), which can be integrated in a caisson breakwater in reinforce concrete.

In particular, the innovative U-OWC (or REWEC) device, developed in the last decades, is a very promising system for the production of electricity from sea waves. The U-OWC is a modified OWC device with the introduction of a U-shaped duct. Although, this new structural element does not determine any significant structural changes with respect to OWC, it is able to determine substantial modifications in the hydrodynamics of the plant due to the interaction between waves and pneumatic absorption chamber. This allows for the realization of the natural resonance for assigned wave conditions, guaranteeing considerable increases in the energy performance of the U-OWC device, in comparison with traditional OWC. The first full-scale realization of the U-OWC in the prototype of the Port of Civitavecchia, has proven the high capability of the U-OWC to absorb a very high percentage of the incoming wave energy [17].

In the present paper, a modified caisson breakwater with incorporated a U-OWC device has been tested in a relevant environment at the NOEL laboratory of Reggio Calabria (Italy). A small-field experiment (scale 1:8) on U-OWC caisson has been pursued in the NOEL laboratory. During the experimental activity, a large number of sea states with extreme sea conditions have been recorded. In the present work, an intensive investigation on wave loads acting on the modified U-OWC structure has been pursued, based on the PROVERBS approach developed on traditional vertical structures. In particular, "quasi-standing" loads produced by irrotational sea waves and "impulsive" loads, due to breaking waves acting on the U-OWC model, has been identified and the related wave pressure distributions have been obtained from experimental data.

At present, there are no studies concerning the mechanics of wave impacts with the new kind of breakwater modified with U-OWC technology. Some initial studies involve the analyses on wave pressure distributions on classical OWC device integrated into breakwater [18]. For this purpose, the aim of the present study is to investigate the nature of such impacts, even when the pneumatic chambers in the U-OWC breakwater are active for wave energy conversion. Moreover, no models have yet been developed to study the wave loads acting on such U-OWC caissons. Therefore, as a first attempt, Goda's [1] model was applied to estimate the wave pressure distributions acting on these U-OWC breakwaters. Finally, these theoretical distributions have been compared with those obtained from the measurement during the experiment at NOEL laboratory.

2. The Small-Scale Field Experiment on a U-OWC Breakwater in the NOEL Laboratory

2.1. Set-up of the Experimental Campaign

The U-OWC, called also REWEC3-Resonant Wave Energy Converter–realization 3, is a modified Oscillating Water Column (OWC) wave energy converter. The device can be incorporated in a classical caisson breakwater for harbor protection. The "active" part of the modified breakwater consists on a pneumatic chamber for absorbing (capturing) the incoming wave energy. In the case of the U-OWC or REWEC3, the pneumatic chamber in connected to the open field via an addition U-duct (see Figure 1). A description of U-OWC working principle for the exploitation of the incident wave energy is provided in the following. The pneumatic chamber contains water in the lower part and air in the upper part. Then, the air pocket is connected to the atmosphere trough a duct, where an air turbine coupled to an electrical generator may convert the absorbed wave energy (pneumatic power) into electricity. Under the action of sea waves, an alternate motion in the water column (and in the air pocket) is induced inside the pneumatic chamber, and, consequently, in the air duct connecting the chamber to atmosphere. As proven at small- and full-scale, the eigenperiod of the U-OWC plant can be tuned to be equal to the period of the most energetic sea state at the considered location, allowing for the improvement of the energetic performances with respect to traditional OWC devices.

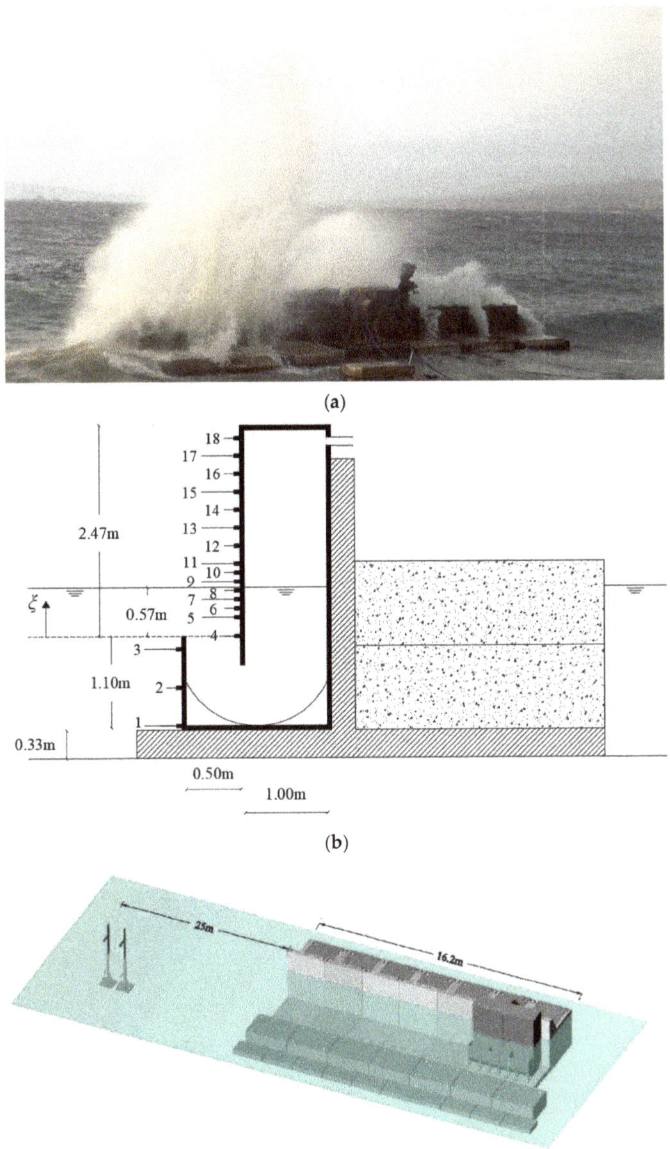

Figure 1. (**a**) The small-scale U-OWC breakwater at NOEL laboratory (off the beach at Reggio Calabria, eastern coast of the Strait of Messina) during a sea storm with the realization of an impact load on the structure; (**b**) Cross- of the U-OWC caisson equipped with pressure transducers (1–18); (**c**) Three-dimensional overview of the small-scale experiment at NOEL laboratory.

A small-scale model (1:8 scale) of a U-OWC caisson breakwater has been installed in 2014 off the beach of Reggio Calabria in the Eastern coast of the Strait of Messina (Southern Italy), at the Natural Ocean Engineering Laboratory (NOEL) and the experimental activities were carried out in 2015 (Figure 1). A peculiarity of the laboratory is that a local wind from NNW often generates sea states consisting of pure wind waves that represent a small-scale model of Mediterranean or Ocean storms, in the Froude dynamic similarity.

The structure, completely made of steel, has been located in front of the reinforced concrete caisson (at the wave-beaten side, see Figure 1b) and structurally connected to the part in reinforce concrete. The "active" part is in steel and it is composed by the vertical U-duct and by the pneumatic chamber, connected to the atmosphere via an orifice. This model has a height of 2.00 m with respect to the mean water level, and the depth of the natural bottom is approximately 2 m. The width of the U-conduit is 0.5 m, while the pneumatic chamber is 1 m wide. The variation of the water depth due to the tide is typically within ± 0.15 m. The opening of the vertical duct is located at 0.57 m below the mean water level. On the top of the chamber, there is a tube with a horizontal axis, connecting the chamber to the atmosphere.

For the small-scale field experiment, the cross-section of the central U-OWC caisson was equipped with a vertical row of 18 pressure transducers allowing the measurement of the wave pressure, and thus of the horizontal wave force, acting on the U-OWC breakwater. Pressure transducers were numbered from 4 to 18, from the opening of the vertical duct to the top of the U-OWC (see Figure 1b). Then, three transducers (from 1 to 3) were located along the external wall from the opening to the bottom. The center distance of the pressure transducers is: 0.21 m from 11 to 18, 0.15 m from 4 to 11, while 0.45 m from 1 to 4.

2.2. Characterization of Incident Wave Field during Experiment

The free surface elevation and wave pressures of the incident waves were calculated through a set of instruments, located in the undisturbed wave field, far from the structure. They were placed on two thin vertical piles positioned 25 m from the U-OWC breakwater, each equipped with a couple of an ultrasonic probe and a pressure transducer (see Figure 1c). The accuracy of wave pressure transducers is 0.5% SB (scale bottom error).

During the experiment, a set of more than 250 records of pure wind waves were recorded from March to October 2015. The duration of each record was 5 min, with a sampling rate of 10 Hz. As shown by Peregrine [5] and Boccotti et al. [7], 10 Hz is a sufficient sampling rate for the identification and characterization of impact wave loads on vertical seawall, also in the presence of impulsive forces due to breaking sea waves. The significant wave height H_s ranges between 0.2 m and 1.0 m, the peak period varies in the range 1.6–8.0 s. The dominant wave direction θ had a range of $[-9°, 15°]$ and was positive when clockwise with respect to the normal outward from the structure.

In addition, the narrow-bandedness parameter ψ of Boccotti [10] for surface waves, defined as the absolute value of the quotient between the minimum and the maximum of the autocovariance function, ranged, during the experiment, among 0.60 and 0.76 (that are the typical values relative to wind wave spectra).

3. Pressure Distributions on the U-OWC Breakwater at NOEL Laboratory

3.1. The PROVERBS Wave Load Classification for a U-OWC Breakwater

Based on the experience gained by the European project PROVERBS on traditional vertical breakwaters, the phenomenon of impulsive forces was investigated on a small-scale field model of U-OWC breakwater. Thus, the aim of this work is to apply the approach already developed by PROVERBS [8] for conventional breakwaters also to this new technology. In this regards, the methodology adopted is reported in the Flow Chart of Figure 2.

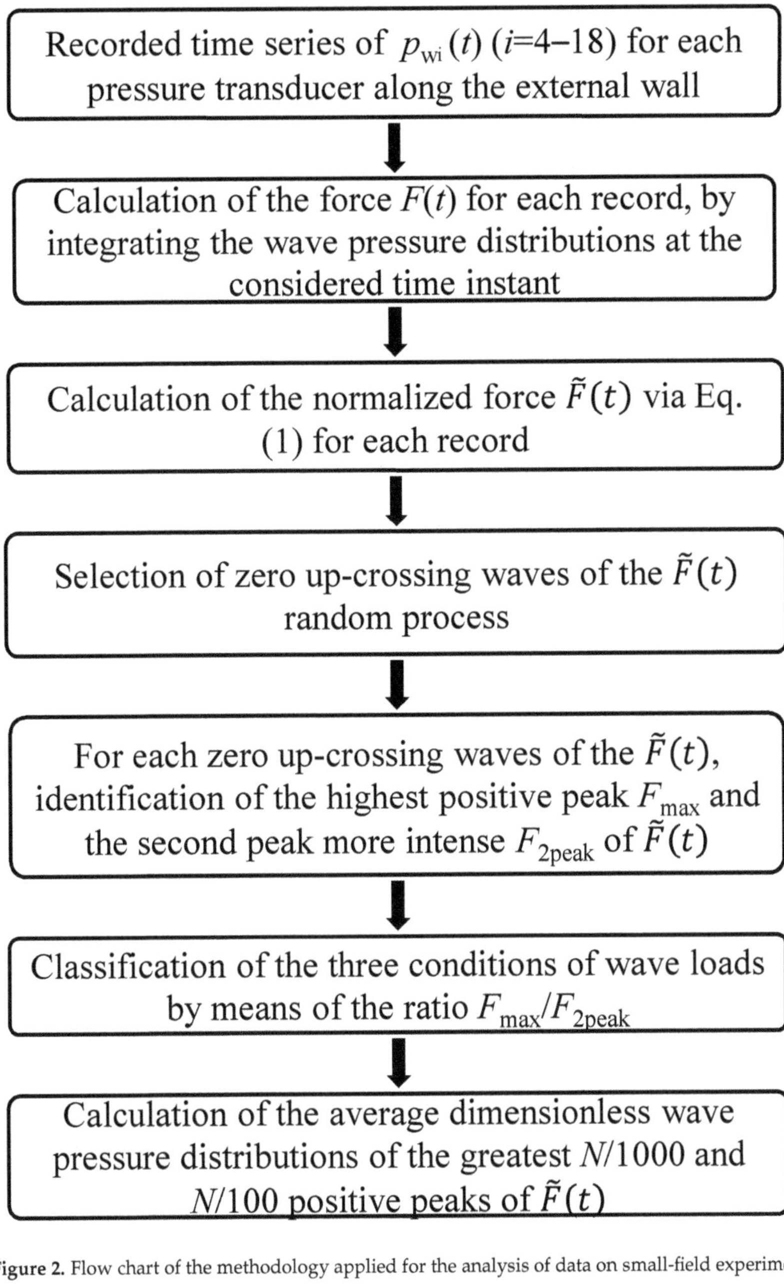

Figure 2. Flow chart of the methodology applied for the analysis of data on small-field experiment.

When the waves are perfectly reflected, a "Quasi-Standing (QS)" wave pattern results, characterized by two double positive peaks of equal intensity, when a high wave crest impacts to the breakwater. When the breaking phenomena is realizing on the structure or in front of it, the wave pressure in the time domain shows an asymmetrical trend with two peaks with different intensity, with the first one greater than the second [3]. The impulsive wave impacts in these configurations are distinguished in "Slightly Breaking Waves (SBW)", where the first higher positive load is greater than 1–2.5 times the subsequent second and in "Impact Loads (IL)", where the highest positive load is greater than 2.5 times the second.

In the present work, PROVERBS methodology has been applied to the investigation of records on the small-scale field model of U-OWC breakwater installed at the NOEL laboratory of the University "Mediterranea" of Reggio Calabria in order to identify the wave load configurations acting on the modified coastal structure.

3.2. The Pressure Distribution and the Relative Wave Forces on the U-OWC Breakwater

In this section, the post-processing of data of the small-scale experiment on the model of U-OWC breakwater (1:8 scale) is pursued in order to investigate and classify impact wave loads acting against the modified breakwater.

Firstly, the time series of wave pressure recorded by the array of eighteen pressure transducers, placed as shown in Figure 1c, have been considered. By these measurements, the wave pressure distributions along the U-OWC breakwater have been calculated for each time instants by considering pressure transducers from 4 to 18. Then, the force $F(t)$ per unit of length at the mid-section of the U-OWC breakwater is obtained by integrating the wave pressure distribution recorded. In detail, the wave loads acting on the external wall of the pneumatic U-OWC chamber are calculated by integrating the pressure distributions of

Then, the normalized force $\widetilde{F}(t)$ for each record was obtained by dividing the measured force $F(t)$ by its own standard deviation σ_F, and it was calculated for each recorded sea state:

$$\widetilde{F}(t) = \frac{F(t)}{\sigma_F}. \tag{1}$$

The positive peaks of the $\widetilde{F}(t)$ random process was determined for every 5-min records. For each peak of $\widetilde{F}(t)$, the corresponding dimensionless wave pressure distributions were obtained, by dividing the measured wave pressure $p_{w_i}(t)$ (i = 4–18) by its own standard deviation σ_{p_w}.

The average pressure distribution of the greatest $N/1000$ and $N/100$ of the positive force peaks (N = total number of waves of the $F(t)$ process) is calculated.

In this paper, the positive force peak refers to the largest positive value of a wave of the $F(t)$ process, and the number of waves refers to the number of zero up-crossing waves of the random wave-force-process, $F(t)$. In addition, it is assumed ξ_i/d (i = 4–18) the dimensionless coordinates of the pressure transducers along the external wall, related to the depth (a = 0.57 m) of the upper opening of the U-duct.

Firstly, the zero up-crossing waves of the random wave force process, $\widetilde{F}(t)$, of the whole dataset of the experiment have been considered. These have been 5000 sea waves.

Considering the wave loads classification of the PROVERBS project, the pressures distributions on the external wall of the U-OWC model have been obtained for the three classes of wave loads. In this regards, in the time series of the wave force process F, more than one positive peak can occur (see different behaviors in Figures 3 and 4). Thus, the highest positive peak is denoted by $F_{\max}$, while the second peak more intense is indicated by F_{2peak}.

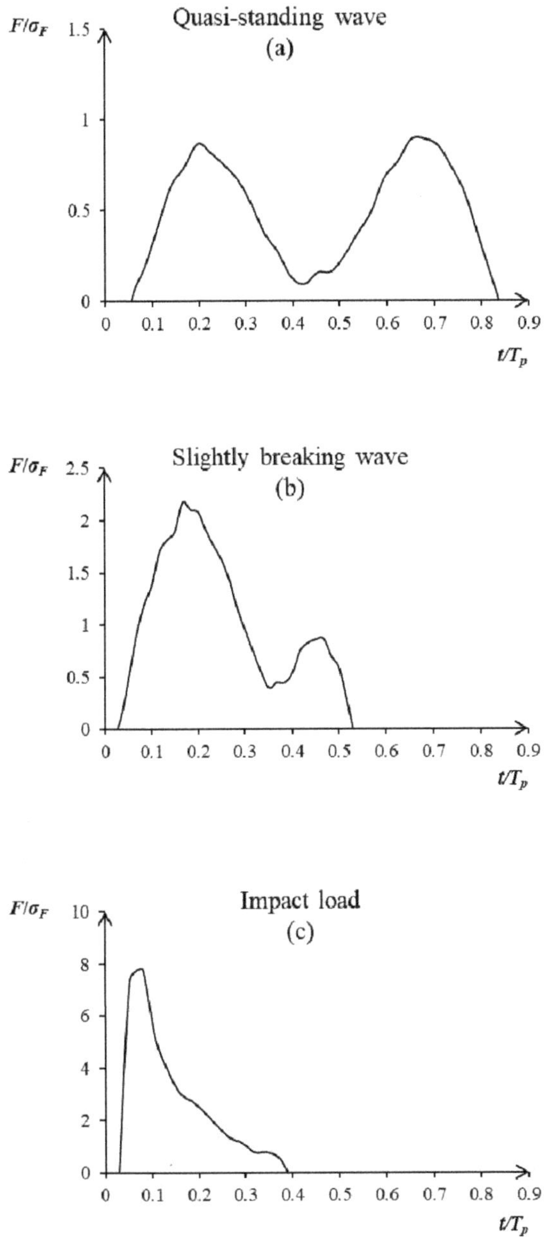

Figure 3. Example of the three conditions of wave loads F on the U-OWC breakwater versus time instant t (related to the peak period T_p): (**a**) a quasi-standing wave; (**b**) a slightly breaking wave $F_{\max}/F_{2peak} = 2.48$ and (**c**) an impact load $F_{\max}/F_{2peak} = 10.18$, recorded at NOEL.

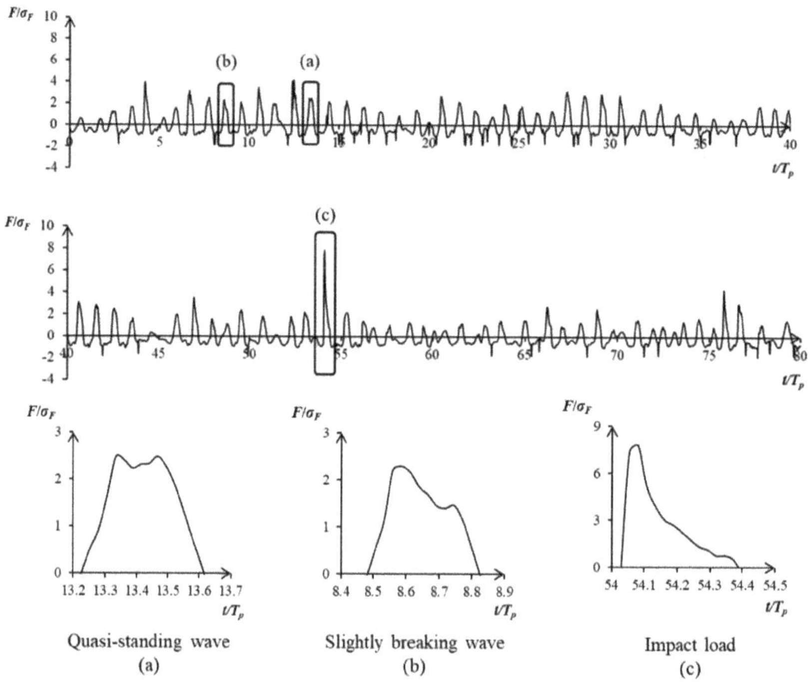

Figure 4. Example of complete time series of the wave force, with the occurrence of the three cases of wave loads: (**a**) Quasi-Standing: (**b**) Slightly Breaking; (**c**) Impact Load.

When F_{max} is comparable to F_{2peak}, a "Quasi-Standing Wave (QSW)" is realized (Figure 3a); if $1 < F_{\max}/F_{2peak} < 2.5$, a "Slightly Breaking Wave (SBW)" occurs (Figure 3b); for $F_{\max}/F_{2peak} > 2.5$, an "Impact Load (IL)" is considered (Figure 3c). Moreover, the peak duration decreases significantly, when the peak intensity increases greatly, as shown in Figure 3.

After, the average dimensionless wave pressure distributions of the greatest $N/1000$ and $N/100$ positive peaks of $\widetilde{F}(t)$ have been calculated. They have been given for the three classes of wave loads in Figure 4, where dots denote the measured wave pressures on the front wall of the U-OWC model.Comparisons from Figure 5 show that the total wave loads are almost double from "Quasi-Standing Wave (QSW)" to "Slightly Breaking Wave (SBW)", for both the $N/1000$ and $N/100$ maximum dimensionless wave pressures. Then, the $N/1000$ maximum wave pressures of the "Impact Load (IL)" are approximately three times greater than that of the "Slightly Breaking Wave (SBW)" and five times greater than that of the "Quasi-Standing Wave (QSW)". An additional behavior is outlined, when a predominance of impulsive component in the total wave forces, due to wave breaking against structure, is realized at the U-OWC. As a matter of fact, the increase in the total force on the modified structure, in the presence of impulsive loads, is essentially due to the increase in wave pressures below the mean-water-level (MWL), while that for "Slightly Breaking Wave (SBW)" and "Impact Load (IL)" is, respectively, approximately 2.5 and 7 times greater than the quasi-static ones. On the other hand, above the mean-water-level, the wave pressures for both the SBW and the IL are approximately 50–60% higher than those that are quasi-static.

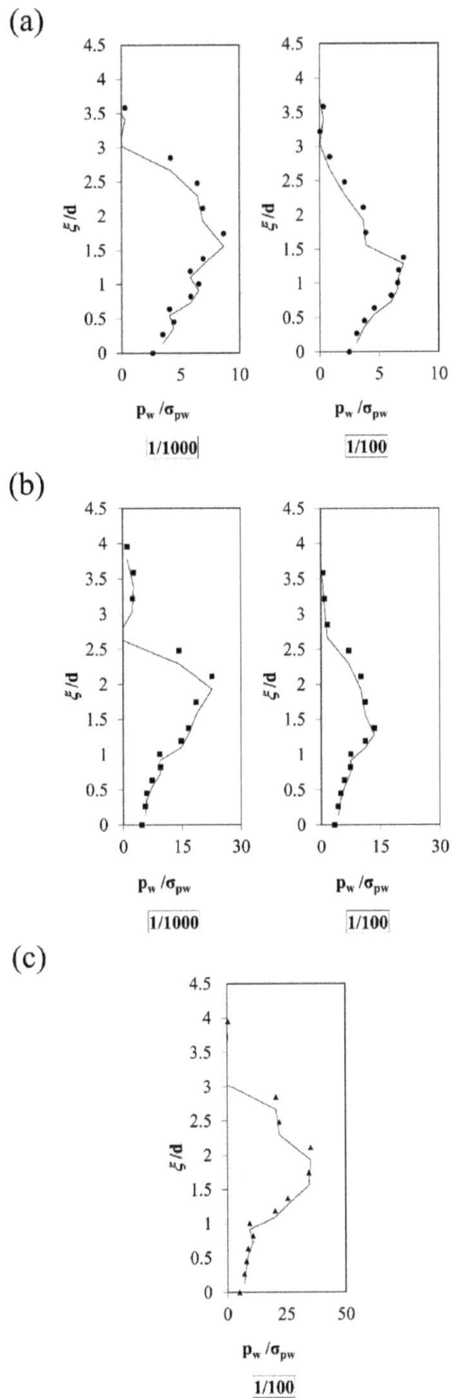

Figure 5. Average wave pressure distribution of the greatest $N/1.000$ and $N/100$ positive force peaks $\widetilde{F}(t)$ of "Quasi-Standing waves" (**a**), of "Slightly Breaking waves" (**b**) and of "Impact Loads" (**c**).

4. Goda's Model Applied to the U-OWC Breakwater

4.1. Goda's Model Applied for the Small-Scale U-OWC Breakwater Installed at the NOEL Laboratory

Ln the literature, Goda's model [1] is one of the most effective method, widely adopted [7] to determine the pressure distribution on a seawall or a vertical breakwater, in the case either of breaking or non-breaking incident sea waves.

In particular, Goda's model is applied with reference to the configuration of the U-OWC breakwater tested during the small-scale field experiment campaign arranged at NOEL laboratory (see Figure 1).

The model for the pressure distribution proposed by Goda is shown in Figure 6 and it provides the maximum elevation of the free surface, $\eta_{\max}$, which is equal to

$$\eta_{\max} = \frac{3}{4}(1+\cos\theta)H \tag{2}$$

and the maximum (positive) wave pressures under a wave crest, which are given, with reference to the scheme of Figure 6, by

$$p_{w1}^{(+)} = \frac{1}{2}(1+\cos\theta)(\alpha^{I}+\alpha^{II}\cos^2\theta)\gamma\, H \tag{3}$$

$$p_{w}^{(+)} = \alpha^{III} p_{w1}^{(+)} \tag{4}$$

$$p_{w2}^{(+)} = \left(1-\frac{h_M}{\eta_{\max}}\right)p_{w1}^{(+)} \tag{5}$$

$$p_{w3}^{(+)} = (1-\alpha^{IV})p_{w1}^{(+)} + \alpha^{IV} p_{w}^{(+)} \tag{6}$$

where γ is the water density, h_M is the quote of the top of U-OWC pneumatic chamber with respect to the mean water level, H is the height of the largest wave in the design sea state and θ is the dominant direction of propagation of incoming waves with respect to the orthogonal to the U-OWC breakwater. In detail, when the irregular sea waves approaches orthogonally to U-OWC caissons, θ is equal to 0.

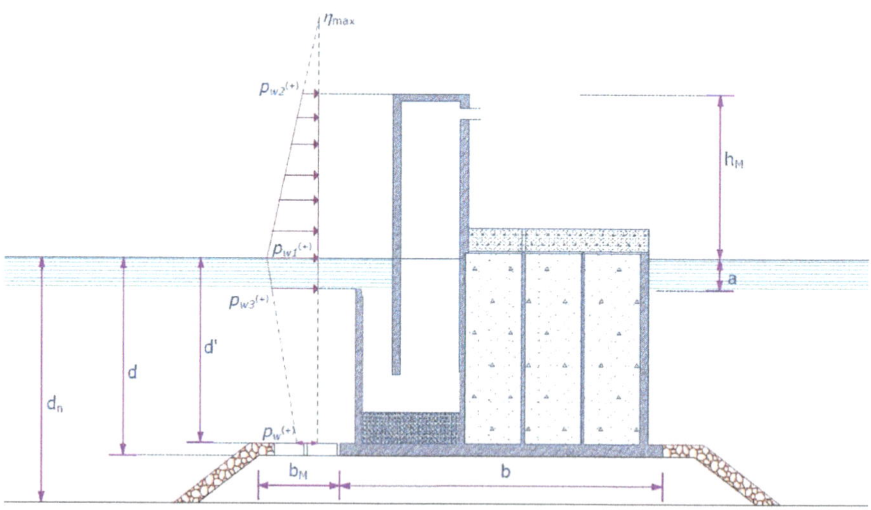

Figure 6. Scheme of Goda's model applied to U-OWC breakwater.

More specifically, we assumed the design wave height H equal to $1.8H_s$, where H_s is the significant height of the incident waves. In the present experiment, the value of H_s have been calculated on the time series data of the free surface displacement obtained from the two ultrasonic probes in the undisturbed wave field (see Figure 1c). In addition, the wave period T of the design wave is assumed equal to the mean wave period of each recorded sea state.

Moreover, in Equations (3)–(6), the coefficients α^I, α^{II}, α^{III} and α^{IV} are empirical are they are defined by Goda's model [1] as follows

$$\alpha^I = 0.6 + \frac{1}{2}\left[\frac{2kd_n}{\sinh(2kd_n)}\right]^2 \tag{7}$$

$$\alpha^{II} = Min\left[\left(\frac{d'_n - d'}{3d'_n}\right)\left(\frac{H}{d'}\right)^2, 2\frac{d'}{H}\right] \tag{8}$$

$$\alpha^{III} = 1 - \frac{d}{d_n}\left[1 - \frac{1}{\cosh(kd_n)}\right] \tag{9}$$

$$\alpha^{IV} = \frac{a}{d'} \tag{10}$$

where k is the wave number relative to the water depth d_n and the mean wave period, with the associated wave length L. Then, d, d' and d_n are the water depths illustrated in Figure 6, while d'_n is defined as the water depth of the natural bed at a distance of $5H_s$ from the U-OWC breakwater.

Following Goda's model [1], the wave pressure $p_{w1}^{(+)}$ on the mean water level, is expressed by Equation (3) for the case of non-breaking waves, which is identified by conditions that

$$\frac{d_n}{H_s} \geq 2.4 \text{ and } \frac{d_n}{L} \geq 0.12. \tag{11}$$

If relations (11) are not satisfied, it is necessary to take into account an additional impulsive wave pressure contribution. In this case, the wave pressure $p_{w1}^{(+)}$ is defined by Goda [1] by the following expression

$$p_{w1}^{(+)} = \frac{1}{2}(1 + \cos\theta)(\alpha^I + \alpha * \cos^2\theta)\gamma\, H \tag{12}$$

where

$$\alpha * = Max\left\{\alpha^{II}, \alpha_I\right\}, \tag{13}$$

with α^{II} expressed by Equation (8) and coefficient α_I calculated by relation that

$$\alpha_I = \alpha_{IH}\alpha_{IB} \tag{14}$$

being

$$\alpha_{IH} = \min\left\{\frac{H}{d'}; 2.0\right\} \tag{15}$$

$$\alpha_{IB} = \begin{cases} \frac{\cos\delta_2}{\cosh\delta_1} & : \delta_2 \leq 0, \\ \frac{1}{\cosh\delta_1 \cosh^{1/2}\delta_2} & : \delta_2 > 0, \end{cases} \tag{16}$$

$$\delta_1 = \begin{cases} 20\,\delta_{11} : \delta_{11} \leq 0, \\ 15\,\delta_{11} : \delta_{11} > 0, \end{cases} \tag{17}$$

$$\delta_2 = \begin{cases} 4.9\,\delta_{22} : \delta_{22} \leq 0, \\ 3.0\,\delta_{22} : \delta_{22} > 0, \end{cases} \tag{18}$$

$$\delta_{11} = 0.93\left(\frac{b_M}{L} - 0.12\right) + 0.36\left(0.4 - \frac{d\prime}{d_n}\right), \tag{19}$$

$$\delta_{22} = -0.36\left(\frac{b_M}{L} - 0.12\right) + 0.93\left(0.4 - \frac{d'}{d_n}\right) \tag{20}$$

In Equations (19) and (20), b_M is the length between the base of the U-OWC breakwater and the head of the foundation berm. L is the wave length at water depth d_n.

4.2. Comparison between Goda's Model and Experimental Pressure Distributions

Here, Goda's formulas have been applied to calculate the pressure distributions on the breakwater modified with the U-OWC technology.

The $N/1000$ average pressure distributions (N = number of total waves for the three class), for the largest $N/1000$ positive force peaks $\widetilde{F}(t)$, represent the most intense $N/1000$ waves of force process recorded during the small-scale experiment for the three classes of quasi-standing waves, slightly breaking waves and impact loads.

In Figures 7–9, the comparison between the experimental distributions and the theoretical pressure distributions using Goda's model is shown.

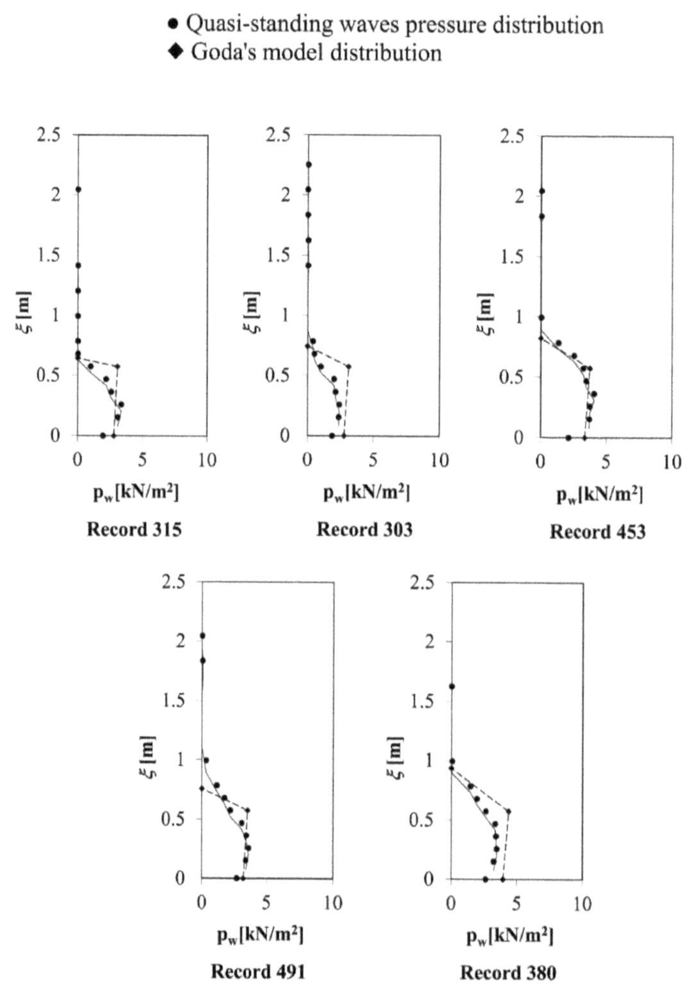

Figure 7. Pressure distributions of each of the greatest N/1.000 positive force peaks $\widetilde{F}(t)$ for "Quasi-Standing waves" versus Goda's model.

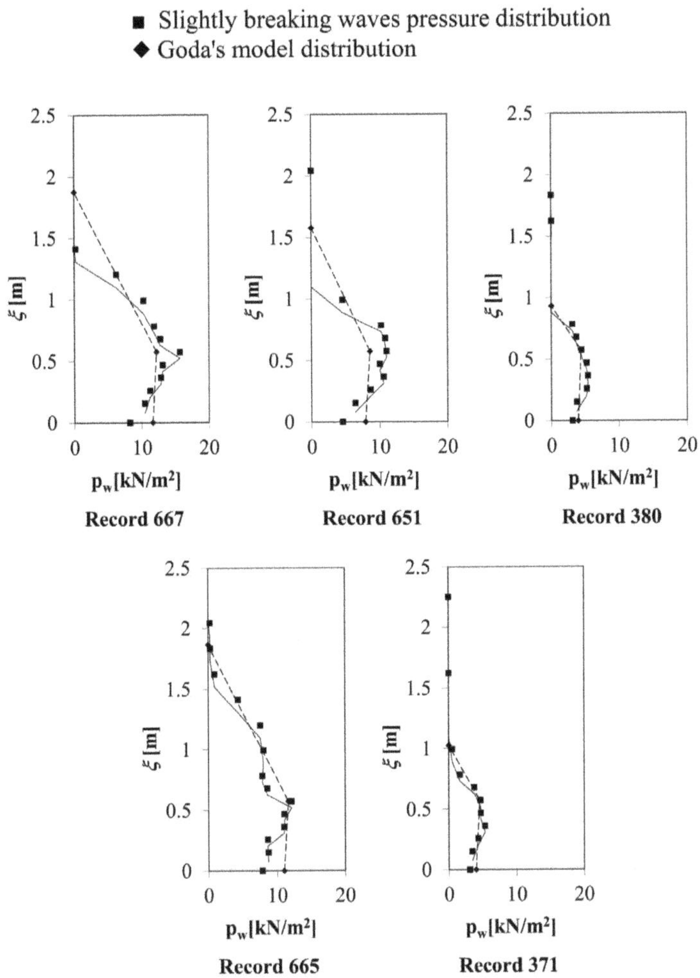

Figure 8. Pressure distributions of each of the greatest N/1.000 positive force peaks $\widetilde{F}(t)$ for "Slightly Breaking waves" versus Goda's model.

Figures 7 and 8 shows the comparison between the theoretical and the experimental pressure distributions corresponding, to the five highest quasi-standing and slightly breaking waves, respectively. In these cases, Goda's model approximates quite well the experimental wave force.

Finally, Figure 9 represents the pressure distributions of the five greatest positive peaks of the force process $\widetilde{F}(t)$ for the class of impact loads. In all three records, the theoretical schemes underestimate significantly the impulsive force obtained by the experimental pressure distributions.

Therefore, the pressure distribution for the impact loads shows experimentally an important limitation to predict impulsive pressures on the U-OWC model with the existing Goda's approach, while this theoretical scheme is in good agreement with experimental results for "quasi-standing" and "slightly breaking" waves.

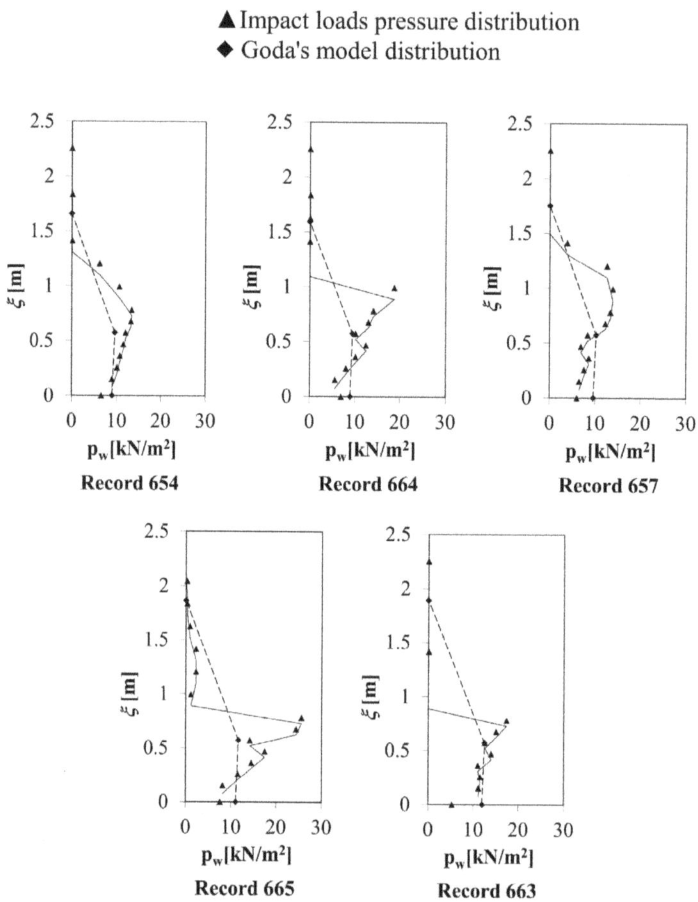

Figure 9. Pressure distributions of each of the greatest N/1.000 positive force peaks $\widetilde{F}(t)$ for "Impact Load waves" versus Goda's model.

5. Conclusions

In this paper, the analysis of the impact wave loads acting on the modified U-OWC structure during a field experiment in the NOEL laboratory has been pursued.

Based on the PROVERBS methodology for traditional breakwater, the phenomenon of impact forces was investigated on the small-scale U-OWC model. The occurrence of both "quasi-standing wave" loads due to non-breaking waves and "impulsive wave loads" with different wave impacts exerted by a wave breaking against the U-OWC model, has been observed. The experimental average pressure distributions for the three classes of loads have been calculated. It has been observed that the total force acting on a U-OWC breakwater doubles during "Slightly Breaking Wave (SBW)" impacts compared to quasi-static conditions; for "Impact Loads (IL)" the impulsive loads due to wave breaking are prevalent and remarkable, becoming almost five times greater than static wave loads.

Finally, the applicability of Goda's model for the prediction of wave pressure distributions on the U-OWC model has been verified, under different wave load conditions, showing that the model can provide efficient estimate in the case of "Quasi-Static Wave" and "Slightly Breaking Wave", but it is not accurate for the prevision of Impact Loads on U-OWC modified breakwater.

Author Contributions: Conceptualization, A.R., V.L, V.F. and F.A.; methodology, A.R., V.F., B.T., V.L. and F.A.; software, B.T, V.L. and A.R.; validation, A.R., B.T. and V.F.; writing—original draft preparation, A.R. and B.T.; writing—review and editing, A.R., B.T., V.L., V.F. and F.A.; supervision, F.A. and A.R. All authors have read and agreed to the published version of the manuscript.

Funding: This research received no external funding.

Informed Consent Statement: Informed consent was obtained from all subjects involved in the study.

Data Availability Statement: The data presented in this study are available on request from the corresponding author. The data are not publicly available because they were acquired as part of a research project owned by the NOEL laboratory.

Conflicts of Interest: The authors declare no conflict of interest.

References

1. Goda, Y. *Random Seas and Design of Maritime Structures*; Advanced Series on Ocean Engineering; World Scientific: Singapore, 2000; pp. 1–433.
2. Boccotti, P. *Wave Mechanics for Ocean Engineering*; Elsevier Science: Oxford, UK, 2000; pp. 1–485.
3. Boccotti, P. *Wave Mechanics and Wave Loads on Marine Structures*; Butterworth-Heinemann; Elsevier: Amsterdam, The Netherlands, 2014; pp. 1–311.
4. Romolo, A.; Arena, F. Mechanics of Nonlinear Random Wave Groups interacting with a Vertical Wall. *Phys. Fluids* **2008**, *20*, 036604. [CrossRef]
5. Romolo, A.; Arena, F. Three-dimensional non-linear standing wave groups: Formal derivation and experimental verification. *Int. J. Non-Linear Mech.* **2013**, *57*, 220–239. [CrossRef]
6. Romolo, A.; Arena, F.; Laface, V. A generalized approach to the mechanics of three-dimensional nonlinear ocean waves. *Probabilistic Eng. Mech.* **2014**, *35*, 96–107. [CrossRef]
7. Bagnold, R.A. *Interim Report on Wave-Pressure Research. Technical Report*; Institution of Civil Engineers: London, UK, 1939.
8. Kirkgoz, M.S. Shock pressure of breaking waves on vertical walls. *J. Waterw. Port Coast. Ocean. Div.* **1982**, *108*, 81–95, WW1, ASCE New York. [CrossRef]
9. Oumeraci, H.; Klammer, P.; Partenscky, H. Classification of breaking wave loads on vertical structures. *J. Waterw. Port Coast. Ocean Eng.* **1993**, *119*, 381–397. [CrossRef]
10. Peregrine, D.H. Water-wave impact on walls. *Annu. Rev. Fluid Mech.* **2003**, *35*, 23–43. [CrossRef]
11. Cuomo, G.; Allsop, W.; Bruce, T.; Pearson, J. Breaking wave loads at vertical seawalls and breakwaters. *Coast. Eng.* **2010**, *57*, 424–439. [CrossRef]
12. Boccotti, P.; Arena, F.; Fiamma, V.; Romolo, A.; Barbaro, G. A small scale field experiment on wave forces on upright breakwaters. *J. Waterw. Port Coast. Ocean. Eng.* **2012**, *138*, 97–114, ASCE. [CrossRef]
13. Oumeraci, H.; Kortenhaus, A.; Allsop, W.; de Groot, M.; Crouch, M.; Vrijling, H.; Voortman, H. *Probabilistic Design Tools for Vertical Breakwaters*; Balkema: Lisse, The Netherlands, 2001; pp. 1–373.
14. Cooker, M.J.; Peregrine, D.H. Pressure impulse theory for liquid impact problems. *J. Fluid Mech.* **1995**, *297*, 193–214. [CrossRef]
15. Wood, D.J.; Peregrine, D.H.; Bruce, T. Wave impact on a wall using pressure impulse theory. I: Trapped air. *J. Waterw. Port Coast. Ocean Eng.* **2000**, *126*, 182–190. [CrossRef]
16. Chena, X.; Hofland, B.; Molenaar, W.; Capel, A.; Van Gent, M.R.A. Use of impulses to determine the reaction force of a hydraulic structure with an overhang due to wave impact. *Coast. Eng.* **2019**, *147*, 75–88. [CrossRef]
17. Arena, F.; Romolo, A.; Malara, G.; Fiamma, V.; Laface, V. Response of the U-OWC Prototype Installed in the Civitavecchia Harbour. In Proceedings of the 37th International Conference of Offshore, Mechanics and Arctic Engineering, Madrid, Spain, 17–22 June 2018. OMAE2018-78762.
18. Pawitan, K.A.; Dimakopoulos, A.S.; Vicinanza, D.; Allsop, W.; Bruce, T. A loading model for an OWC caisson based upon large-scale measurements. *Coast. Eng.* **2019**, *145*, 1–20. [CrossRef]

Journal of
Marine Science and Engineering

MDPI

Article

Application of SWASH to Compute Wave Overtopping in Ericeira Harbour for Operational Purposes

Anika Manz [1,*], Ana Catarina Zózimo [1,*] and Juan L. Garzon [2,*]

[1] Laboratório Nacional de Engenharia Civil (LNEC), 1700-111 Lisboa, Portugal
[2] CIMA—Centre for Marine and Environmental Research, Universidade do Algarve, 8005-139 Faro, Portugal
* Correspondence: a67280@ualg.pt (A.M.); aczozimo@lnec.pt (A.C.Z.); jlhervas@ualg.pt (J.L.G.)

Abstract: This work aimed at testing the capability of the numerical model SWASH to be implemented in the prototype of the overtopping and flooding forecast system HIDRALERTA for Ericeira harbour. In contrast to the neural network NN_OVERTOPPING2, which is currently implemented in HIDRALERTA, SWASH is able to estimate the flood extension and wave propagation along the domain, which makes it a possible improvement to NN_OVERTOPPING2. The one-dimensional version of the SWASH model was implemented to simulate overtopping at two different profiles (antifer and tetrapods) and calibrated for three storms in 2019 by comparing the simulated overtopping discharge to NN_OVERTOPPING2 results. For the calibration, the Manning coefficient was used to represent the friction of the armour layer. Then, for operational purposes, four expressions to calculate the Manning coefficient were developed based on: the relative crest freeboard, the wave steepness, the incident wave angle and the type of armour layer. The expressions showed small errors between the calculated and calibrated Manning coefficients and highlighted the importance of the incident wave angle to obtain an accurate calibration. Despite an underestimation of the overtopping discharge in some cases, the SWASH model was found to provide overall good results when applied with calculated Manning coefficients and suitable to be implemented in HIDRALERTA.

Keywords: wave overtopping; early warning system; SWASH; Manning coefficient; HIDRALERTA; To-SEAlert

1. Introduction

Adverse sea conditions can result in wave-induced overtopping and flooding, having a negative impact on society, the environment and economy [1,2]. In many regions, coastal communities and socio-economic activities rely on the capability of coastal structures to minimise the consequences of wave-induced flooding. Yet, these structures are rarely designed for zero overtopping discharges [3]. To reduce the risks caused by these hazards, forecast and early warning systems (EWSs) have been identified as an important tool to detect emergency situations and to initiate the necessary safety measures [4]. While the United Nations calls to strengthen these prediction systems, they are still in their infancy. In Europe, only some efforts have been made to implement EWSs (e.g., [5,6]) and in many countries such as Portugal, highly exposed to coastal hazards, no fully operational national flood forecast and early warning system exists yet.

The HIDRALERTA system [7–11] aims to contribute to filling this gap. HIDRALERTA is a wave overtopping and flooding forecast system with early warning and risk assessment capabilities. It runs in real-time mode and enables the identification of emergency situations in coastal and harbour areas with 72 h of anticipation. The system uses a neuronal network system, NN_OVERTOPPING2 [12], to compute mean overtopping discharge q at several cross-sections of the structure in harbour areas. NN_OVERTOPPING2 uses as input conditions the results of the numerical wave propagation model DREAMS [13] and tide levels from the XTIDE model [14]. Warnings are triggered when pre-set thresholds for q are exceeded and forecasts are sent daily to decision-makers [2,7,8].

While wave overtopping has been traditionally estimated through physical modelling or with (semi-) empirical formulas (e.g., [15,16]), advances in computer hardware and numerical methods in recent years have made it possible to use numerical models to obtain accurate estimations of physical processes such as wave propagation and transformation in shallow areas and overtopping over the defense structures [17].

A recent approach to model wave overtopping is based on the dispersive NLSW (Non Linear Shallow Water) equations, which include a non-hydrostatic pressure term, as well as a resolution of the vertical flow and its structure. The SWASH model (an acronym for Simulating WAves till SHore) [18] numerically simulates non-hydrostatic, free-surface, rotational flows in one or two horizontal dimensions. As the governing equations rely on the NLSW equations and include non-hydrostatic pressure, they can describe complex and rapidly changing flows in detailed topo-bathymetries that are often found in coastal flooding events. Previous studies have demonstrated the good skills of the model in simulating wave propagation in shallow conditions and wave overtopping discharge [17,19–23].

This study was developed within the frame of the To-SEAlert project, which aims at increasing the reliability and accuracy of HIDRALERTA. The main objective of this study was to investigate the suitability of the numerical model SWASH to be integrated into the HIDRALERTA system for the Ericeira harbour. The SWASH model is considered a robust process-based model with high accuracy of simulating wave propagation in shallow water [16]. Given that any point in the numerical domain can be specified in the output of the SWASH model, it is possible to estimate the inland incursion of overtopping discharges or, in the case of two-dimensional models, the flooded area, as well as the wave propagation along the domain, which the neural network tool NN_OVERTOPPING2, the current approach implemented in HIDRALERTA, is not capable of. In order to achieve this objective, a one-dimensional model was implemented for two profiles of Ericeira breakwater, where the model was applied to simulate overtopping during storm conditions. Its results for mean overtopping discharge were compared to NN_OVERTOPPING2. The main efforts were devoted to calibrating the model by varying the Manning coefficient that represents the friction of the armour layer of the breakwater and to get the best agreement with the overtopping estimations obtained from NN_OVERTOPPING2. For predictive purposes, expressions for the determination of the Manning coefficient were derived, allowing to simulate wave-induced flooding during storms for a given type of armour layer. These expressions are of capital importance for the successful implementation of a numerical model into a real-time forecast system, since its parameterisation has to be previously and seamlessly defined based on some oceanic and site-specific conditions of the case.

2. Materials and Methods

2.1. Study Site

Ericeira harbour is located on the west coast of Portugal. It is sheltered by a 430 m long breakwater, oriented to the south-west, with a quay in the rear side. For the simulation of wave overtopping, two profiles were chosen. Profile-A has an armour layer of antifer cubes and is located at the head of the breakwater. It has an orientation of 262° N. Profile-T consists of tetrapods and is in the vicinity of the quay of the breakwater and has an orientation of 309° N (Figure 1).

The two profiles differ in their slope angle, their crest freeboard R_c and armour freeboard A_c and in the width of the crest G_c (Table 1).

Table 1. Structure characteristics of the breakwater profiles referred to the Portuguese hydrographic zero (ZH).

	Slope	R_c (m)	A_c (m)	G_c (m)
Profile-A	1:2.0	9.03	10.85	5.79
Profile-T	1:1.5	8.98	10.20	5.28

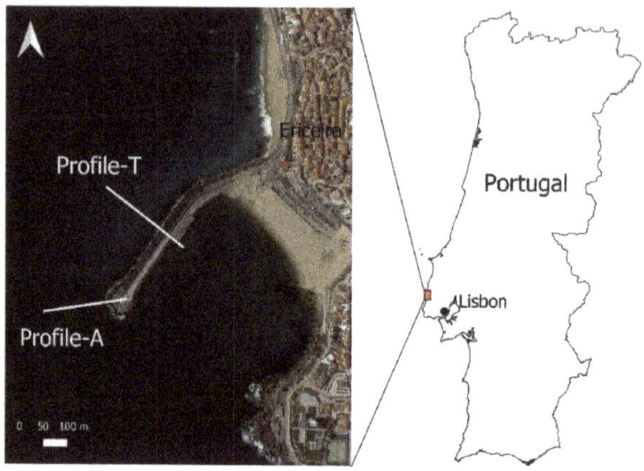

Figure 1. Location of Ericeira harbour and the breakwater profiles.

The simulations were performed for three consecutive storms that hit the study area between the 15th and 23rd of December 2019 (Figure 2). The first storm occurred between the 15th and 17th and remained unnamed, with wave directions between 280° and 300° N, a maximum peak period T_p of 16.7 s and a maximum significant wave height H_s of 5.57 m. The Elsa storm occurred from 19th to 21st of December and had wave directions between 260° and 280° N, maximum T_p of 13.6 s and maximum H_s of 7.93 m. The Fabien storm hit the study area between 22nd and 23rd of December and had wave directions between 257° and 290° N, maximum T_p of 16.7 s and maximum H_s of 7.14 m.

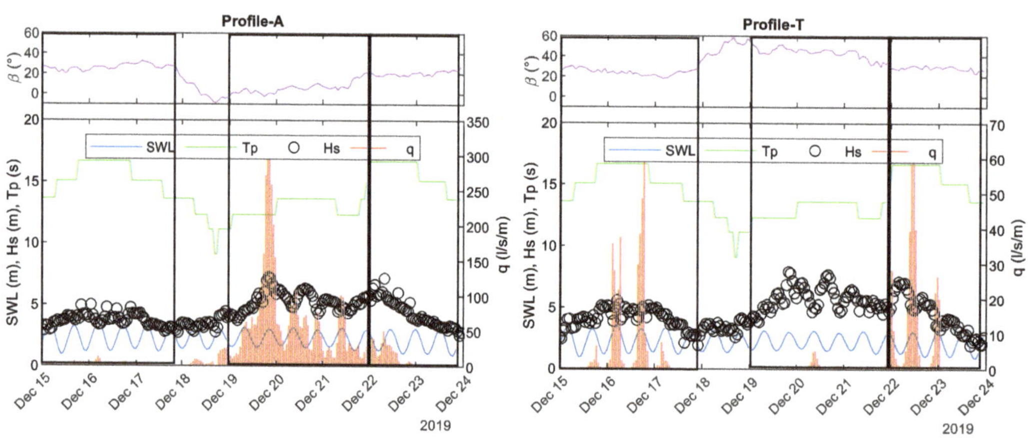

Figure 2. Time series of the simulated storms of Profile-A and Profile-T (from left to right). The boxes indicate the first storm (unnamed), the Elsa storm and the Fabien storm. SWL indicates the Still Water Level, q is the mean overtopping discharge and β (°) is the angle of wave attack, which is the angle between the incident wave direction and the perpendicular to the profile.

The oceanic offshore conditions (waves, wind and surge) during these storms were obtained from ERA-5 hindcast data downloaded from the Copernicus Climate Change Service (C3S) Climate Data Store [24]. This information was then introduced in the numerical framework developed for HIDRALERTA. This framework uses the XTIDE model [14] for the astronomical tide data, and the models SWAN [25] and DREAMS [13] to propagate the wave conditions from offshore to the vicinity of the structure. To force the SWASH models, water levels and wave regimes were extracted at the points located at the offshore boundary of the simulated profiles (341 m distance of the breakwater with a depth of 14.47 m below the Portuguese hydrographic zero, ZH, for Profile-A and 335 m distance of the breakwater with a depth of 9.5 m ZH for Profile-T). In addition, to compare with the performance of the actual HIDRALERTA system, oceanic conditions were extracted at the toe of the structure and used on NN_OVERTOPPING2 to compute the overtopping discharge. The overtopping discharges were calculated with a 1-h time step.

2.2. SWASH Model Setup

Simulations were performed in a one-dimensional mode for a computational period corresponding to 500 waves, with an additional spin-up period of 15% of the computational period. While the SWASH user manual recommends between 500 and 1000 waves for a steady-state resolution, the minimum was applied for a low computational cost. An initial timestep of 0.008 s and an automatic time step control was applied with a maximum Courant number of 0.5 and a minimum Courant number of 0.1. The number of vertical layers used for the simulations was determined by the maximum frequency of each case, which resulted in one vertical layer for all simulations. Considering that the SWASH user manual sets the default Manning friction coefficient to 0.019 $s/(m^{1/3})$ for wave simulations over large distances, a minimum value of 0.02 $s/(m^{1/3})$ was imposed for the calibration of the model for the armour layers of the breakwater, as the tetrapods and antifer cubes likely cause higher friction than found in the approximation area.

Profile-A had a domain length of 392 m, where 341 m corresponded to the approximation area and 51 m to the breakwater. The length of the numerical domain was 419 m for profile-T, where 334.5 m corresponded to the approximation area, 48 m to the breakwater and 36.5 m to the lee side of the structure (Figure 3). The length of the SWASH computational domains was a consequence of the depth needed to comply with the model requisites at the offshore boundary in terms of the relation between water depth and significant wave height and between water depth and mean wave period. The bathymetry was constructed with data acquired from EMODnet (150 m grid spacing) and DGTerritorio (LiDAR survey of 2011, 2 m spacing). The profiles had a constant grid spacing of 0.5 m. Mean overtopping discharge, q, was defined at the red markers in Figure 3 by simulating the instantaneous overtopping discharge at each time step using the DISCH command in SWASH. In order to obtain the mean overtopping discharge, the sum of discharges was divided by the computational time (or the amount of instantaneous discharges that was received in the output).

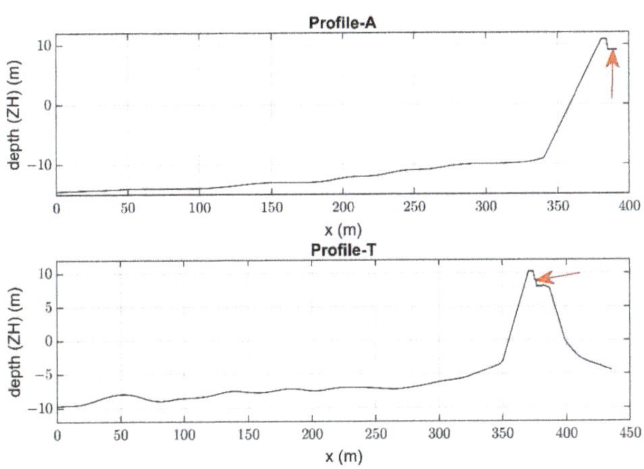

Figure 3. Bathymetries used for the simulations in SWASH, with the cross-shore distance referred to the wavemaker boundary and red arrows indicating where overtopping results were extracted.

At the offshore boundary, a Jonswap wave spectrum defined the shape of the irregular waves, with a peak enhancement parameter $\gamma = 3.3$, and a weakly-reflective boundary was imposed. A Sommerfeld radiation boundary condition was applied at the end of the domain to prevent the reflection of outgoing waves that could give rise to instabilities within the numerical domain. The boundary conditions were chosen based on several existing overtopping studies [17,20]. For the non-hydrostatic pressure term, a Keller Box scheme with ILU preconditioner was used to increase the stability of the model.

2.3. Model Calibration

A possible approach to simulate wave overtopping over a breakwater is to define the breakwater in terms of the porosity of its material. However, previous studies showed that the use of a porosity term may lead to wave dissipation without wave run-up and overtopping [26,27]. Alternatively, a permeable breakwater can be treated as impermeable terrain using a bottom friction coefficient, which represents the effect of comprehensive energy dissipation as a consequence of roughness [26]. So, in this study, the friction and rugosity of the breakwater was included in the form of a Manning coefficient, which is also a simpler approach to implement in operational systems like HIDRALERTA. As the literature does not provide a friction value or a Manning coefficient for tetrapods or antifer cubes, the model was calibrated by varying Manning values in order to obtain discharges similar to the ones estimated by HIDRALERTA by using the NN-OVERTOPPING2 tool. This way, the effect of porosity was accounted for in the calibration of the Manning coefficient. The neural network NN_OVERTOPPING2 was developed as part of the CLASH European project and based on a large database of physical model tests for a wide range of coastal structure types. Therefore, the mean overtopping discharge of the antifer and tetrapod armoured breakwater for the calibration was provided by the NN_OVERTOPPING2 neural network. NN_OVERTOPPING2 uses a roughness and permeability coefficient, provided in [28], depending on the type of armour layer.

Following the results of [21], who found a correlation of the friction coefficient with the wave steepness S_{op} (wave height H/wavelength λ) and the relative crest freeboard R_c/H_s, the cases were chosen to account for a wide range of S_{op} and R_c/H_s conditions, where the crest freeboard R_c defines the height of the superstructure relative to SWL. Given that it is not possible to account for wave direction in one-dimensional simulations and that NN_OVERTOPPING2, in contrast, does account for wave direction in overtopping estimations, the cases were also sorted according to their incident wave angles β above and below 15°. The main reason for this is that oblique wave attack reduces the overtopping discharge estimated by NN_OVERTOPPING2 and that SWASH can replicate this in the form of a higher Manning friction coefficient, resulting in lower discharges. The threshold of 15° was chosen based on the studies of [29], who found that there is no significant difference in overtopping for wave attack over an armoured breakwater between 0 and 15°.

Twenty-four simulations were performed for profile-T, which included sea states from the three different storms, where β varied between 18° and 49°. The values for wave steepness ranged from 0.02 to 0.05 and for dimensionless crest freeboards between 1.0 and 1.9. Profile-A was almost aligned with the dominant wave direction that occurred during the Elsa storm (β lower than 15°, representing normal wave attack). Twelve cases were simulated from this storm with incident wave angles below 15°, wave steepness values ranging from 0.02 to 0.05 and dimensionless crest freeboards between 1.1 and 2.1.

2.4. Manning Coefficient Expression Development and Validation

Following the study from [21], the relation of both wave steepness and relative crest freeboard with overtopping discharges and the Manning coefficients obtained in the calibration process ($n_{calibrated}$) were investigated. Then, for predictive purposes, empirical equations to obtain $n_{calculated}$ as a function of the hydrodynamic conditions and the geometry of the structure were developed. To develop these equations, the Matlab CFtool was utilised to find the best fitting using least square regression. To evaluate the accuracy of the empirical equations, discharges computed with $n_{calibrated}$ were confronted with the discharges computed with $n_{calculated}$ along with the discharges provided by NN_OVERTOPPING2 by using the Root-Mean-Square-Error (RMSE).

For Profile-A, based on the chosen cases (Table 2), one equation was developed:

- Equation (1)—defining n_A as a function of R_c/H_s and S_{op}, where n_A is the Manning coefficient of an armour layer of antifer cubes. This equation does not account for wave obliquity and can only be applied under normal wave attack conditions, i.e., incident angles lower than 15°.

Table 2. Cases chosen for SWASH simulations for profile-A with wave characteristics from the Elsa storm.

Case No	Date and Time	R_c/H_s	S_{op}	H_s (m)	T_p (s)	SWL (m)	Incident Wave Angle β (°)
1	19 December 2019, 18:00	1.10	0.05	6.03	12.33	2.39	2.00
2	20 December 2019, 01:00	1.10	0.04	6.19	13.64	1.97	5.00
3	20 December 2019, 03:00	1.20	0.04	6.07	13.64	1.68	5.00
4	20 December 2019, 07:00	1.20	0.03	5.20	13.64	2.70	8.00
5	19 December 2019, 15:00	1.30	0.05	5.90	12.33	1.55	2.00
6	19 December 2019, 10:00	1.30	0.04	4.68	12.33	2.77	3.00
7	19 December 2019, 11:00	1.40	0.04	4.85	12.33	2.39	0.40
8	21 December 2019, 08:00	1.40	0.03	4.46	12.33	2.59	4.00
9	18 December 2019, 22:00	1.50	0.04	4.45	11.14	2.45	4.00
10	19 December 2019, 06:00	1.50	0.03	4.15	12.33	2.76	3.00
11	19 December 2019, 00:00	1.80	0.04	4.03	11.14	1.80	2.00
12	19 December 2019, 02:00	1.80	0.03	4.10	12.33	1.58	1.00
13	18 December 2019, 10:00	2.10	0.03	3.28	12.33	2.28	1.00
14	18 December 2019, 05:00	2.10	0.02	2.95	13.64	2.79	11.00

For Profile-T, firstly, the 24 simulations for the time period between the 15th and 23rd of December 2019 were considered. Secondly, given that the SWASH model does not take wave direction into account for one-dimensional simulations, the importance of the wave angles β was investigated as well. The 24 simulations were separated into two incident wave climates. Wave climate 1 was characterised by lower wave steepness and β between 15° and 30°. Wave climate 2 showed higher values for wave steepness and waves approached the structure with β between 30° and 50° (Table 3). This way, it was possible to investigate whether or not the performance of the expression to define the Manning coefficient depends on the specific wave conditions that were used to build the expression. Thus, for Profile-T, three equations for the calculation of the Manning coefficient were developed for different ranges of wave attack angles:

- Equation (2)—defining $n_{T,oblique}$ as a function of R_c/H_s and $\cos(\beta)$, where $n_{T,oblique}$ is the Manning coefficient for a tetrapod armour layer. This equation accounts for obliquity and is only applicable to oblique wave attack with incident angles greater than 15°. For that, the 24 cases were considered.
- As an alternative to Equation (2), the following Equations can be used, depending on the incident wave angle:
- Equation (3)—defining $n_{T\ oblique(15\text{–}30)}$ as a function of R_c/H_s and S_{op}, where $n_{T\ oblique(15\text{–}30)}$ is the Manning coefficient for a tetrapod armour layer for wave attack between 15° and 30°, based on the data from wave climate 1.
- Equation (4)—defining $n_{T\ oblique(30\text{–}50)}$ as a function of R_c/H_s and S_{op}, where $n_{T\ oblique(30\text{–}50)}$ is the Manning coefficient for a tetrapod armour layer for wave attack between 30° and 50°, based on the data from wave climate 2.

Table 3. Cases chosen for SWASH simulations for Profile-T with wave characteristics from the three storms. Separation of cases for the tetrapod profile for wave climate 1 (framed) and wave climate 2 (not framed).

Case No	Date and Time	R_c/H_s	S_{op}	H_s (m)	T_p (s)	SWL (m)	Incident Wave Angle β (°)
1	16 December 2019. 05:00	1.00	0.03	5.57	16.69	3.50	23.00
2	16 December 2019. 06:00	1.10	0.03	5.23	16.69	3.41	23.00
3	20 December 2019. 06:00	1.10	0.05	6.02	13.64	2.35	45.00
4	20 December 2019. 20:00	1.10	0.04	5.83	13.64	3.06	42.00
5	19 December 2019. 18:00	1.20	0.05	5.43	12.33	2.39	49.00
6	21 December 2019. 07:00	1.20	0.04	5.64	13.64	2.21	45.00
7	16 December 2019. 17:00	1.20	0.03	4.81	16.69	3.21	21.00
8	22 December 2019. 12:00	1.20	0.03	4.84	16.69	3.03	32.00
9	21 December 2019. 14:00	1.40	0.05	5.10	12.33	2.06	32.00
10	21 December 2019. 13:00	1.40	0.04	4.70	12.33	2.49	34.00
11	17 December 2019. 04:00	1.40	0.03	4.27	15.09	2.92	18.00
12	22 December 2019. 22:00	1.40	0.03	4.53	16.69	2.53	30.00
13	21 December 2019. 20:00	1.50	0.04	4.51	13.64	2.14	35.00
14	15 December 2019. 22:00	1.50	0.03	5.02	16.69	1.25	27.00
15	16 December 2019. 07:00	1.50	0.02	3.96	16.69	3.06	25.00
16	15 December 2019. 14:00	1.60	0.03	4.11	15.09	2.35	27.00
17	16 December 2019. 15:00	1.60	0.02	3.93	16.69	2.54	23.00
18	15 December 2019. 15:00	1.70	0.03	3.66	15.09	2.80	26.00
19	17 December 2019. 08:00	1.70	0.02	3.56	15.09	2.83	21.00
20	23 December 2019. 00:00	1.70	0.02	3.53	16.69	3.06	31.00
21	15 December 2019. 03:00	1.80	0.03	3.28	13.64	3.15	29.00
22	15 December 2019. 06:00	1.80	0.02	3.24	13.64	3.07	30.00
23	23 December 2019. 03:00	1.90	0.03	3.63	15.09	2.14	30.00
24	18 December 2019. 06:00	1.90	0.02	3.07	13.64	3.08	42.00

3. Results

3.1. Profile-A

For Profile-A, the mean overtopping discharge simulated by SWASH was, in general, in agreement with the values obtained by NN_OVERTOPPING2 (see Appendix A, Table A1). Only a limited number of cases were found where the discharge computed by SWASH was much lower than the one estimated by NN_OVERTOPPING2 (3, 43, 11 and 10 l/s/m for cases 13, 1, 11 and 5, respectively). However, it is important to note that the overtopping discharges estimated by NN_OVERTOPPING2 ranged from 3 to 160 l/s/m (Appendix A, Table A1). The discrepancies between q simulated by SWASH and q simulated by NN_OVERTOPPING2 yielded an RMSE of 12.11 l/s/m.

When confronting the calibrated Manning coefficients with the dimensionless numbers, an inverse correlation between the Manning coefficient and R_c/H_s and S_{op} was observed, where the Manning coefficient increased when both parameters decreased (Figure 4a). Similarly, the overtopping discharge tended to decrease with increasing R_c/H_s for a specific S_{op} (Figure 4b).

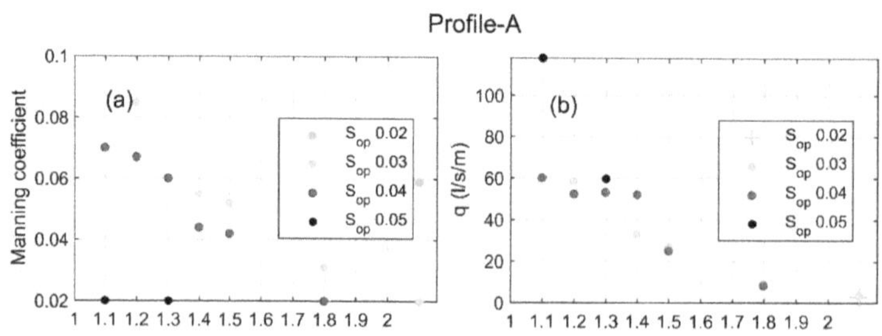

Figure 4. Profile-A. Relation between the Manning coefficient and R_c/H_s (**a**) and between q and R_c/H_s (**b**).

Based on these correlations, an empirical equation of the Manning coefficient for an armour layer of antifer cubes as a function of R_c/H_s and S_{op} was developed for wave attack angles smaller than 15°. The equation of the Manning coefficient n_A (s/m$^{1/3)}$) was given as follows:

$$n_A = a1 + a2 \times R_c/H_s + a3 \times S_{op} + a4 \times (R_c/H_s)^2 + a5 \times R_c/H_s \times S_{op} + a6 \times S_{op}{}^2 + a7 \times (R_c/H_s)^3 + a8 \times (R_c/H_s)^2 \times S_{op} + a9 \times R_c/H_s \times S_{op}{}^2 + a10 \times S_{op}{}^3 \quad (1)$$

where a1 = 1.149; a2 = −1.402; a3 = −22.96; a4 = 0.5556; a5 = 22.77; a6 = 166.5; a7 = −0.0453; a8 = −8.814; a9 = 73.37; a10 = −3105.

The range of application for Equation (1) was between 1.1 and 2.1 for R_c/H_s and between 0.02 and 0.05 for S_{op} (see Appendix B, Figure A1). The Manning coefficients calculated with the equation showed good agreement with the Manning coefficients that were calibrated in the previous sections for each case (Figure 5). According to these results, RMSE between the 14 calibrated and calculated Manning coefficients was 0.002 s/(m$^{1/3}$).

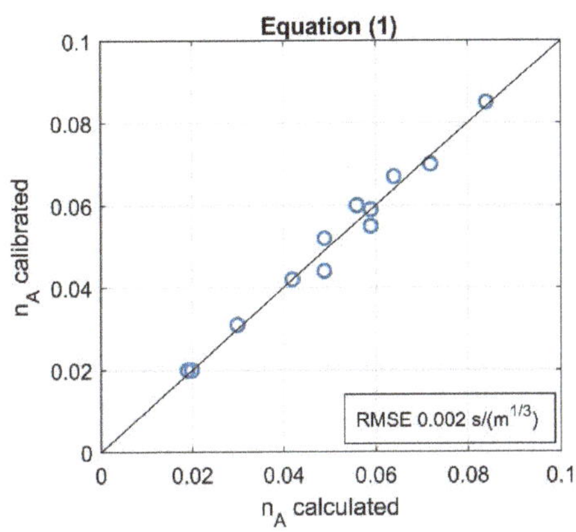

Figure 5. Profile-A. Comparison of calibrated Manning coefficients (s/(m$^{1/3}$)) with the ones calculated with Equation (1).

When the calculated Manning coefficient was used to simulate the overtopping discharge with SWASH for the 14 cases considered for Profile-A, only slight deviations in q with respect to the values obtained with a calibrated Manning coefficient were observed (Figure 6a). In fact, in most of the cases, n_A calibrated and n_A calculated were equal and therefore resulted in the same value for q. For the cases where the calculated Manning coefficient did not fully agree with the calibrated Manning coefficient, the differences in q did not exceed 3 l/s/m in most cases. Only case 3, 6 and 7 showed larger differences in q of 5.1, 9.1 and 4 l/s/m, respectively. When q obtained with the calculated Manning simulations was compared with the values of NN_OVERTOPPING2, the differences were larger. For instance, cases 1, 5 and 11 showed deviations in q of >10 l/s/m and case 6 differed in about 9 l/s/m (Figure 6b), while the maximum discharge estimated by NN_OVERTOPPING2 was 160 l/s/m. The RMSE between q simulated by SWASH using the calculated Manning coefficient and calibrated Manning coefficient was 3.13 l/s/m. In comparison, the RMSE between q simulated by NN_OVERTOPPING2 and q simulated by SWASH using the calculated Manning coefficient was 12.48 l/s/m.

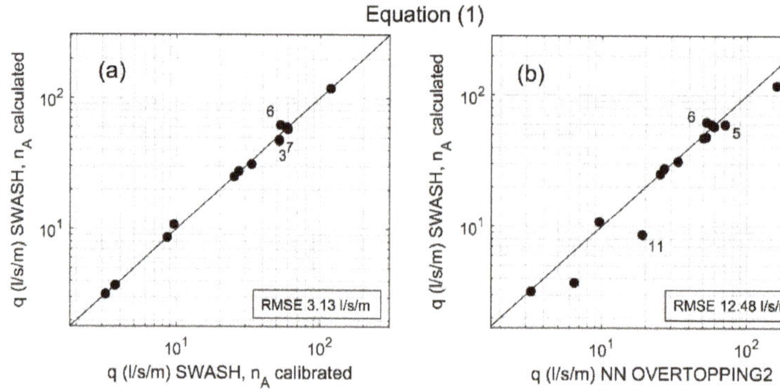

Figure 6. Profile-A. Comparison of q estimated by SWASH using the calculated Manning coefficient by Equation (1) with q estimated by SWASH using the calibrated coefficient (**a**) and with q estimated by NN_OVERTOPPING2 (**b**).

3.2. Profile-T

After calibration, the estimations of mean overtopping discharge by SWASH for Profile-T fairly matched the ones obtained by NN_OVERTOPPING2 in most of the 24 cases (see Appendix A, Table A2), that had values ranging from 0.13 to 59 l/s/m. For most cases, the difference between q simulated by SWASH and by NN_OVERTOPPING2 did not exceed 3 l/s/m. For a few cases (8, 15, 20), however, q computed by SWASH was much lower than the one estimated by NN_OVERTOPPING2 (with deviations up to 44 l/s/m). For those cases, during the calibration process, the Manning coefficient could not be decreased further to improve the result for q as it had already reached the established lower limit of 0.02 s/(m$^{1/3}$). The remaining cases showed a maximum difference of 3 l/s/m. The RMSE between q simulated by NN_OVERTOPPING2 and by SWASH was of 10.49 l/s/m.

For the development of the Manning coefficient expression, no clear correlation between the Manning coefficient and S_{op} or R_c/H_s was found for the 24 cases simulated with incident wave angles higher than 15° (Figure 7a). However, an inverse relation between q and R_c/H_s could be observed, where for a specific wave steepness, q decreased with increasing R_c/H_s (Figure 7b).

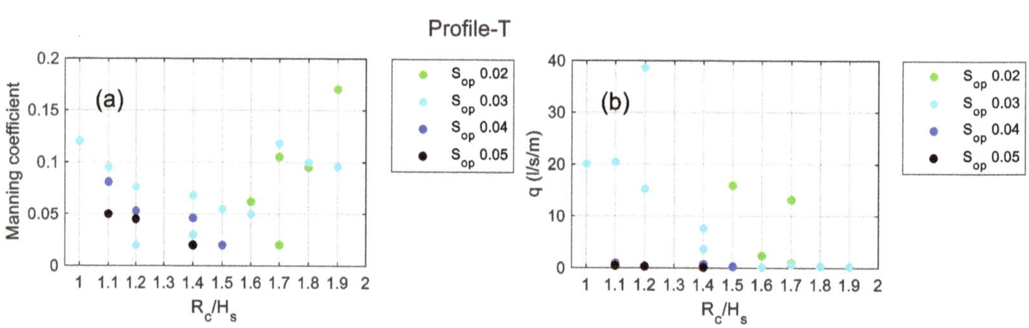

Figure 7. Profile-T. Relation between the Manning coefficient and R_c/H_s (**a**) and q and R_c/H_s (**b**).

During these storms, the incident wave angle was related with S_{op} (Figure 8). Cases with high incident waves angles (low cos(β)) had mostly higher values of wave steepness, while cases with lower incident wave angles tended to have a lower steepness.

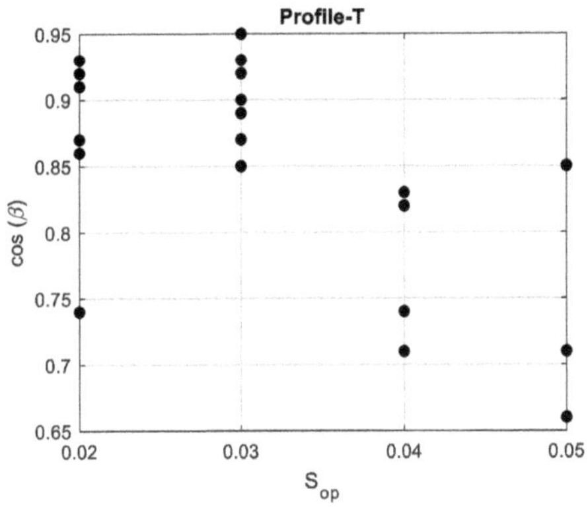

Figure 8. Profile-T. Relation between cos(β) and S_{op}.

Thus, three empirical equations of the Manning coefficient for an armour layer of tetrapods were developed: Equation (2), based on all 24 cases and accounting for the oblique wave attack between 15° and 50°, was developed as a function of the dimensionless crest freeboard R_c/H_s and cos(β). Since the latter was highly related to the wave steepness, S_{op} was not included in the equation and the cosine of the incident wave angle was considered as it displayed higher variability than wave steepness (only four values). This higher variability contributed to the development of the empirical equation. In order to investigate whether the calculated Manning coefficients could achieve a better agreement with the calibrated Manning coefficients when the test cases are separated into two different wave climates, two more equations were developed that can be used alternatively. Equation (3) was based on the 12 simulations of wave climate 1 and accounted for β between 15° and 30°, and Equation (4) was based on the 12 simulations of wave climate 2 and accounted for β between 30° and 50°. Both equations were developed as a function of wave steepness and the dimensionless crest freeboard. The equations of the Manning coefficients $n_{T\,oblique}$, $n_{T\,oblique(15\text{–}30)}$ and $n_{T\,oblique(30\text{–}50)}$ (s/(m$^{1/3}$)) were given as follows:

$$n_{T\,oblique} = a1 + a2 \times Rc/Hs + a3 \times \cos(\beta) + a4 \times (Rc/Hs)^2 + a5 \times Rc/Hs \times \cos(\beta) + a6 \times \cos(\beta)^2 \tag{2}$$

where a1 = 1.655; a2 = −0.738; a3 = −2.827; a4 = 0.3895; a5 = −0.4065; a6 = 2.114

$$n_{T\,oblique(15\text{-}30)} = a1 + a2 \times Rc/Hs + a3 \times Sop + a4 \times (Rc/Hs)^2 + a5 \times Rc/Hs \times Sop + a6 \times (Rc/Hs)^3 + a7 \times (Rc/Hs)^2 \times Sop \tag{3}$$

with a1 = −12.96; a2 = 15.26; a3 = 460.3; a4 = −4.398; a5 = −548.2; a6 = −0.03234; a7 = 162.9

$$n_{T\,oblique\,(30\text{-}50)} = a1 + a2 \times Rc/Hs + a3 \times Sop + a4 \times (Rc/Hs)^2 + a5 \times Rc/Hs \times Sop + a6 \times Sop^2 + a7 \times (Rc/Hs)^3 + a8 \times (Rc/Hs)^2 \times Sop + a9 \times Rc/Hs \times Sop^2 + a10 \times Sop^3 \tag{4}$$

being a1 = 2.617; a2 = −11.47; a3 = 188.8; a4 = 7.496; a5 = 66.13; a6 = −5562; a7 = −1.312; a8 = −49.98; a9 = 686.7; a10 = 3.756 × 10^4.

As these equations were developed under different regimes—they have different ranges of applicability (see Appendix B, Figure A2). While Equation (2) could be used for R_c/H_s ranging between 1.0 and 1.9 and cos(β) ranging between 0.6 and 1.0, Equations (3) and (4) had lower ranges of applicability. For the former, R_c/H_s was between 1.0 and 1.8 and S_{op} between 0.02 and 0.03, and for the latter, R_c/H_s varied between 1.1 and 1.9 and S_{op} between 0.02 and 0.05.

The Manning coefficients that were calculated with Equation (2) showed differences from the calibrated Manning coefficients. While a few values were close or equal, others differed by 0.02 s/(m$^{1/3}$) or more (Figure 9a). The RMSE between the 24 calibrated n_T and calculated $n_{T\ oblique}$ was 0.02 s/(m$^{1/3}$), with the largest discrepancies shown in the lowest values. When separating in wave climates, Equation (3) showed better skills to predict the calibrated Manning coefficients than the coefficients obtained for the same cases by Equation 2 (Figure 9b). A RMSE of 0.012 s/(m$^{1/3}$) was found between the calibrated and calculated Manning coefficients with Equation (3). The calculated Manning coefficients obtained with Equation (4) also showed a good agreement with the calibrated Manning coefficients (Figure 9c). The RMSE calculated between the calibrated and calculated Manning coefficients with Equation (4) was 0.006 s/(m$^{1/3}$) exhibiting the lowest value among the three derived equations.

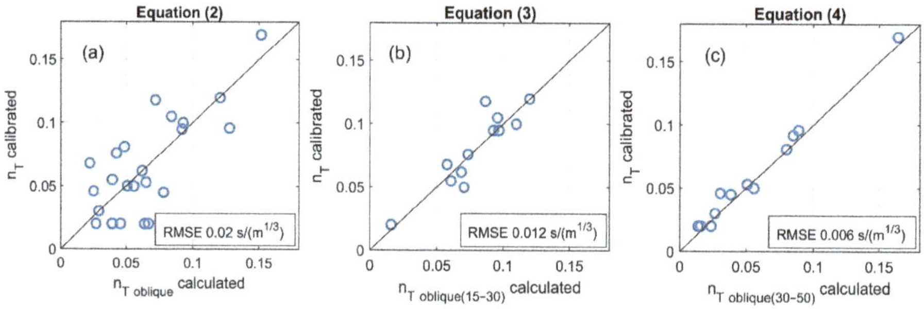

Figure 9. Comparison of calibrated Manning coefficients (s/(m$^{1/3}$)) with the ones calculated with Equations (2) (**a**), (3) (**b**) and (4) (**c**).

To verify the performance of Equations (2)–(4), the calculated Manning coefficients were applied to simulate overtopping discharges with the SWASH model for the same cases considered during the calibration phase. When comparing the overtopping discharges computed with the calibrated (n_T calibrated) and the calculated ($n_{T\ oblique}$) Manning coefficients by Equation (2), it was found that for the cases of lower discharge, q simulated with $n_{T\ oblique}$ was generally overestimated when compared to q simulated with n_T calibrated (Figure 10a). The RMSE between q simulated by SWASH with the calibrated Manning coefficients and with the calculated Manning coefficients was 1.53 l/s/m. Furthermore, it could be seen that q simulated with $n_{T\ oblique}$ showed slightly larger deviations when compared to the values given by NN_OVERTOPPING2. Generally, overtopping was overestimated in the range of lower discharges, and underestimated in the range of higher discharges (Figure 10b). The RMSE between q simulated by SWASH with $n_{T\ oblique}$ and q computed by NN_OVERTOPPING was 10.62 l/s/m.

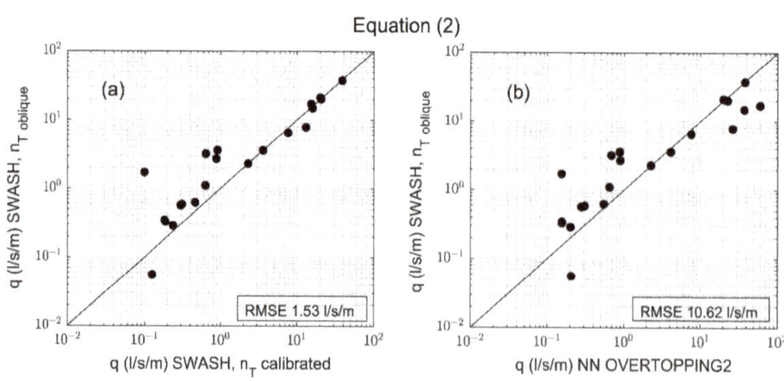

Figure 10. Comparison of q estimated by SWASH using the calculated Manning coefficient by Equation 2 with (**a**) q estimated by SWASH using the calibrated coefficient and (**b**) with q estimated by NN_OVERTOPPING2.

The overtopping discharges obtained with the calculated Manning coefficient from Equation (3) ($n_{T\ oblique(15–30)}$) showed a close fit with n_T calibrated (RMSE 1.3 l/s/m) (Figure 11a). In fact, only one case (case 2 in Table 3) resulted in an important deviation (4 l/s/m), while the rest showed deviations of less than 2 l/s/m. The comparison of q computed by NN_OVERTOPPING2 and SWASH with the $n_{T\ oblique(15–30)}$ revealed that both discharges were similar (Figure 11b). Only one case showed a difference of 12 l/s/m (case 15). The RMSE between q estimated by SWASH with the calculated Manning coefficients and the q of NN_OVERTOPPING2 was 6.28 l/s/m. Regarding Equation (4), the results for q obtained by the simulations with $n_{T\ oblique(30–50)}$ matched closely the discharge obtained with n_T calibrated (RMSE = 0.59 l/s/m) (Figure 11c). The discharges for this wave climate were generally low (most cases had discharges of < 1 l/s/m). The highest deviation found between q simulated with a calibrated Manning and q simulated with a calculated Manning was of 1.9 l/s/m. Comparing the SWASH simulated discharges with $n_{T\ oblique(30–50)}$, with the values of NN_OVERTOPPING2, a larger RMSE was found (13.46 l/s/m) (Figure 11d). The two cases that were responsible for this large error were cases 8 and 20 in Table 3, with discharges of 60 and 26 l/s/m, respectively, where SWASH showed deviations of 45 and 13 l/s/m in regard to q computed by NN_OVERTOPPING2.

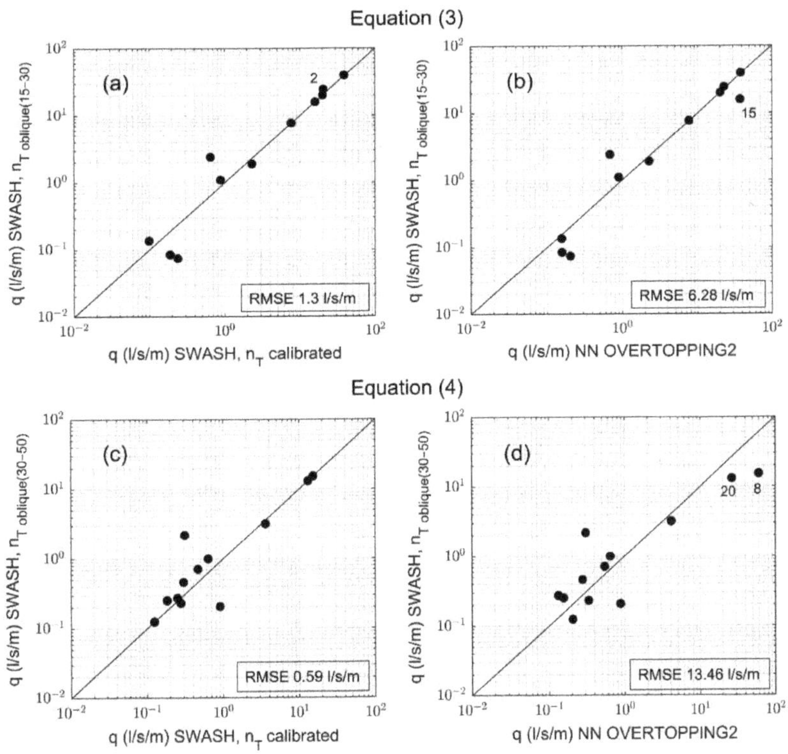

Figure 11. Comparison of q estimated by SWASH using $n_{T\ oblique(15–30)}$ with (**a**) q estimated by SWASH using the n_T calibrated and with (**b**) q estimated by NN_OVERTOPPING2, and of q estimated by SWASH using $n_{T\ oblique(30–50)}$ with (**c**) q estimated by SWASH using n_T calibrated and with (**d**) q estimated by NN_OVERTOPPING2.

4. Discussion

The overtopping simulations at Ericeira harbour demonstrated that the SWASH model is capable of providing similar results as the neural network tool NN_OVERTOPPING2 when the value for bottom friction of the armour layer is calibrated, even for different profile morphologies and typologies. However, in some cases where the Manning coefficient was limited to 0.02 s/(m$^{1/3}$), the discharge was lower in the SWASH simulations than in the NN_OVERTOPPING2 tool. A similar issue was observed by [30], who found that the model clearly underestimated the overtopping discharge for complex structure types. They suggested that the velocity of overtopped water masses is underestimated because the velocity differences in front of the structure between bottom and surface were not well described by the model in simulations with one vertical layer. Furthermore, previous studies have shown that the accuracy of the SWASH model can depend on the model settings. For example, according to [17], the estimation of mean overtopping discharge appears to be sensitive to the non-dimensional parameter L/dx (wavelength at the toe of the structure over grid size). Based on calculations performed by [31], however, this sensitivity could not be confirmed (Appendix C, Figures A3 and A4).

In addition, for the evaluation of the performance of SWASH, it must be recalled that the results obtained by SWASH were not validated with real data, but with another tool for the estimation of overtopping. In order to confirm if it is indeed the SWASH model that underpredicts, or possibly NN_OVERTOPPING2 that overpredicts, real data are necessary to compare the results and draw a conclusion. While the neural network is a commonly used tool to perform the overtopping predictions, the SWASH model shows some advantages in the use for operational systems like HIDRALERTA, as mentioned in Section 1, namely the capability to estimate the flood extension or the flooded area by implementing a 2D model. Especially for studies where the wave propagation plays an important role, SWASH, in contrast to NN_OVERTOPPING2, is able to extract wave characteristics travelling through the domain at any given point.

During the model calibration for Profile-A, a RMSE of 12.11 l/s/m was obtained between the simulated mean overtopping discharge by SWASH and the one by NN_OVERTOPPING2. It must be noted here, however, that the discharges computed by NN_OVERTOPPING2 were generally very high (maximum discharge 160 l/s/m). The impact of the inaccuracy of q predictions depends on the kind of receptor that is considered in the Early Warning System. An overtopping discharge of 3 l/s/m may not cause severe erosion or damage at the structure but can be harmful for pedestrians.

The model calibration for Profile-T included a large variety of cases with different wave characteristics, but the values for q estimated by NN_OVERTOPPING2 could be matched closely in most cases by SWASH. The RMSE between the values of both models was 10.49 l/s/m, although as observed at Profile-A, only a few cases, where q was underestimated by SWASH by > 10 l/s/m, were responsible for this high error.

For the development of the Manning coefficient expressions, the results partially confirmed the findings of [21], who found a correlation between the Manning coefficient, R_c/H_s and S_{op}. However, this correlation could only be observed when the angle of wave attack was considered. At this point, it must be noted that the work of [21] was performed under controlled conditions, while in this study, real storm events were used, which makes the model calibration more demanding. The fact that the simulations not only included large ranges of wave periods and wave heights, but also incident wave angles between 15° and 50° degrees can make it difficult to find these relations between R_c/H_s and S_{op}. Many studies have shown that wave obliquity has an impact on overtopping estimation (e.g., [16,29]). As the developed expressions are based exclusively on the wave conditions during real storm events in Ericeira, the ranges of applicability presented in Appendix B are of great importance to the use of these expressions. Once the input conditions for the SWASH model lie outside of these ranges, the expressions will not produce reliable results and are therefore not applicable.

Due to the relatively high root-mean-square error of the overtopping simulations based on Equation (2), the simulated cases were analysed and separated into two different wave conditions according to their wave characteristics. Thus, it was possible to investigate whether also the behaviour of the Manning coefficient was related to these wave characteristics.

The equations that resulted from the separation of the simulations of this profile resulted in very small errors for the calculation of the Manning coefficient in comparison to the ones obtained by Equation (2). At this point, it must be noted that the relevance of the RMSE error estimations reduces, once the number of calibration data and the amount of coefficients in the expression become similar, as in Equations (3) and (4), which are based on only a few test cases. The reduction in RMSE was 0.008 $s/(m^{1/3})$ for Equation (3) and 0.014 $s/(m^{1/3})$ for Equation (4). This showed that, by dividing the cases according to their wave direction and characteristics, the performance of the equations could be improved but will have to be validated with more data in future studies. Although each equation consequently had a smaller range of applicability, their combined use can overcome this constraint. This indicates that there might not only be a correlation between the Manning coefficient, Rc/Hs and Sop, but also a possible dependency of the Manning coefficient on the incident wave angle. This dependency can be explained by the fact that an oblique wave attack reduces the overtopping discharge estimated by NN_OVERTOPPING2 and the way that the calibrated SWASH model can replicate this is with an increased Manning friction coefficient, causing a decrease in overtopping discharge. Overall it can be said that these findings suggest to implement Equation (3) and (4), rather than Equation (2), as they showed a better performance with smaller errors and will most likely deliver more reliable results.

For the implementation of SWASH into an operational system like HIDRALERTA, the model run time has to be considered. The simulations were performed on a Windows operating system with Intel Core i5-1035G1 (CPU 1,0 GHz, 4 cores). SWASH was run serial, using one processor, and took approximately 10 min for each simulation. NN_OVERTOPPING2, however, shows a lower model run time of only a few seconds.

5. Conclusions

This study was conducted to investigate the capabilities of SWASH to be integrated into the HIDRALERTA Early Warning System by calibrating two one-dimensional models for storm conditions at the Ericeira harbour prototype and developing an expression for the definition of the Manning coefficient of different armour layers of the breakwater based on the sea wave conditions.

The calibration of the one-dimensional model for Profile-A (antifer cubes) and Profile-T (tetrapods) for the Ericeira harbour demonstrated that, in this study, the SWASH model performance for the estimation of mean overtopping discharge is strongly governed by the Manning coefficient. With a calibrated Manning coefficient, the model is capable of providing similar results than with NN_OVERTOPPING2 at a low computational effort. Nevertheless, in some cases, the SWASH model still underpredicted the overtopping discharge.

The simulations revealed that the variables R_c/H_s and S_{op}, which were considered in this study, are important variables defining the Manning coefficient and that, for the development of an expression for the Manning coefficient, the angle of wave attack must be considered. Thus, the incident wave angles must be included as an independent variable in the formula in the form of the cosine of the incident angle. Alternatively, the simulations that are used to develop the Manning coefficient expressions can be separated, based on their incident wave angles. Therefore, it is possible to account for wave obliquity as it cannot specifically be included in one-dimensional simulations in SWASH. For the implementation into HIDRALERTA, the ranges of application of the developed expression are fundamental, as they will not produce reliable results in situations where the values of the wave conditions lie outside of these ranges. In those cases, the results for mean overtopping discharge will solely be based on the estimations by NN_OVERTOPPING2.

As a final conclusion and in the framework of this study, it can be said that the SWASH model is capable of providing reasonable results and it has the potential to be implemented in the Early Warning System HIDRALERTA. However, the type of receptor that is considered in the risk assessment plays an important role when assessing the performance of the SWASH model for the Ericeira prototype. The discrepancies of simulated mean overtopping discharge with the ones estimated by NN_OVERTOPPING2 are significant in the case of a risk assessment for pedestrians and have to be improved if the SWASH model was to be implemented in HIDRALERTA. Generally, after calibration, SWASH mostly delivers reliable results at a low computational cost (for its one-dimensional version). Nevertheless, in order to make a final statement on the performance of SWASH, the results have to be compared against field data. In addition, SWASH is capable of modelling the wave propagation as well as the overtopping process, allowing for a future definition of the extension of the flooded area, which cannot be accomplished with tools like NN_OVERTOPPING2. The development of expressions to seamlessly calculate the Manning coefficient helped to build the frame for the SWASH implementation in HIDRALERTA system as an alternative to the neural network NN_OVERTOPPING2.

Author Contributions: Conceptualisation, methodology, writing—original draft preparation, A.M.; review, editing and supervision, A.C.Z. and J.L.G.; project administration, A.C.Z. All authors have read and agreed to the published version of the manuscript.

Funding: This research was supported by the Portuguese Foundation of Science and Technology (FCT) under the project LA/P/0069/2020 granted to the Associate Laboratory ARNET and UID/00350/2020 CIMA and by the research project BSafe4Sea (Ref. PTDC/ECI-EGC/31090/2017).

Institutional Review Board Statement: Not applicable.

Informed Consent Statement: Not applicable.

Acknowledgments: The authors acknowledge the research projects To-SEAlert (Ref. PTDC/EAM-OCE/31207/2017), BSafe4Sea (Ref. PTDC/ECI-EGC/31090/2017) and EW-COAST (Ref. ALG-LISBOA-01-145-FEDER-028657, with financial support of FCT under the project LA/P/0069/2020 granted to the Associate Laboratory ARNET and UID/00350/2020 CIMA). Infraestrutura Nacional de Computação Distribuída (INCD) is also acknowledged for the access to their computational resources, as well as Copernicus program.

Conflicts of Interest: The authors declare no conflict of interest.

Appendix A

Results for mean overtopping discharge obtained by NN_OVERTOPPING2 and by SWASH and values of the calibrated Manning coefficients

Table A1. Estimated overtopping Profile-A.

Case No	q (l/s/m) NN_OVERTOPPING2	q (l/s/m) SWASH	Manning Coefficient Calibrated (s/(m$^{1/3}$))
1	160.42	117.91	0.02
2	59.64	60.14	0.07
3	50.33	52.34	0.07
4	57.42	58.50	0.08
5	70.88	59.72	0.02
6	52.72	53.20	0.06
7	52.37	52.11	0.04
8	33.38	33.24	0.06
9	25.19	25.14	0.04
10	26.85	27.02	0.05
11	19.01	8.60	0.02
12	9.61	9.62	0.03
13	6.49	3.74	0.02
14	3.26	3.18	0.06

Table A2. Estimated overtopping Profile-T.

Case No	q (l/s/m) NN_OVERTOPPING2	q (l/s/m) SWASH	Manning Coefficient Calibrated (s/(m$^{1/3}$))
1	20.14	20.14	0.12
2	22.74	20.52	0.10
3	0.54	0.46	0.05
4	0.88	0.91	0.08
5	0.31	0.31	0.05
6	0.28	0.30	0.05
7	38.26	38.64	0.08
8	59.93	15.33	0.02
9	0.21	0.13	0.02
10	0.65	0.63	0.05
11	7.58	7.60	0.07
12	4.09	3.60	0.03
13	0.33	0.28	0.02
14	0.17	0.14	0.06
15	37.48	15.94	0.02
16	0.15	0.10	0.05
17	2.26	2.34	0.06
18	0.68	0.64	0.12
19	0.90	0.88	0.11
20	26.44	13.14	0.02
21	0.16	0.19	0.10
22	0.20	0.22	0.10
23	0.13	0.25	0.10
24	0.16	0.18	0.17

Appendix B

Ranges of application for the developed equations.

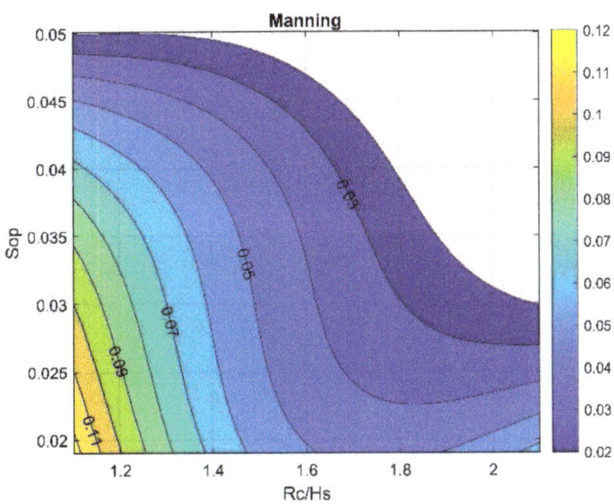

Figure A1. Profile-A. Range of application of Equation (1).

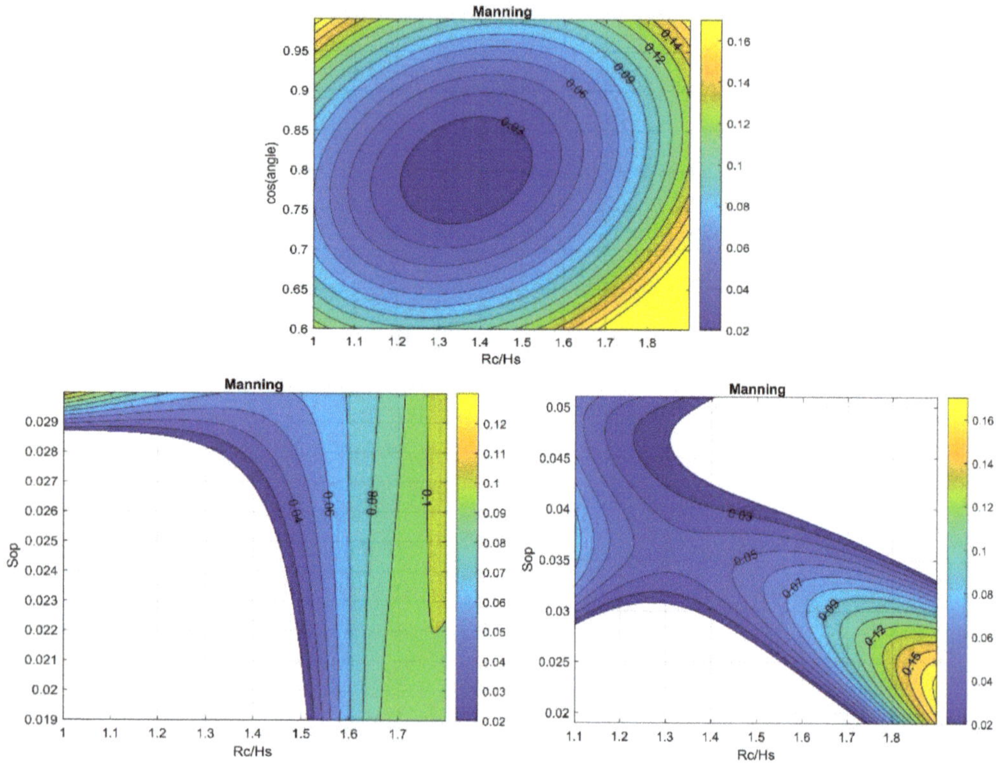

Figure A2. Profile-T. Range of application of Equations 2 (**top**), 3 (**bottom left**) and 4 (**bottom right**).

Appendix C

Correlations of mean overtopping discharge estimations and L/dx

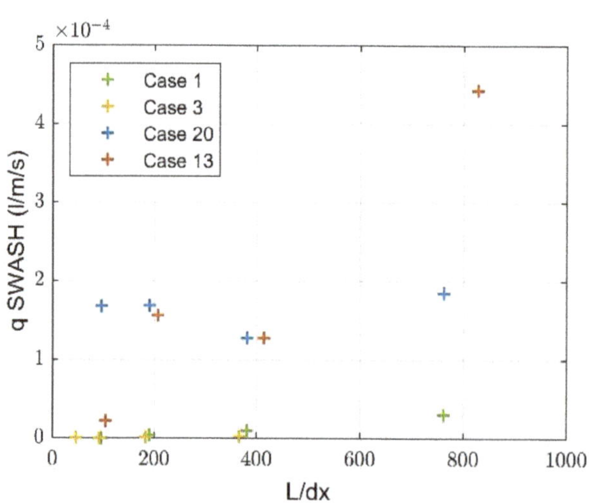

Figure A3. Relation of the mean overtopping discharge versus L/dx based on the calculations of [31].

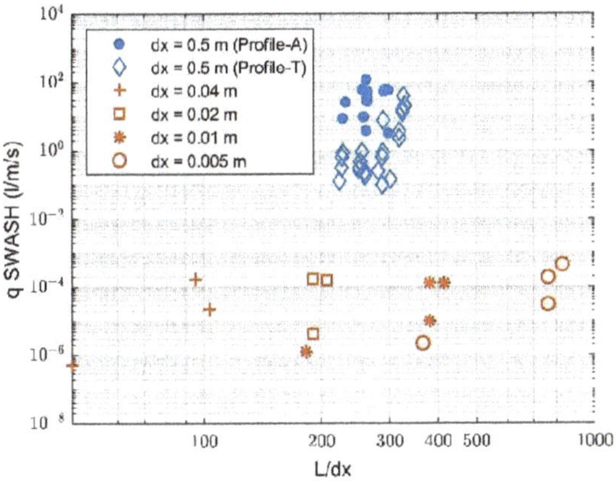

Figure A4. Ratio of the mean overtopping discharge versus L/dx based on the calculations of [31] and the cases of this study.

References

1. Tavares, A.O.; Barros, J.L.; Freire, P.; Santos, P.P.; Perdiz, L.; Fortunato, A.B. A Coastal Flooding Database from 1980 to 2018 for the Continental Portuguese Coastal Zone. *Appl. Geogr.* **2021**, *135*, 102534. [CrossRef]
2. Fortes, C.; Reis, M.T.; Poseiro, P.; Capitão, R.; Santos, J.; Pinheiro, L.; Craveiro, J.; Rodrigues, A.; Sabino, A.; Ferreira Silva, S.; et al. HIDRALERTA Project—A Flood Forecast and Alert System in Coastal and Port Areas. In Proceedings of the IWA – World Water Congress & Exhibition, Lisbon, Portugal, 21–26 September, 2014. [CrossRef]
3. Pillai, K.; Etemad-Shahidi, A.; Lemckert, C. Wave Overtopping at Berm Breakwaters: Review and Sensitivity Analysis of Prediction Models. *Coast. Eng.* **2017**, *120*, 1–21. [CrossRef]
4. Lavell, A.; Oppenheimer, M.; Diop, C.; Hess, J.; Lempert, R.; Li, J.; Muir-Wood, R.; Myeong, S.; Moser, S.; Takeuchi, K.; et al. Climate Change: New Dimensions in Disaster Risk, Exposure, Vulnerability, and Resilience. In *Managing the Risks of Extreme Events and Disasters to Advance Climate Change Adaptation*; Field, C.B., Barros, V., Stocker, T.F., Dahe, Q., Eds.; Cambridge University Press: Cambridge, UK, 2012; pp. 25–64. [CrossRef]
5. van Dongeren, A.; Ciavola, P.; Martinez, G.; Viavattene, C.; Bogaard, T.; Ferreira, O.; Higgins, R.; McCall, R. Introduction to RISC-KIT: Resilience-Increasing Strategies for Coasts. *Coast. Eng.* **2018**, *134*, 2–9. [CrossRef]
6. Gracia, V.; García-León, M.; Sánchez-Arcilla, A.; Gault, J.; Oller, P.; Fernández, J.; Sairouní, A.; Cristofori, E.; Toldrà, R. A new generation of early warning systems for coastal risk. the icoast project. *Coast. Eng. Proc.* **2014**, *1*, 1–8. [CrossRef]
7. Poseiro, P. *Forecast and Early Warning System for Wave Overtopping and Flooding in Coastal and Harbour Areas: Development of a Model and Risk Assessment*; IST-UNL: Lisbon, Portugal, 2019.
8. Fortes, C.J.E.M.; Reis, M.T.; Pinheiro, L.; Poseiro, P.; Serrazina, V.; Mendonça, A.; Smithers, N.; Santos, M.I.; Barateiro, J.; Azevedo, E.B.; et al. The HIDRALERTA System: Application to the Ports of Madalena Do Pico and S. Roque Do Pico, Azores. *Aquat. Ecosyst. Health Manag.* **2020**, *23*, 398–406. [CrossRef]
9. Pinheiro, L.; Fortes, C.J.E.M.; Reis, M.T.; Santos, J.; Soares, C.G. Risk Forecast System for Moored Ships. *Vicce Virtual Int. Conf. Coast. Eng.* **2020**, *36*, 37. [CrossRef]
10. Santos, M.I.; Pinheiro, L.; Fortes, J.C.E.M.; Reis, M.T.; Serrazina, V.; Azevedo, E.B.; Reis, F.V.; Salvador, M. Simulation of Hurricane Lorenzo at the Port of Madalena Do Pico, Azores, by Using the HIDRALERTA System. In Proceedings of the MARTECH 5th International Conference on Maritime Technology and Engineering, Lisbon, Portugal, 16–19 November 2020.
11. Zózimo, A.C.; Ferreira, A.M.; Pinheiro, L.; Fortes, C.J.E.; Baliko, M. Implementação Do Sistema HIDRALERTA Para a Zona Costeira Da Costa Da Caparica. In Proceedings of the X Congresso sobre Planeamento e Gestão das Zonas Costeiras dos Países de Expressão Portuguesa, Rio de Janeiro, Brazil, 6–10 December 2021.
12. Coeveld, E.M.; Van Gent, M.R.A.; Pozueta, B. *Neural Network Manual for NN_Overtopping Program*; Technical Report, TU Delft: Delft, The Netherlands, 2005.
13. Fortes, C.J.E. Transformações Não-Lineares de Ondas Marítimas Em Zonas Portuárias. Análise Pelo Método Dos Elementos Finitos. Ph.D. Thesis, Laboratorio Nacional de Engenharia Civil, Lisboa, Portugal, 2002.
14. Flater, D. Xtide. Available online: https://flaterco.com/xtide (accessed on 15 December 2021).
15. Tonelli, M.; Petti, M. Numerical Simulation of Wave Overtopping at Coastal Dikes and Low-Crested Structures by Means of a Shock-Capturing Boussinesq Model. *Coast. Eng.* **2013**, *79*, 75–88. [CrossRef]

16. EurOtop. *Manual on Wave Overtopping of Sea Defences and Related Structures. An Overtopping Manual Largely Based on European Research, but for Worldwide Application.*; Van der Meer, J.W., Allsop, N.W.H., Bruce, T., De Rouck, J., Kortenhaus, A., Pullen, T., Schüttrumpf, H., Troch, P., Zanuttigh, B., Eds. EurOtop. 2018. Available online: http://www.overtopping-manual.com (accessed on 15 December 2021).
17. Suzuki, T.; Altomare, C.; Veale, W.; Verwaest, T.; Trouw, K.; Troch, P.; Zijlema, M. Efficient and Robust Wave Overtopping Estimation for Impermeable Coastal Structures in Shallow Foreshores Using SWASH. *Coast. Eng.* **2017**, *122*, 108–123. [CrossRef]
18. Zijlema, M.; Stelling, G.; Smit, P. SWASH: An Operational Public Domain Code for Simulating Wave Fields and Rapidly Varied Flows in Coastal Waters. *Coast. Eng.* **2011**, *58*, 992–1012. [CrossRef]
19. Suzuki, T.; Verwaest, T.; Veale, W.; Trouw, K.; Zijlema, M. A numerical study on the effect of beach nourishment on wave overtopping in shallow foreshores. *Coast. Eng. Proc.* **2012**, *1*. [CrossRef]
20. Suzuki, T.; Altomare, C.; Verwaest, T.; Trouw, K.; Zijlema, M. Two-dimensional wave overtopping calculation over a dike in shallow foreshore by swash. *Coast. Eng. Proc.* **2014**, *1*, Structures.3. [CrossRef]
21. Zhang, N.; Zhang, Q.; Wang, K.-H.; Zou, G.; Jiang, X.; Yang, A.; Li, Y. Numerical Simulation of Wave Overtopping on Breakwater with an Armor Layer of Accropode Using SWASH Model. *Water* **2020**, *12*, 386. [CrossRef]
22. Celli, D.; Pasquali, D.; De Girolamo, P.; Di Risio, M. Effects of Submerged Berms on the Stability of Conventional Rubble Mound Breakwaters. *Coast. Eng.* **2018**, *136*, 16–25. [CrossRef]
23. Liang, B.; Wu, G.; Liu, F.; Fan, H.; Li, H. Numerical Study of Wave Transmission over Double Submerged Breakwaters Using Non-Hydrostatic Wave Model. *Oceanologia* **2015**, *57*, 308–317. [CrossRef]
24. Hersbach, H.; Bell, B.; Berrisford, P.; Biavati, G.; Horányi, A.; Muñoz Sabater, J.; Nicolas, J.; Peubey, C.; Radu, R.; Rozum, I.; et al. *ERA5 Hourly Data on Single Levels from 1959 to Present*; Copernicus Climate Change Service (C3S) Climate Data Store (CDS): Reading, UK, 2018. [CrossRef]
25. Swan Team. *Swan User Manual, 40.51.*; Department of Civil Engineering and Geosciences, Delft university of Technology: Delft, The Netherlands, 2006.
26. Pés, V.M. *Applicability and Limitations of the SWASH Model to Predict Wave Overtopping*; TU Delft: Delft, The Netherlands; Universitat Politecnica de Catalunya: Barcelona, Spain, 2013.
27. Salas Pérez, M. *Overtopping over a Real Rubble Mound Breakwater Calculated with SWASH*; Universitat Politecnica de Catalunya: Barcelona, Spain; Delft University of Technology: Delft, The Netherlands, 2014.
28. TAW. *Technisch Rapport Golfoploop En Golfoverslag Bij Dijken (Technical Report on Wave Run-up and Wave Overtopping at Dikes—In Dutch)*; Technical Advisory Committee on Flood Defence: Brussels, Belgium, 2002.
29. Galland, J.C. Rubble mound breakwater stability under oblique waves: And experimental study. *Coast. Eng. Proc.* **1994**, *1*, 24. [CrossRef]
30. Vanneste, D.F.A.; Altomare, C.; Suzuki, T.; Troch, P.; Verwaest, T. Comparison of numerical models for wave overtopping and impact on a sea wall. *Coast. Eng. Proc.* **2014**, *1*, 5. [CrossRef]
31. Manz, A. *Application of SWASH to Determine Overtopping during Storm Events in the Port of Ericeira and Its Introduction into HIDRALERTA System*; Universidade do Algarve: Faro, Portugal, 2021.

Journal of
Marine Science and Engineering

MDPI

Article

Coastal Defences on Low-Lying Reef Flats: A Laboratory Study of Seawall Shape and Position

Goélanne Rossignol [1,†] and Damien Sous [1,2,*,†]

1 Université de Toulon, Aix Marseille Université, CNRS, IRD, Mediterranean Institute of Oceanography (MIO), 83130 La Garde, France
2 Universite de Pau et des Pays de l'Adour, E2S UPPA, SIAME, Anglet, 64000 Pau, France
* Correspondence: sous@univ-tln.fr
† These authors contributed equally to this work.

Abstract: The present study reports on a series of laboratory experiments aiming to understand and to optimize the deployment of seawall defences over low-lying coral reef flats. The study is performed in a purely cross-shore configuration, with varying wave forcing, water depth, and seawall shapes and positions along the reef flat. The observations reveal the individual and combined effects of the external parameters. A higher seawall efficiency is reached for low waves, shallow water depth, seaward position, and a stepped seawall shape. A dimensionless formulation is proposed to predict the short wave transmission coefficient dependency on the seawall freeboard, incoming wave height, berm width, and seawall shape.

Keywords: reef flat; seawall; wave transmission; optimal shape; laboratory experiments

1. Introduction

Mainly found along tropical and subtropical coastlines, coral reefs provide a major form of shelter for low-lying coastal areas against marine hazards, including submersion and wave damages during tropical storms, cyclones, and tsunamis. The sustainability of coastal coral reefs is now threatened by the effects of global warming, ocean acidification, plastic debris, and human actions [1–3]. The potential degradation (smoothing, patchiness, and flattening) of shallow water coral colonies, combined with the rise of the sea level, is expected to increase the exposure of coastlines [4] leading to higher risks of damages to natural and engineered structures, and flooding and salinization of coastal aquifers [5]. In many situations, hard engineering structures will remain the most relevant solution to protecting the shore from the ocean's attack. Optimizing the deployment of coastal defences is therefore essential to prevent further damages while avoiding excessive costs and consumption of materials (generally locally made coral concrete [6]).

The defence design is currently a crucial issue for many low-lying coral islands such as closed coral atolls, where in the absence of well-developed reef passes, channel-basin systems have been directly dug into the reef flat to ensure boat access for all island supplies. A typical illustration of such a system at Reao island, French Polynesia, is displayed in Figure 1. The seawalls here play a direct role on wave buffering, but they also affect, by controlling the transmitted energy and the water mass, the intensity and stability of the return flow in the channel, which is a major threat for boat navigation and berthing. The wave conditions at the time of the picture were typical of well-developed swell conditions, but not a particularly energetic event, according to local fishermen (no wave data were available in this remote area). One notes that the submersion of most defences, including the 2.4 m high stepped wall in the red frame.

Figure 1. Channel-basin system in the Reao fringing reef French Polynesia—*Courtesy of A.Para, ODEWA.*

While a vast body of literature exists on wave force and stability issues in the design of breakwaters [7–15], together with the parameterization of overtopping for simple seawall shapes [15–17], a few studies have been dedicated to the shape optimization [18]. Most existing seawalls can be classified as vertical, sloping, stepped, or curved types (these latter include the so-called "non-overtopping" design) [19–24]. Furthermore, the seawall effectiveness is strongly dependent on the local deployment and hydrodynamic conditions, making direct comparisons delicate [18]. The main trends can be summarized as follows.

- Vertical seawalls maximize wave reflection but can be exposed to direct flip-through impact, inducing high pressure peaks and structural damages [25]. Other typical drawbacks are strong scouring at the seawall toe [26], and jet and aerosol production.
- Sloping seawalls reduce wave reflection compared to the vertical seawall, lowering the scouring risk and impact pressure [27], but they are less efficient in limiting overtopping (the ramp effect).
- Stepped seawalls have the same advantages as sloping seawalls (less impact and scouring), but are expected to favor wave dissipation.
- Curved seawalls are generally designed to promote destructive interaction between the incoming and reflected waves, and to reduce impact pressure [28]. While curved

walls are expected to show better performance, they are more complex to build and maintain [24].

Compound designs have also been proposed, such as the combination of a stepped/curved profile at the lower/upper part of the seawall profile, potentially providing the advantages of both shapes (wave breaking and reflection) and minimizing the disadvantages (ramping and impact) [19,20].

In addition to the sparse knowledge of the shape effect, the specific configuration of seawall systems deployed on coral reef flat remains undocumented. On natural reef systems, being barrier or fringing reefs, the wave dynamics show remarkable features when compared to the more classical case of sand beaches. The steep slope observed on forereefs induces more violent plunging wave breaking [29–33], leading to an increased wave setup, while the shallowness and the roughness of the reef flat system are associated with an increased degree of dissipation and spectral energy transfers [34–43]. Waves and water levels are then intrinsically and non-linearly connected all across the reef system. Starting at the breaking point, the main trends are an attenuation of short waves (SW), an increase of the mean water level, and the development of long infragravity (IG) waves [44]. However, the coupling between the driving physical processes (reflexion, breaking, friction, and non-linear transfers) controlling wave transformation are strongly non-linear [37], and are dependent on the local hydrodynamical and morphological features. This complex coupled evolution of waves and the water level is therefore a major obstacle for the prediction of the exposure level of any defence placed on the reef flat.

The aim of the present study is to decipher, under controlled conditions, the hydrodynamical functioning of a seawall deployed on a fringing reef flat. To address this issue, direct field studies should be ruled out, considering their cost and environmental impact, while numerical models able to reach sufficient resolution to decipher the shape effect [18] are still unusable at the full reef system scale. The approach followed here is to use laboratory wave flume experiments to understand the seawall effect on the wave field, focusing on the role played by shape and position. Section 2 describes the experimental setup, including the instrumentation and methods. Section 3 presents the results, while Section 4 provides a discussion and outlines the prospects.

2. Materials and Methods

2.1. Experimental Setup

Experiments were carried out in the CASH (Canal Aero-Sedimento- Hydrodynamics) wind-wave flume Seatech engineering school, University of Toulon [45]. The flume has a length of 6m and a width of 0.5m. It was equipped with an idealized fringing reef bathymetry (Figure 2A) made of (i) a 2.5 m length flat horizontal bottom, (ii) a 0.5 m length linear sloping forereef (slope of 0.35), (iii) a 2.5 m length and 0.18m height horizontal reef flat, and (iv) a dissipative beach. The reef bed, including the forereef, the reef flat, and the dissipative beach, was made of a plastic LEGO® baseplate covered by a regular layout of 1.6cm height plastic cubes arranged in alternating gridlines (Figure 2B,C), showing an overall roughness standard deviation of 0.6 mm.

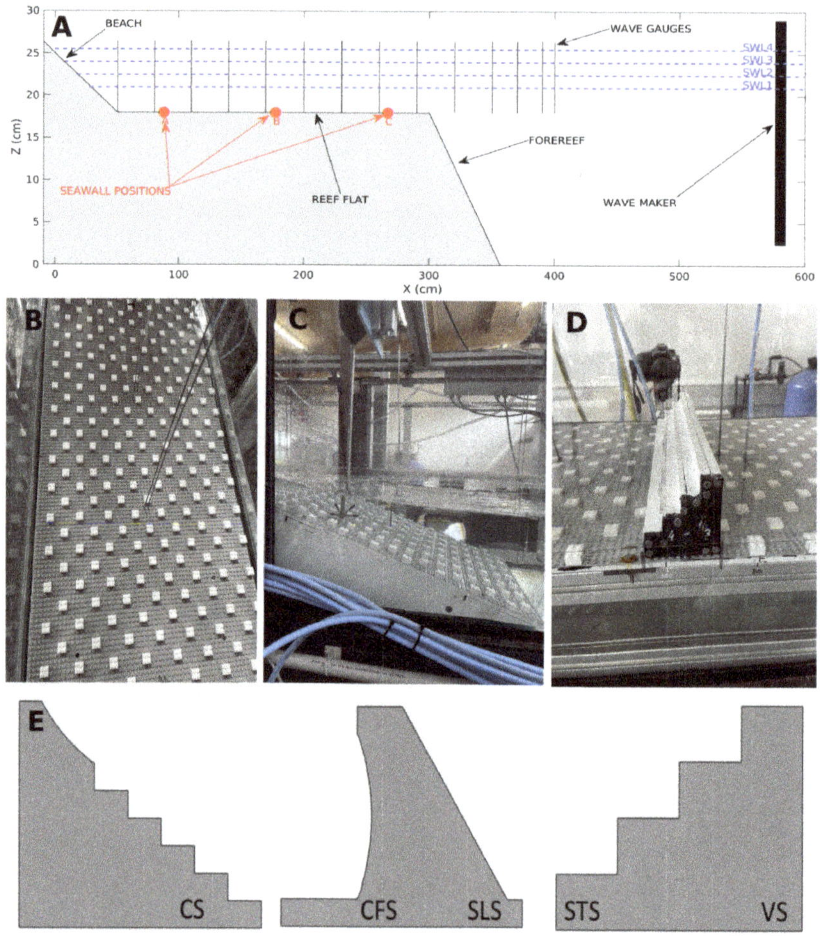

Figure 2. Experimental setup. (**A**): Bed profile and instrumentation. (**B**) and (**C**): Reef flat and forereef covered in plastic cubes. (**D**): Stepped seawall on the reef flat. (**E**): Tested seawall shapes.

2.2. Test Cases

2.2.1. Wave and Depth Conditions

Irregular waves, based on a JONSWAP spectrum ($\gamma = 6$), were produced using a piston wave maker in the flat bottomed section of the flume. Each run lasted 5 min. Four experimental water depth cases and five experimental wave cases with different wave characteristics were performed, aiming to cover a wide range of seawall exposure (Table 1). For a given wave forcing, the actual significant height of the generated wave depends on the water depth (deeper conditions at the wave maker produces slightly higher waves).

Table 1. Hydrodynamic conditions for the test cases.

	Reef Flat Depth (cm)	Total Depth (cm)
SWL1	3	21
SWL2	4.5	22.5
SWL3	6	24
SWL4	7.5	25.5
	H_s (cm)	**T_p (s)**
W1	1.98–2.5	0.8
W2	3.2–4.28	1
W3	4.43–5.77	1
W4	4.74–6.37	1.2
W5	7.22–8.66	1.4

2.2.2. Seawall Shapes and Positions

For each wave/depth condition, five seawall shapes and three seawall positions were systematically tested (Figure 2D,E):

- Curved front seawall (CFS),
- Stepped seawall (STS),
- Sloping seawall with a slope of 45° (SLS).
- Vertical seawall (VS).
- Compound seawall with a stepped part at the toe and a curved part at the top of the structure (CS).

Three asymmetric seawalls were actually built (Figure 2E) and flipped to expose the selected face to the incoming wave. All seawall shapes were built in PVC with a 8 cm crest, and filled with lead beads to ensure stability. The smooth PVC surface ensured that the study focused on the effect of shape only, without additional surface-roughness dissipation. Each seawall was tested at three different positions, defined from the shoreward end of the flume: near the reef crest ($X = 2.68$ m), in the middle of the reef flat ($X = 1.79$ m), and near the shoreline ($X = 0.89$ m) (Figure 2A).

2.3. *Instrumentation and Data Processing*

Free surface elevation measurements were performed using 14 resistor-type wave gauges arranged as shown in Figure 2, with an acquisition frequency of 100 Hz. Data processing was carefully conducted in order to maintain the original wave characteristics while separating the short waves (SW) and infragravity waves (IG). Each elevation timeseries was processed using a Fourier transform. The resulting spectral density of variance was integrated on the short waves (SW) band of $0.5F_p < F < 1.5F_p$ to obtain the corresponding significant wave height H_s. The lower limit at $0.5F_p$ was selected because a clear energy drop was observed between the short and long (infragravity) bands for all wave and depth conditions.

2.4. *Estimation of Seawall Efficiency*

Estimating the efficiency of a seawall in a realistic reef flat context remains a challenge, owing to the fact that the seawall is integrated within a natural system already affected by a strong wave dissipation and fluctuation of the free surface at various time scales (SW, IG, and wave setup). The seawall implementation on the reef flat may not only act as a local filter of wave energy, but is expected to affect several processes acting in wave transformation (reflexion, breaking, and harmonic transfers). Therefore, the assessment of seawall efficiency requires a quantification of the defence-induced modification applied on the reef system as a whole. Moreover, in many situations observed in Polynesian atolls, seawalls are exposed to regular submersion during typical wave events, as illustrated by Figure 1, without the need to reach extreme wave conditions. In this context, the defence will act as a partly submerged breakwater. In addition, a focus point of the present study was to characterize the remaining wave agitation downstream of the seawall. For these reasons, instead of assessing the seawall effect in terms of the overtopping discharge [46–48], the seawall efficiency was here quantified by the global transmission coefficients K_t (Equation (1)), defined as the ratio between the deep water and the shoreline wave heights. Higher transmission coefficients were then associated with a lower defence efficiency.

The SW transmission coefficient is given by:

$$K_t = \frac{H_s}{H_o} \quad (1)$$

where H_o is the deep-water wave height calculated from the linearly deshoaled free surface measurement average of the two wave gauges closest to the wave maker, and H_s is the shoreline SW height. In order to distinguish the seawall effect from the free reef functioning, normalized transmission coefficients K_{tn} were defined as the ratio between transmission coefficients with the seawall K_t and without the seawall K_{tr}.

$$K_{tn} = \frac{K_t}{K_{tr}} \quad (2)$$

The estimation of the transmission coefficient can be affected by secondary wave reflection at the wave maker, through possible changes in H_o due to differential wave height modulation. For given depth and wave forcing conditions, the average fluctuation of H_o is 2.9%. This remains weak compared to the effects of seawall shape and position on K_t highlighted hereafter, and should therefore not drastically affect the main interpretations.

2.5. Scaling Strategy

The laboratory experiments did not aim to provide a comprehensive scaling of the fine wave impact dynamics at the real scale, but rather, to assess the overall hydrodynamical defence effect under controlled conditions. The focus was placed on the relevancy of conditions for wave propagation over the reef flat ($0.07 < kh < 0.25$), wave shape ($0.2 < ka < 0.4$), and breaking features (surf similarity parameter $1 < \xi < 2$, with $xi = tan\beta/\sqrt{H/L_o}$, β being the forereef slope, H being the breaker height, and L_o being the deep water wave length). The Froude scaling indicates that the scaling ratio between nature and laboratory was $\frac{1}{30}$ for the vertical scale and $\frac{1}{60}$ for the horizontal scale. Therefore, the real-scale depths are 6.3 to 7.65 m at the toe of the forereef, and 0.9 to 2.25 m over the reef flat, while the real-scale offshore wave heights range from 0.6 to 2.6 m. Note that the waves are generated at the wave maker under intermediate depth conditions. The experimental setup being focused on shallow wave dynamics, from the breaking over the forereef to the beach, the transformation from deep water up to the forereef is out of the scope of the present study. As for most wave flume experiments, the Reynolds number similitude is impossible to achieve.

3. Results

3.1. Overview

3.1.1. No-Seawall Cases

Figure 3 depicts the cross-shore profiles of significant wave height in SW bands, respectively, for four illustrative cases with low (SWL1) and high (SWL4) water levels and weak (W1) and strong (W5) wave forcing. Note that the zero for the y-axis corresponds to the still water level for each case. We first focus on the description of the reference no-seawall tests (see the black dashed lines in Figure 3). The aim of the analysis is not to precisely decipher each physical process acting in the wave dynamics on the reef system, which has already been detailed in dedicated studies [34–37,49,50], but rather, to assess the relevancy of the present reef system in reproducing the common hydrodynamics observed on typical fringing reef systems before exploring the effect of the seawall. As expected in such environments [37,51,52], the SW profiles show a strong dependency to both deep water wave forcing and depth. For weak wave and shallow reef flat conditions (Figure 3A), wave breaking occurs over the upper forereef. The attenuation rate, due to the combined effects of friction and breaking dissipation and non-linear energy transfers, is moderate: half of the incident wave height is attenuated at around $X = 230$ cm, and 18% remains at the shore. Increasing the water depth (Figure 3C) leads to a strong decrease of wave dissipation. No breaking-induced inflexion of the wave height profile is observed, indicating that most of the attenuation should be attributed to frictional dissipation. This observation is confirmed with the estimation of the breaking parameter γ, defined as the ratio between the local wave height and depth, which remains below 0.3 throughout the reef system. The remaining wave height at the shore is about 54 % of the incident wave height. Increasing the wave energy has a drastic effect on the SW profile ((Figure 3B,D). The breaking point is clearly identified, with a nearly constant position over the forereef (about $X = 320$, see Figure 3B,D) regardless of the water depth due to the very strong bottom slope [49]. Offshore of the breaking point ($X > 320$ cm), the SW height profile displays spatial modulations due to the combined effects of shoaling and partial reflection. Onshore of the breaking point, the wave attenuation is first governed by breaking dissipation over the forereef (260 cm $< X <$ 320 cm). Decreasing the water depth (Figure 3b vs. Figure 3d) leads to a stronger dissipation rate in the surf zone, due to a more violent plunging breaking. The remaining SW heights at the shore are 8% and 21% of the offshore wave height for the shallow and deep cases, respectively.

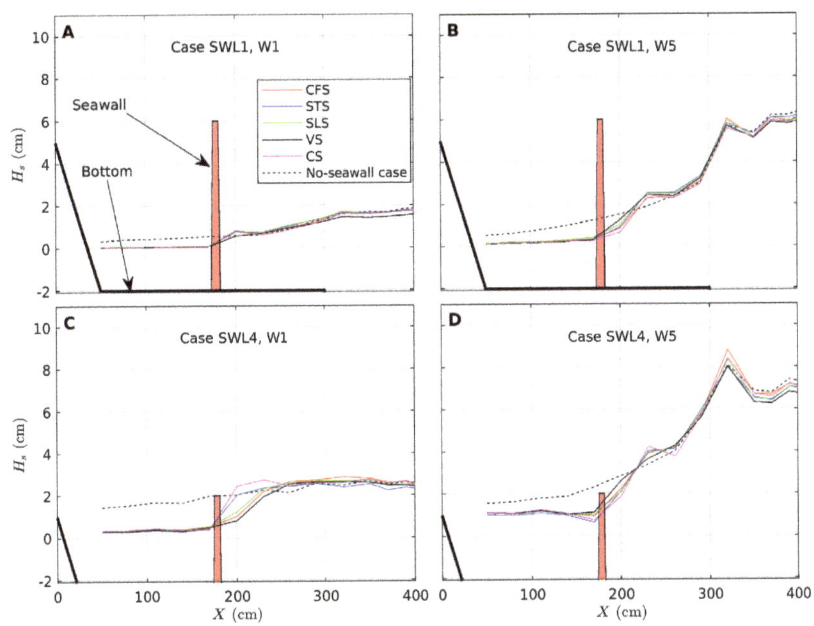

Figure 3. Cross-shore profiles of SW in the no-seawall reference cases (black dashed line) and seawall cases at position (**B**) (colored solid lines). (**A–D**): SWL1/W1, SWL1/W5, SWL4/W1, and SWL4/W5 cases, respectively. The zero for the y-axis corresponds to the still water level for each case. The red dot indicates the seawall position, when present. Note that the breaking point in (**A**,**B**,**D**) is located at around $X = 320$ cm, i.e., near the mid forereef slope, which is not represented here, see Figure 2 for complete bottom profile.

3.1.2. Seawall Cases

In each studied case, the seawall has a drastic effect on the wave dynamics. Illustrative cases with the seawall in Position B are depicted in solid lines in Figure 3, with colors indicating the seawall shape. First, it is worthwhile to mention that the seawalls do not only show a local buffering effect, but they also strongly affect the upstream wave field due to reflection and complex wave–structure interaction. This is particularly visible under strong wave conditions (Figure 3B,D) where a spatial modulation of the wave height is observed, leading to a decrease of wave height just upstream of the obstacle when compared to the no-seawall case. This observation reinforces the need to use a global approach at the reef scale to understand the seawall effect. In the small wave and shallow depth case (Figure 3A), an abrupt drop is observed on the SW profile. Regardless of the seawall type, most of the incoming energy is filtered and only residual agitation is measured downstream of the seawall ($H_s < 0.1$ cm). The wave damping decreases with increasing incoming wave height or water depth. The distinct effects of depth and wave condition are observed when comparing Figure 3B,C with Figure 3A, respectively. While forced by different conditions (small depth/large wave vs. large depth/small waves), both the SWL1/W5 and SWL4/W1 cases show similar levels of agitation downstream of the obstacle (H_s about 0.3 cm at the shore). Combining large waves and deeper conditions (Figure 3D) leads to a higher transmission of SW energy toward the shore ($H_s = 1$ cm).

3.2. *Transmission Coefficients*

Figure 4 depicts the transmission coefficient K_t for each considered case: the water depth (x-axis), wave condition (rows), seawall position (columns), and seawall shape (colored markers). The reference no-seawall conditions are shown in black dots. The free (no-seawall) reef system shows the expected increase of SW dissipation for shallow depths and large waves. In most cases, the transmission coefficients are significantly lowered by the implementation of the defence. The seawall effect first shows a marked dependency on depth: the larger the depth, the greater the transmission coefficient. For a small depth, K_t remains below 0.07 regardless of the wave forcing, or the seawall shape of position. An increasing depth leads to an increase of K_t, with a further dependency on wave forcing and seawall position. For small waves and shallow depths, the seawalls show similar performances regardless of their positions. Increasing wave forcing and water depth leads to a differential behavior depending on the seawall position: the more shoreward the seawall, the weaker the transmission coefficient. Overall, the most unfavorable case is reached when combining a large depth, high waves, and an offshore seawall position, for which the transmission coefficient is quite close to the one obtained in the no-seawall case.

Finer information is gained on the seawall shape effect by computing the normalized transmission coefficient K_{tn}^{sw} (Equation (2)); see Figure 5. Strong variations are observed, with K_{tn}^{sw} ranging from 9 to 88%. The previous trends are confirmed: the wave transmission over the seawall is promoted by large waves, a large depth, and an offshore-located seawall. An additional sensitivity to the seawall shape is observed. The best overall performance in terms of SW attenuation is reached for STS and CS, which shows a 5 to 25% lower K_{tn}^{sw} than the SLS and CFS types. VS show variable efficiency, with the best performances reached at low water depths.

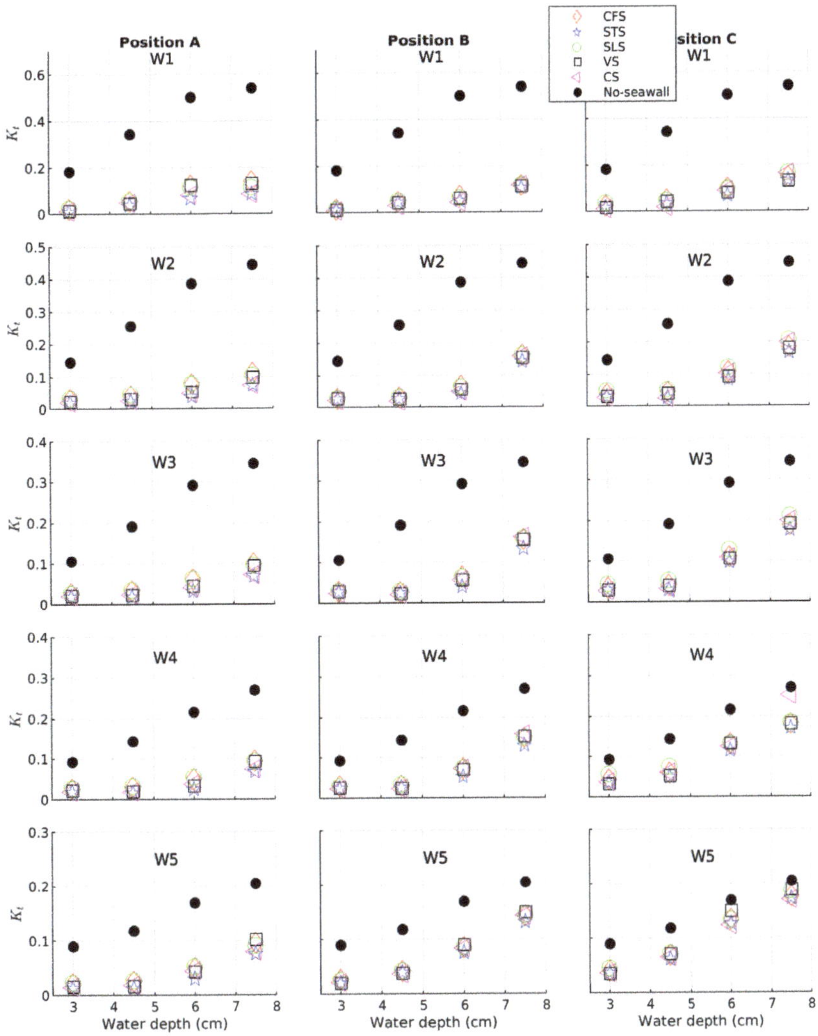

Figure 4. SW transmission coefficient vs. reef flat water depth. Water depth, wave condition, seawall position, and seawall shape are depicted with the x-axis, rows, columns, and colored markers, respectively. Note the difference in y-axis scale.

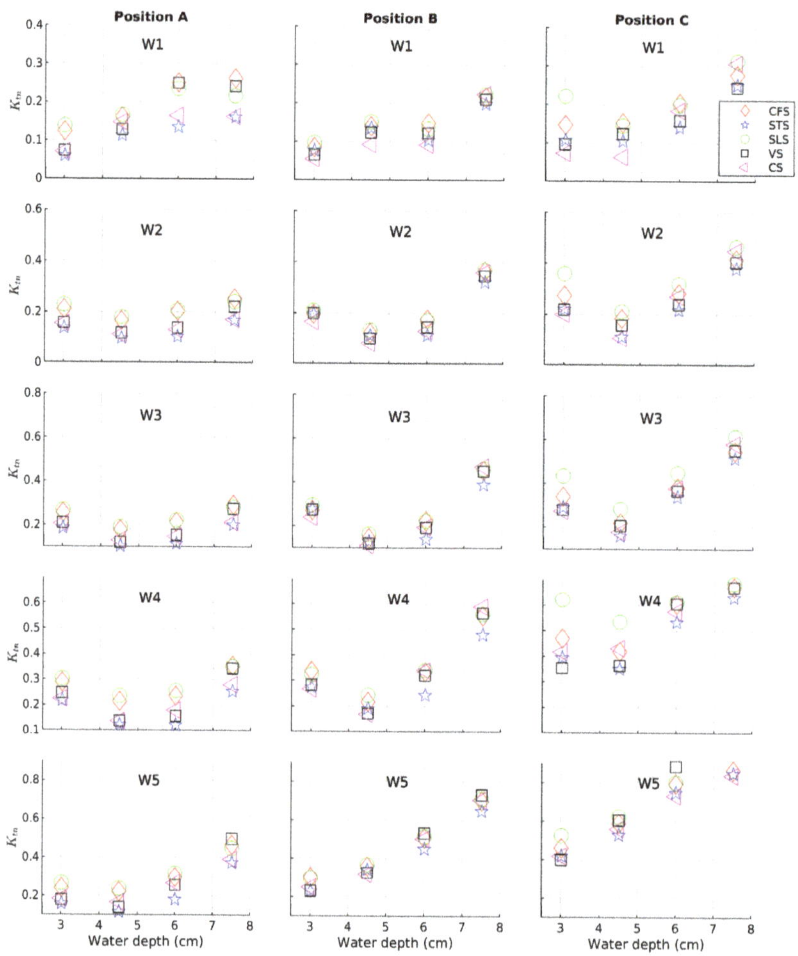

Figure 5. Normalized SW transmission coefficient vs. reef flat water depth. Water depth, wave condition, seawall position, and seawall shape are depicted in x-axis, rows, columns, and colored markers, respectively. Note the difference in y-axis scale.

3.3. K_t Relationship to External Parameters

Following previous works carried out on overtopping issues [46,48,53], we propose here an empirical formulation aiming to predict the transmission coefficient as a function of the external parameters of the system. We first define the dimensionless freeboard R^* as

$$R^* = \frac{R_c}{H_0} \tag{3}$$

where R_c is freeboard. Two reductions factors γ_l and γ_s are formed to account for the effect of the berm width and the seawall shape, respectively, defined as:

$$\gamma_l = \sqrt{\frac{X}{\lambda}} \tag{4}$$

where X and λ are the cross-shore distances between the seawall and the reef crest, and the wave length on the reef flat. The best predicting model is obtained for

$$K_t = 0.017(R^*\gamma_s\gamma_l)^{-0.5} \quad (5)$$

The shape factor γ_s is fixed at 0.85, 0.9, 1, 1.1, and 1.3 for SLS, CFS, VS, CS, and STS, respectively, based on an optimization of the linear determination coefficient between K_t and $(R^*\gamma_s\gamma_l)^{-0.5}$.

Figure 6 depicts the relationship between K_t and $(R^*\gamma_s\gamma_l)^{-0.5}$. The linear determination coefficient is 0.85, confirming the validity of the parameterization for the whole dataset.

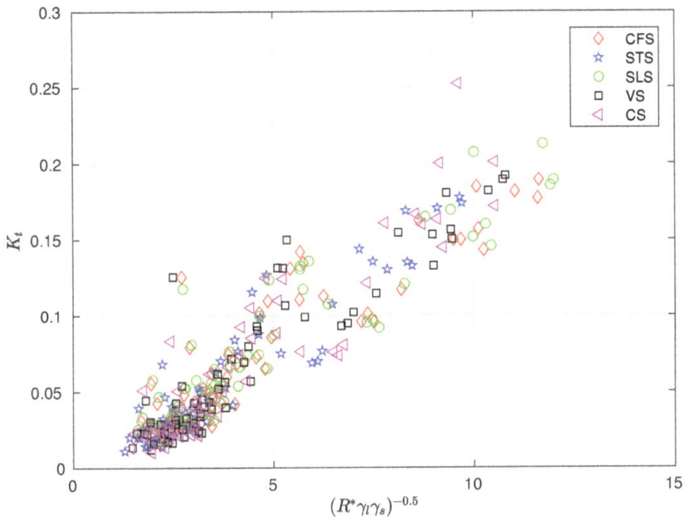

Figure 6. Transmission coefficient vs. R^*. The symbols denote the seawall type.

4. Discussion

Building upon previous research on seawall–wave interactions, this study presents a detailed quantification of wave attenuation by a single seawall defence deployed on an archetypal reef flat. The study has been carried out in controlled laboratory experiments, which currently remains the only possible approach, allowing for a careful assessment of the seawall effect at the complete reef system scale. The seawall efficiency is addressed here in terms of the wave transmission coefficient K_t, in order to cover a large range of submersion conditions forced by irregular wave fields and varying mean water levels. The main findings can be summarized as follows:

- The seawall efficiency, measured in terms of the transmission coefficient, is intrinsically linked to its level of exposure, and therefore, to the coupled dynamics of wave and water depth across the reef system.
- Higher efficiencies are reached for a low water level and a low incoming wave height.
- As a consequence, the more onshore the seawall, the stronger the wave attenuation.
- The experimental results confirm the recent numerical observations [18] showing that that a stepped seawall shape is more efficient than vertical, sloping, and curved seawall shapes.
- The additional compound seawall shape tested here shows good performance, but it does not surpass the results reached with a stepped seawall.
- A simple monotonic K_t prediction formula is proposed to represent the overall effect of the main external factor controlling the seawall efficiency: the freeboard, the incoming wave height, the berm width, and the seawall shape.

The observations presented here result from bulk observations over a complete irregular wave run, ensuring robust statistics. The detailed dynamics of each individual wave impact are not isolated by the measurements. However, the overall results are in agreement with the qualitative observations performed during the experiments. In particular, it has been observed that the most reflective shapes (vertical and curved) are actually the most efficient in reflecting most individual wave impacts. However, placed in an irregular wave field, this does not systematically result in lower overtopping and energy transmission due to the complex local interactions between the incident and reflected wave fronts just upstream of the seawall, regularly leading to the opposite effect. The sloping seawall showed much lower reflection, but also a clear ramp effect, inducing an increase in overtopping and wave transmission for nearly all conditions. The stepped and compound seawalls showed visually more regular but weaker overtopping than the vertical and curved seawalls. The stepped/compound seawalls do not establish a impassable obstacle, but they are very efficient in buffering the wave energy by increased breaking-induced dissipation, whatever the depth and wave conditions.

In the general context of global sea level rises and increases in the vulnerability of low-lying coral islands, the present database can be used for operational engineering purposes, but it may also be helpful in calibrating numerical wave models for which the representation of wave dynamics in such complex environments remains a major challenge. The present study was totally focused on hydrodynamics, and did not consider wave load and stability issues. However, it is worthwhile to mention that the most effective configurations, in terms of shape and position, are certainly those for which the loads are minimized.

Author Contributions: Conceptualization, D.S.; methodology, G.R. and D.S.; validation, G.R. and D.S.; investigation, G.R. and D.S; resources, D.S.; data curation, G.R.; writing—original draft preparation, G.R.; writing—review and editing, D.S.; visualization, D.S.; supervision, D.S.; project administration, D.S.; funding acquisition, D.S. All authors have read and agreed to the published version of the manuscript.

Funding: This research was funded by LE MINISTERE DE L'EQUIPEMENT DE POLYNESIE FRANCAISE, grant number 20/21/MAPA/MGT.

Institutional Review Board Statement: Not applicable.

Informed Consent Statement: Not applicable.

Data Availability Statement: The data presented in this study are available on request from the corresponding author. The data are not publicly available due to MINISTERE DE L'EQUIPEMENT DE POLYNESIE FRANCAISE's ownership.

Conflicts of Interest: The authors declare no conflict of interest.

References

1. Comeau, S.; Cornwall, C.; DeCarlo, T.M.; Doo, S.; Carpenter, R.; McCulloch, M. Resistance to ocean acidification in coral reef taxa is not gained by acclimatization. *Nat. Clim. Chang.* **2019**, *9*, 477–483. [CrossRef]
2. Cheal, A.J.; MacNeil, M.A.; Emslie, M.J.; Sweatman, H. The threat to coral reefs from more intense cyclones under climate change. *Glob. Chang. Biol.* **2017**, *23*, 1511–1524. [CrossRef] [PubMed]
3. Lartaud, F.; Meistertzheim, A.; Reichert, J.; Ziegler, M.; Peru, E.; Ghiglione, J. Plastics: An additional threat for coral ecosystems. In *Perspectives on the Marine Animal Forests of the World*; Springer: Berlin/Heidelberg, Germany, 2020; pp. 469–485.
4. Fernando, H.; McCulley, J.; Mendis, S.; Perera, K. Coral poaching worsens tsunami destruction in Sri Lanka. *Eos Trans. Am. Geophys. Union* **2005**, *86*, 301–304. [CrossRef]
5. Paldor, A.; Michael, H. Storm surges cause simultaneous salinization and freshening of coastal aquifers, exacerbated by climate change. *Water Resour. Res.* **2021**, *57*, e2020WR029213. [CrossRef]
6. Ehlert, R.A. Coral concrete at bikini atoll. *Concr. Int.* **1991**, *13*, 19–24.
7. Sainflou, G. Essai sur les digues maritimes verticales. *Ann. Ponts Chauss.* **1928**, *98*, 5–48.
8. Allsop, N.; McKenna, J.; Vicinanza, D.; Whittaker, T. New design formulae for wave loadings on vertical breakwaters and seawalls. In Proceedings of the 25th ICCE, Conference on Coastal Engineering, Orlando, FL, USA, 2–6 September 1996; ASCE: New York, NY, USA, 1996.

9. Allsop, N.; Vicinanza, D. Wave impact loadings on vertical breakwaters: Development of new prediction formulae. In Proceedings of the 11th International Harbour Congress, Antewerp, Belgium, 17–21 June 1996.
10. Klammer, P.; Kortenhaus, A.; Oumeraci, H. Wave impact loading of vertical face structures for dynamic stability analysis-prediction formulae. In Proceedings of the 25th Conference on Coastal Engineering, Orlando, FL, USA, 2–6 September 1996.
11. Kortenhaus, A.; Oumeraci, H. Classification of wave loading on monolithic coastal structures. In Proceedings of the Coastal Engineering, Copenhagen, Denmark, 22–26 June 1998.
12. Oumeraci, H.; Kortenhaus, A.; Allsop, W.; de Groot, M.; Crouch, R.; Vrijling, H.; Voortman, H. *Probabilistic Design Tools for Vertical Breakwaters*; CRC Press: Boca Raton, FL, USA, 2001.
13. Takahashi, S. *Design of Vertical Breakwaters*; PHRI Reference Document nr. 34; PARI: Yokosuka, Japan, 2002.
14. Cuomo, G.; Allsop, W.; Bruce, T.; Pearson, J. Breaking wave loads at vertical seawalls and breakwaters. *Coast. Eng.* **2010**, *57*, 424–439. [CrossRef]
15. Goda, Y. *Random Seas and Design of Maritime Structures*; World Scientific Publishing Company: Singapore, 2010.
16. Thomas, R.S.; Hall, B. *Seawall Design*; Butterworth-Heinemann: Oxford, UK, 2015.
17. Tsinker, G. *Handbook of Port and Harbor Engineering: Geotechnical and Structural Aspects*; Springer: Berlin/Heidelberg, Germany, 2014.
18. Dang, B.L.; Nguyen-Xuan, H.; Wahab, M.A. Numerical study on wave forces and overtopping over various seawall structures using advanced SPH-based method. *Eng. Struct.* **2021**, *226*, 111349. [CrossRef]
19. CERC. *Shore Protection Manual*; US Army Coastal Engineering Research Center: Montgomery, AL, USA, 1973; Volume 1.
20. Manual, E. *Design of Coastal Revetments, Seawalls, and Bulkheads*; Department of the Army, Corps of Engineers: Washington, DC, USA, 1995.
21. Marukami, K.; Irie, I.; Kamikubo, Y. Experiments on a non-wave overtopping type of seawall. In Proceedings of the Coastal Engineering, Orlando, FL, USA, 2–6 September 1996.
22. Pullen, T.; Allsop, N.; Bruce, T.; Kortenhaus, A.; Schüttrumpf, H.; Van der Meer, J. *EurOtop Wave Overtopping of Sea Defences and Related Structures: Assessment Manual*; Boyens Medien GmbH: Groningen, The Netherlands, 2007.
23. Stagonas, D.; Ravindar, R.; Sriram, V.; Schimmels, S. Experimental evidence of the influence of recurves on wave loads at vertical seawalls. *Water* **2020**, *12*, 889. [CrossRef]
24. Hosseinzadeh, N.; Ghiasian, M.; Andiroglu, E.; Lamere, J.; Rhode-Barbarigos, L.; Sobczak, J.; Sealey, K.S.; Suraneni, P. Concrete seawalls: A review of load considerations, ecological performance, durability, and recent innovations. *Ecol. Eng.* **2022**, *178*, 106573. [CrossRef]
25. Martin-Medina, M.; Abadie, S.; Mokrani, C.; Morichon, D. Numerical simulation of flip-through impacts of variable steepness on a vertical breakwater. *Appl. Ocean. Res.* **2018**, *75*, 117–131. [CrossRef]
26. Zapata, M.U.; Zhang, W.; Marois, L.; Hammouti, A.; Van Bang, D.P.; Nguyen, K. Two-phase experimental and numerical studies on scouring at the toe of vertical seawall. *Eur. J. Mech.-B/Fluids* **2022**, *93*, 13–28. [CrossRef]
27. Salauddin, M.; Pearson, J. Experimental study on toe scouring at sloping walls with gravel foreshores. *J. Mar. Sci. Eng.* **2019**, *7*, 198. [CrossRef]
28. Anand, K.V.; Sundar, V.; Sannasiraj, S.A. Dynamic pressures on curved front seawall models under random wave. *J. Hydrodyn.* **2010**, *22*, 521–527. [CrossRef]
29. Young, I.R. Wave transformation over coral reefs. *J. Geophys. Res. Ocean.* **1989**, *94*, 9779–9789. [CrossRef]
30. Smith, E.R.; Kraus, N.C. Laboratory study of wave-breaking over bars and artificial reefs. *J. Waterw. Port Coast. Ocean. Eng.* **1991**, *117*, 307–325. [CrossRef]
31. Gourlay, M.R.; Colleter, G. Wave-generated flow on coral reefs—An analysis for two-dimensional horizontal reef-tops with steep faces. *Coast. Eng.* **2005**, *52*, 353–387. [CrossRef]
32. Yao, Y.; Zhang, Q.; Chen, S.; Tang, Z. Effects of reef morphology variations on wave processes over fringing reefs. *Appl. Ocean. Res.* **2019**, *82*, 52–62. [CrossRef]
33. Lowe, R.; Altomare, C.; Buckley, M.; da Silva, R.; Hansen, J.; Rijnsdorp, D.; Domínguez, J.; Crespo, A. Smoothed Particle Hydrodynamics simulations of reef surf zone processes driven by plunging irregular waves. *Ocean Model.* **2022**, *171*, 101945. [CrossRef]
34. Lowe, R.J.; Falter, J.L.; Bandet, M.D.; Pawlak, G.; Atkinson, M.J.; Monismith, S.G.; Koseff, J.R. Spectral wave dissipation over a barrier reef. *J. Geophys. Res. Ocean.* **2005**, *110*, 1–16. [CrossRef]
35. Monismith, S.G. Hydrodynamics of coral reefs. *Annu. Rev. Fluid Mech.* **2007**, *39*, 37–55. [CrossRef]
36. Péquignet, A.C.; Becker, J.; Merrifield, M.; Boc, S. The dissipation of wind wave energy across a fringing reef at Ipan, Guam. *Coral Reefs* **2011**, *30*, 71–82. [CrossRef]
37. Sous, D.; Tissier, M.; Rey, V.; Touboul, J.; Bouchette, F.; Devenon, J.L.; Chevalier, C.; Aucan, J. Wave transformation over a barrier reef. *Cont. Shelf Res.* **2019**, *184*, 66–80. [CrossRef]
38. Reidenbach, M.A.; Monismith, S.G.; Koseff, J.R.; Yahel, G.; Genin, A. Boundary layer turbulence and flow structure over a fringing coral reef. *Limnol. Oceanogr.* **2006**, *51*, 1956–1968. [CrossRef]
39. Rosman, J.H.; Hench, J.L. A framework for understanding drag parameterizations for coral reefs. *J. Geophys. Res. Ocean.* **2011**, *116*, 1–15. [CrossRef]

40. Monismith, S.G.; Rogers, J.S.; Koweek, D.; Dunbar, R.B. Frictional wave dissipation on a remarkably rough reef. *Geophys. Res. Lett.* **2015**, *42*, 4063–4071. [CrossRef]
41. Rogers, J.S.; Maticka, S.A.; Chirayath, V.; Woodson, C.B.; Alonso, J.J.; Monismith, S.G. Connecting flow over complex terrain to hydrodynamic roughness on a coral reef. *J. Phys. Oceanogr.* **2018**, *48*, 1567–1587. [CrossRef]
42. Davis, K.A.; Pawlak, G.; Monismith, S.G. Turbulence and Coral Reefs. *Annu. Rev. Mar. Sci.* **2020**, *13, 1948189.* [CrossRef]
43. Li, Y.; Zhang, C.; Chen, S.; Sui, T.; Chen, D.; Qi, H. Influence of artificial sandbar on nonlinear wave transformation: Experimental investigation and parameterizations. *Ocean. Eng.* **2022**, *257*, 111540. [CrossRef]
44. Bertin, X.; De Bakker Anouk, V.D.A.; Giovanni, C.; Gael, A.; Fabrice, A.; Philippe, B.; Frederic, B.; Bruno, C.; Crawford Wayne, C.; Mark, D.; et al. Infragravity waves: from driving mechanisms to impacts. *Earth Sci. Rev.* **2018**, *177*, 774–799. [CrossRef]
45. Sous, D.; Forsberg, P.L.; Touboul, J.; Nogueira, G.G. Laboratory experiments of surf zone dynamics under on-and offshore wind conditions. *Coast. Eng.* **2021**, *163*, 103797. [CrossRef]
46. Owen, M. *Design of Seawalls Allowing for Overtopping;* Report EX924; HR Wallingford: Wallingford, UK, 1980.
47. Besley, P. *Wave Overtopping of Seawalls, Design and Assessment Manual;* R&D Technical Report W178; HR Wallingford: Wallingford, UK, 1998.
48. Allsop, W.; Bruce, T.; Pearson, J.; Besley, P. Wave overtopping at vertical and steep seawalls. In Proceedings of the Institution of Civil Engineers; Thomas Telford Ltd. *Marit. Eng.* **2005**, *158*, 103–114. [CrossRef]
49. Buckley, M.L.; Lowe, R.J.; Hansen, J.E.; Van Dongeren, A.R. Dynamics of wave setup over a steeply sloping fringing reef. *J. Phys. Oceanogr.* **2015**, *45*, 3005–3023. [CrossRef]
50. Buckley, M.L.; Lowe, R.J.; Hansen, J.E.; Van Dongeren, A.R. Wave setup over a fringing reef with large bottom roughness. *J. Phys. Oceanogr.* **2016**, *46*, 2317–2333. [CrossRef]
51. Lentz, S.; Churchill, J.; Davis, K.; Farrar, J. Surface gravity wave transformation across a platform coral reef in the R ed S ea. *J. Geophys. Res. Ocean.* **2016**, *121*, 693–705. [CrossRef]
52. Bruch, W.; Cordier, E.; Floc'h, F.; Pearson, S.G. Water Level Modulation of Wave Transformation, Setup and Runup over La Saline Fringing Reef. *J. Geophys. Res. Ocean.* **2022**, *127*, e2022JC018570. [CrossRef]
53. Franco, L.; De Gerloni, M.; Van der Meer, J. Wave overtopping on vertical and composite breakwaters. In Proceedings of the Coastal Engineering, Kobe, Japan, 23–28 October 1994; pp. 1030–1045.

Journal of
Marine Science and Engineering

MDPI

Article

Wave Height Reduction Inside Pohang New Port, Korea, Due to the Construction of a Detached Breakwater

Kyong Ho Ryu, Weon Mu Jeong, Jung-Eun Oh, Won-Dae Baek and Yeon S. Chang *

Maritime ICT R & D Center, Korea Institute of Ocean Science and Technology, Busan 49111, Korea
* Correspondence: yeonschang@kiost.ac.kr; Tel.: +82-51-664-3566

Abstract: The effect of a detached breakwater, which was constructed to improve harbor tranquility inside Pohang New Port, was examined through the comparison of wave data measured before and after the construction of the breakwater. The observation data showed that the wave energy was effectively reduced by the breakwater, although the wave height measured outside the breakwater was higher after its construction. The wave energy was reduced in all of the measured wave-propagating directions, but it was also observed that the breakwater became less effective in protecting against northeastwaves than in protecting against NNE waves. The BOUSS-2D Boussinesq-type wave model was employed to analyze the pattern of wave propagation, showing that, before the breakwater's construction, NE waves could directly enter the port, increasing the wave energy inside the port. After the breakwater's construction, simulations showed that the detached breakwater effectively blocked the waves approaching the port from both the NNE and NE directions, although the wave heights of the waves from the extreme NE direction inside the port increased. Considering that the estimated probability of failing to preserve the port tranquility was only 0.2–0.5% for these extreme NE waves, it was concluded that no secondary structures were necessary, and the existing breakwater was sufficient for the protection of the port.

Keywords: detached breakwater; port protection; harbor tranquility; BOUSS-2D; Pohang New Port

1. Introduction

Since the construction of Pohang New Port inside Yeongil Bay in the southeastern end of the Korean Peninsula in 1985, questions have been raised regarding the harbor's tranquility because downtime occurs frequently, which might lower the port's operation rate. Unexpected downtime due to harbor agitation may interrupt port operation and reduce throughput, and harbors with complex topologies and layouts are prone to experiencing severe agitation and downtime problems that are difficult to predict [1]. Therefore, studies have been performed to determine the factors that affect the tranquility of a harbor (e.g., [2]). For example, knowledge of short-period waves with wave periods less than 30 s, such as wind waves and swells, is an essential prerequisite for the planning and design of harbor protection structures [3], and numerical models have been employed to assess wave agitations induced by short waves in ports [4–6]. On the other hand, it has been suggested that long waves, including infragravity waves, might also be important factors that affect the harbor operation rate even when the harbor offers good protection against short waves [7–9], and low-frequency motions via harbor resonance could also cause agitations inside a harbor [10]. In addition, it was reported that the vertical movement of a vessel at berth was mainly determined by the swell energy inside the port basin, but the horizontal movement was strongly correlated with the low-frequency band energy [11].

In the case of Pohang New Port, it is difficult to determine the exact causes of the frequent downtimes, as its initial design changed due to port expansion, which meant that the combined effects of short waves, infragravity waves, and natural oscillation had to be considered. For example, it was suggested that the contribution of long waves and the natural oscillation period could be also considerable [1]. On the contrary, there have been research results that support the concept that short gravity waves are still the major cause of the downtime in Pohang New Port [12–14], and a detached ~700 m long breakwater with 15 caissons was constructed from October 2018 to December 2020 according to these study results [15,16]. Therefore, the present study was designed to investigate the effectiveness of the detached breakwater in improving the harbor tranquility and thus to confirm the research results that suggested short waves are the main cause for the downtime in Pohang New Port. Previously, a similar experiment was conducted to analyze the effectiveness of a detached breakwater [17]. According to this study, the harbor tranquility was improved after a detached breakwater was constructed. However, the accuracy of the results was questioned [17] because the wave data outside the port were measured at the entrance of Yeongil Bay.

To resolve this problem, in the present study, we employed wave observation data measured at additional stations inside Yeongil Bay, both before and after the construction of the breakwater. The new wave data in this study were measured at three additional stations (W1, W6, and W7 in Figure 1). Therefore, one of the advantages of the present study is that the waves were measured (just) offshore of the detached breakwater, meaning that a direct comparison with the waves inside Pohang New Port increased the accuracy of the results. The most significant contribution of this study is that the harbor tranquility was closely investigated based on field wave data obtained from the observations in both periods before and after the construction of the detached breakwater. Therefore, we were able to examine the effect of the breakwater on the tranquility inside the port under various wave conditions that occurred in the real field. To the best of the authors' knowledge, most of the previous studies that investigated harbor tranquility were based on physical and numerical model experiments, laboratory experiments, or the application of artificial intelligence approaches. Field observation data were used in some previous studies, but only for limited times of the observational periods at daily scales. In this study, the results from the analysis of field data removed the uncertainties that could arise from the simulations that used models and AI approaches. The variety of the wave conditions during the experimental periods could also offer inferences for extreme cases through statistical data. In addition to the analysis of the observation data, a Boussinesq-type wave propagation model was used to conduct numerical experiments to support the observational results by investigating the mechanism during the courses of wave propagation before and after the construction of the breakwater.

The purpose of this study was to analyze the rate of the reduction of wave height inside Pohang New Port due to the construction of a detached breakwater by using improved wave observation data and also by using a phase-resolving wave model that is capable of simulating the effects of wave refraction and diffraction in the complicated topography near the port. In this study, we also aimed to investigate the effect of the breakwater by examining the reduction rate according to the wave propagation direction and thus to provide precedent data in designing future structures in similar environments.

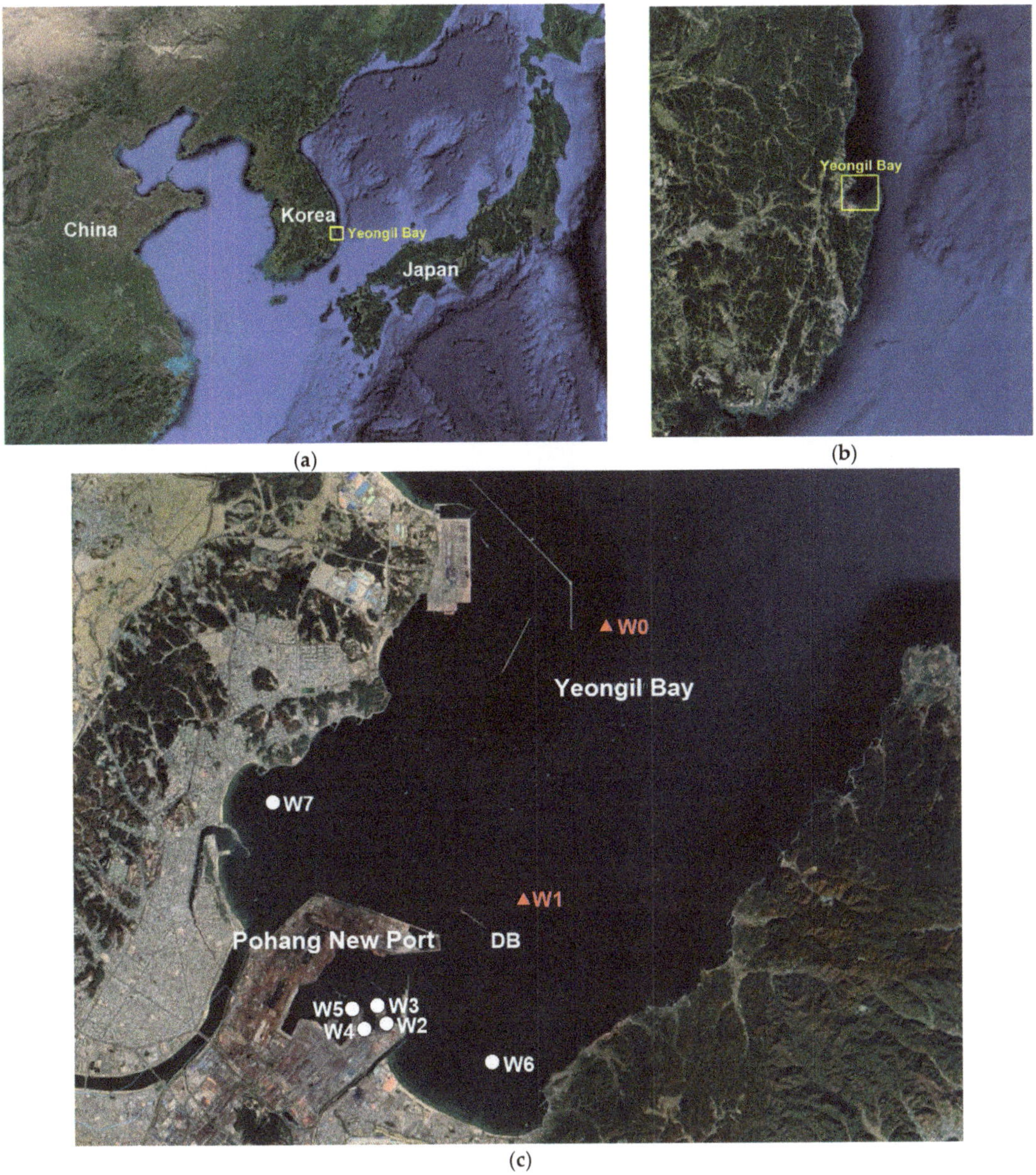

Figure 1. (**a**) Map of the Korean Peninsula with the location of Yeongil Bay, (**b**) map of Yeongil Bay located in the southeast of the Korean Peninsula, and (**c**) magnified map of Yeongil Bay with the locations of wave measurements. W0 and W1 (red triangles): locations where AWAC instruments were installed on the sea bed at the entrance of Yeongil Bay (W0) and just offshore of the detached breakwater (W1); W2–W5 (white circles): locations of the pressure sensors inside Pohang New Port; W6 and W7 (white circles): locations of pressure sensors in the nearshore of Dogu Beach (W6) and Yeongildae Beach (W7).

2. Materials and Methods

2.1. Study Site

Figure 1 shows the map of Yeongil Bay, captured from Google Earth, located in the southeast of the east coast of the Korean Peninsula, which is located in Northeast Asia. Yeongil Bay faces northeast with a mouth of ~10 km. Therefore, the waves that approach from the NE direction could reach the inner part of the bay and even affect the tranquility inside Pohang New Port [18]. As shown in Figure 1c, Pohang New Port was designed to protect the port from waves that mainly approach in the NE direction. However, the frequent downtime remained problematic in terms of the port's operation, which was unusual compared to the cases of other trading ports in the Republic of Korea.

As introduced in Section 1, previous research results have supported the concept that short gravity waves are the major cause for downtime in Pohang New Port, suggesting that harbor tranquility would be improved if the intrusion of short waves into the port was effectively prevented [12,14] and proposing that the target wave height to reach the desired harbor tranquility is 0.3 m [13]. As a measure to maintain the target wave height inside the port, it was suggested that a detached breakwater could be constructed outside the port to block approaching waves. Harbor breakwaters are usually constructed parallel to the shoreline near the edge of the surf zone to reduce the wave energy and protect the shore [19,20]. However, detached breakwaters have also been commonly constructed outside harbors to reduce the energy of waves that propagate to the harbor [21], along with other devices [22]. As a result, a detached, ~700 m long breakwater with 15 caissons was constructed from October 2018 to December 2020 [15,16]. In a previous study, the effect of the detached breakwater was investigated [17], and it was shown that the harbor tranquility was improved after the construction of the breakwater, as the frequency of the measured wave height exceeding the target wave height of 0.3 m clearly decreased. However, questions arose regarding the accuracy of the results of [17], because the wave data outside the port were measured at the entrance of Yeongil Bay. Due to the long distance from the wave station located at the bay entrance (W0 in Figure 1), the comparison with the wave data measured inside the port lowered the credibility of the analysis results. For example, the data in [17] showed that the breakwater was effective against waves in all directions measured at the bay entrance, W0. However, as shown in Figure 1, the course of wave propagation is varied inside Yeongil Bay, and waves from specific directions can successfully intrude into Pohang New Port, which implies that the effectiveness of the detached breakwater depends on the direction of the wave propagation.

2.2. Wave Measurement

Because Yeongil Bay is surrounded by land, the wave climate inside the bay is usually calmer than that outside the bay. However, waves could become harsh enough to cause severe coastline erosion in Yeongildae Beach when severe storm waves approach the site, even though the beach is located in the western corner of the bay [23]. Pohang New Port was constructed in 1985 at the southwest end of Yeongil Bay, as shown in Figure 1c, and the figure also shows the seven locations where waves were measured. An AWAC (Acoustic Wave And Current profiler) was installed on the sea bed at the entrance of Yeongil Bay (W0), as this is where waves were measured outside Pohang New Port in [17]. In this study, wave observations were carried out in W1, where an AWAC was installed on the sea bed ~1.2 km offshore from the head of the detached breakwater in the NE direction to measure the significant wave height (H_{m0}), peak wave period (T_p), and peak wave direction (D_p) every 30 min. Because an AWAC uses acoustic sensors to directly detect the wave surface elevation, it generally provides eave data with higher accuracy compared to a pressure transducer, which indirectly measures the surface elevation by converting the pressure time series using transfer functions [24]. During stormy conditions with strong wind, however, the white-capping of the waves disturbs the acoustic sensors, which may lower the accuracy of the AWAC measurements. In the case of W1, the AWAC was located inside the bay, and the wind energy was reduced as it was blocked by the land surrounding the bay, securing the validity of the AWAC data during most of the observational periods. The AWAC was used to measure the wave data from 21 May 2018. Before then, a Directional Waverider buoy from Datawell Inc. was moored at the same location, W1, to observe the wave conditions for ~9 months from 15 October 2008 to 12 July 2009, and the data during this period were also used in this study. Inside Pohang New Port, pressure sensors and a WTG (Wave and Tide Gauge) were deployed to measure H_{m0} and T_p at the four stations (W2, W3, W4, and W5), and their data were available during the same time periods, with the data from W1 being used for comparison. In addition to these stations inside the port, wave data were also measured using a pressure sensor and a WTG in the nearshore of Dogu Beach (W6) located in the southeast corner of Yeongil bay ~1.4 km in front of the beach. Similarly, another WTG was deployed in the west corner of the bay (W7) ~1.0 km in front of Yeongildae Beach. Due to these wave measurements in W6 and W7, the changes in the wave environment before/after the detached breakwater construction could be observed over the wider area inside Yeongil Bay. The wave data in W6 and W7 were measured from 21 May 2018, and thus, were not available in 2008 and 2009. The locations of the wave sensors, measuring periods, and speculations of the instruments are listed in Table 1.

Table 1. List of observational instruments, locations, installation depths, and measuring frequency.

Station	Locations (Lat., Lon.)	Depth (m)	Instrument	Measuring Period	Measuring Frequency (Hz)
W0	36°05′21.90″ N 129°28′21.20″ E	D.L. (−)24.5	AWAC	21 May 2018–9 May 2020	2
W1	36°02′16.34″ N 129°27′12.03″ E	D.L. (−)22.0	Directional Waverider buoy	15 October 2008–12 July 2009	1.28
			AWAC	21 May 2018–9 May 2020	2
W2	36°00′46.30″ N 129°25′05.60″ E	D.L. (−) 8.5	WTG	15 October 2008–9 May 2020	2
W3	36°01′00.10″ N 129°24′58.00″ E	D.L. (−)11.0	WTG	15 October 2008–9 May 2020	2
W4	36°00′42.70″ N 129°24′45.30″ E	D.L. (−)10.0	WTG	15 October 2008–9 May 2020	2
W5	36°00′58.20″ N 129°24′36.10″ E	D.L. (−) 9.7	WTG	15 October 2008–9 May 2020	2
W6	36°00′21.15″ N 129°26′39.83″ E	D.L. (−)12.5	WTG	21 May 2018–9 May 2020	2
W7	36°03′22.65″ N 129°23′25.86″ E	D.L. (−) 8.5	WTG	21 May 2018–9 May 2020	2

As previously described, the effect of the detached breakwater was examined by separating the periods of wave measurements into two groups—before and after the construction of the breakwater. The 1st~6th caissons were placed between 15 October 2018 and 14 January 2019, and the 12th caisson was placed by 6 August 2019. In the present study, therefore, the period before construction (hereafter, BC) was set to the wave measurements that were conducted before 13 February 2019 when the 1st~6th caissons were placed (before the 7th caisson was placed), and the period after construction (hereafter, AC) was set to after 6 August 2019, when the 12th caisson was placed. The starting time of BC was set to 15 October 2008, when the observation in W1 started, and the ending time of AC was set to 9 May 2020. Therefore, two separate time periods were included in BC, as the first started from 15 October 2008 to 12 July 2009, during which the Directional Waverider buoy was installed in W1; this period lasted for 271 days. This first period of BC is abbreviated as BC01 hereafter. In BC01, the wave measurements were available at five (W1, W2, W3, W4, and W5) out of a total of eight stations. The second period in BC was from 21 May

2018 to 13 February 2019, which lasted for 269 days (hereafter, BC02). In BC02, the data in W2 were of poor quality, with a missing period of 85 days, which was excluded from the analysis. The AC period comprised 278 days from 8 August 2019 to 9 May 2020, and good-quality wave data were available at all measuring stations. The observational periods in BC (hereafter, BC refers the period that includes both BC01 and BC02) and AC are listed in Table 2.

Table 2. List of the observation periods.

	Period	Data Availability
Before construction (BC) of the detached breakwater	15 October 2008–12 July 2009 (271 days, BC01)	Avail.: 5 station N.A.: 3 stations (W0, W6, W7)
	21 May 2018–13 February 2019 (269 days, BC02)	Avail.: 7 station N.A.: 1 station (W2)
After construction (AC) of the detached breakwater	6 August 2019–9 May 2020 (278 days, AC)	Avail.: 8 station N.A.: 0 station

In Figure 2, H_{m0}, measured in W1 and W3, is plotted for both BC and AC to compare the wave height outside and inside Pohang New Port. The value of H_{m0} measured in BC01 is shown in Figure 2a, whereas the data measured in BC02 are shown in Figure 2b. The maximum H_{m0} measured in W1 was 3.4 m in BC on 26 April 2009 and 4.8 m in AC on 22 September 2019, showing that it was higher in AC by ~1.4 m. In BC, there were ~168 h in which there were high wave conditions in which H_{m0} in W1 was greater than 2 m, and in AC, there were ~332 h. Considering that the BC period was longer (540 days) than the AC period (278 days), it was found that the wave conditions were more energetic during AC. In contrast, the maximum H_{m0} measured in W3 was 0.98 m in BC on 26 April 2009 and 0.64 m in AC on 22 September 2019, showing that it was higher in BC by ~0.34 m. In addition, in BC, there were ~183 h in which H_{m0} was higher than 0.5 m in W3, and this was a much shorter time in AC, as it was ~10 h. Considering that the wave condition was generally higher in AC outside the port, the shorter period in which H_{m0} was greater than 0.5 m in AC indicates that the detached breakwater clearly reduced the wave height inside the port. In Figure 2, the waves measured in W1 were used as the data outside the port instead of those measured in W0, and the reason for this is described in Section 3.

2.3. Selection of Wave Events

In order to examine the effect of the detached breakwater more closely, wave events were selected to filter out the calm wave conditions in which the breakwater effect could not be clearly observed because the wave energy was low both outside and inside the port. The wave events were selected by considering two conditions. The first was that H_{m0} should be higher than 1 m in W1. The second condition was that the time in which H_{m0} was greater than 1 m should last for at least 12 h so as to exclude inconsistent data. The number of selected wave events was 38 in BC and 27 in AC, the total time that corresponded to these wave events was 1150 h in BC and 1091 h in AC, and the time required for statistics was secured sufficiently. In Figure 3, time variations in wave height in W1 and W3 and wave direction in W1 are compared between the wave event examples selected in BC and AC. In both cases, the example events were selected as they had the maximum wave heights during each period. Therefore, for BC, a period of two days from 26 April 2009 was chosen as the example event. During this period, the H_{m0} value in W1 increased to a maximum of ~3.4 m, and the H_{m0} value in W3 increased to ~1.0 m. The wave direction range was N15°E~N50°E, showing that the waves generally approached in the NNE and NE directions, as these are the primary wave propagation directions in this region. For AC, a period of four days from 21 September 2019 was chosen as the example event; the H_{m0} value in W1 increased to its maximum value of ~4.8 m, and the H_{m0} value in W3 increased to ~0.6 m. The wave direction range was N40°E~N60°E, showing that the waves generally approached in the NE direction during this storm period.

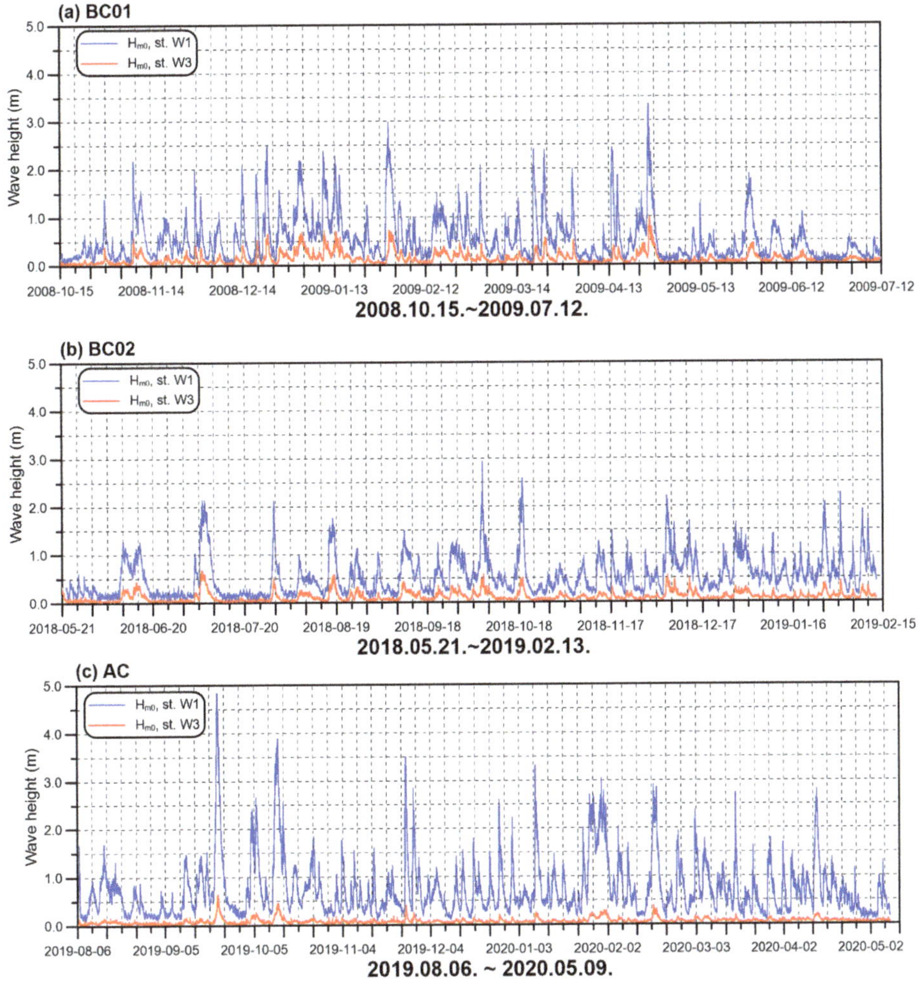

Figure 2. Comparison of H_{m0} in W1 (blue) and W3 (red): (**a**) first period before detached breakwater construction (BC01) in 15 October 2008–12 July 2009, (**b**) before detached breakwater construction (BC02) in 21 May 2018–13 February 2099, and (**c**) after detached breakwater construction (AC).

2.4. Numerical Model Experiment

The BOUSS-2D wave model was employed to conduct numerical experiments to understand the results from the observational data. The observation results showed the breakwater reduced the wave energy inside the port through a comparison of wave height measurements between BC and AC. However, this comparison could not provide the detailed ways in which the reductions actually occurred, which would require additional numerical experiments to understand the processes of wave transformation. The BOUSS-2D wave model was developed based on the non-linear Boussinesq equations [25], and the background of the model is described as follows. The classical form of the Boussinesq equations for wave propagation over water of variable depth was derived by [26], and the equations were restricted to relatively shallow water depths ($h < L/5$ with h = water depth and L = wave length). Nwogu [27] extended the range of applicability of the Boussinesq-type equations to deeper water. Despite this improvement, these equations were based on the assumption that wave heights are much smaller than the water depth. This limited the ability of the equations to describe highly non-linear waves in shallow water and led to a fully non-linear form of equations that are useful for the simulation of asymmetric waves in shallow water, wave-induced currents, wave setup, and wave–current interaction [28]. Then, Nwogu [29] extended the fully non-linear form of Boussinesq equations to the surf zone to consider the turbulence generated by breaking waves and the effect of bottom friction. These modified equations were able to simulate most of the hydrodynamic phenomena in the coastal regions and harbor basins, which included shoaling, refraction, diffraction, reflection, bottom friction, wave–wave and wave–current interactions, wave-induced currents, and wave breaking and runup. The BOUSS-2D model has been successfully applied to simulate wave diffraction in the lee of coastal structures and wave propagation over relatively steep slopes.

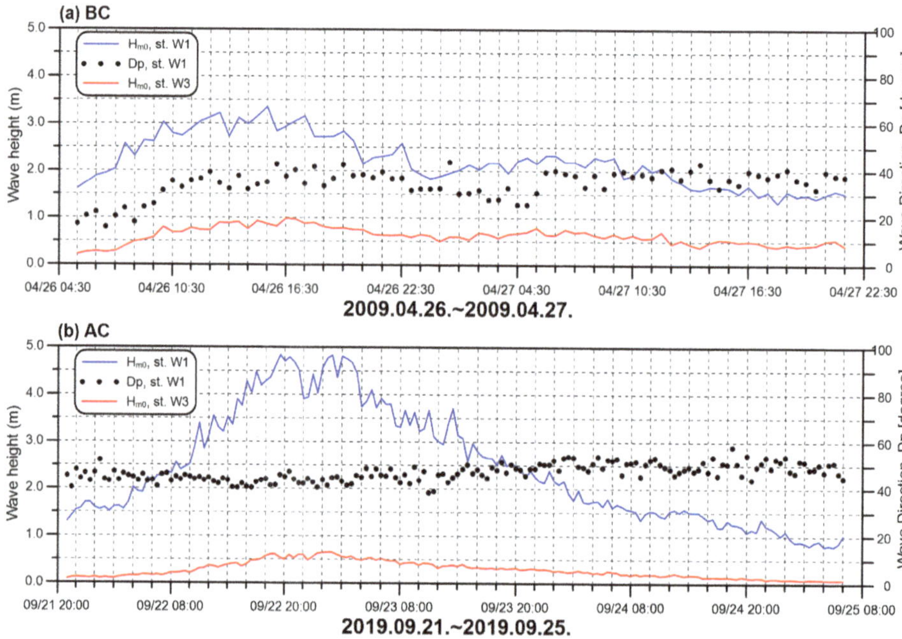

Figure 3. (**a**) H_{m0} (blue line) and D_p (black circles) in W1 and H_{m0} in W3 (red line) for a selected wave event in BC; (**b**) same as (**a**) but an event in AC. These two wave events were also selected for the numerical experiments, as described in Sections 2.3 and 3.2.

BOUSS-2D equations are depth-integrated equations for the conservation of mass and momentum for non-linear waves propagating in shallow and intermediate water depths. They can be considered to be perturbations from shallow-water equations, which are often used to simulate tidal flows in coastal regions. For short-period waves, the horizontal velocities are no longer uniform over depth, and the pressure is non-hydrostatic. The vertical profile of the flow field is obtained by expanding the velocity potential, Φ, as a Taylor series for an arbitrary elevation, z_α, in the water column. For waves with length, L, much longer than the water depth, h, the series is truncated at the second order, resulting in a quadratic variation in the velocity potential over depth:

$$\Phi(x,z,t) = \phi_\alpha + \mu^2(z_\alpha - z)[\nabla\phi_\alpha \cdot \nabla h] + \frac{\mu^2}{2}\left[(z_\alpha + h)^2 - (z+h)^2\right]\nabla^2\phi_\alpha + O\left(\mu^4\right) \quad (1)$$

where $\phi_\alpha = \Phi(x, z_\alpha, t)$, $\nabla = (\partial/\partial x,\ \partial/\partial y)$, and $\mu = h/L$ is a measure of frequency dispersion. The horizontal and vertical velocities are obtained from the velocity potential as:

$$u(x,z,t) = \nabla\Phi = u_\alpha + (z_\alpha - z)[\nabla(u_\alpha \cdot \nabla h) + (\nabla \cdot u_\alpha)\nabla h] + \frac{1}{2}\left[(z_\alpha + h)^2 - (z+h)^2\right]\nabla(\nabla \cdot u_\alpha) \quad (2)$$

$$w(x,z,t) = \frac{\partial\Phi}{\partial z} = -[u_\alpha \cdot \nabla h + (z+h)\nabla \cdot u_\alpha] \quad (3)$$

where $u_\alpha = \nabla\Phi|_{z_\alpha}$ is the horizontal velocity at $z = z_\alpha$. Given a vertical profile for the flow field, the continuity and Euler (momentum) equations can be integrated over depth, reducing the three-dimensional problem to two dimensions. For more detailed information regarding the BOUSS-2D model, readers are referred to the manual in [25].

In this study, the BOUSS-2D model was set up in a rectangular grid system using the staggered method by calculating the time-varying surface elevation and horizontal velocities at the grid nodes. The computational domain around Pohang New Port is shown in Figure 4. A rectangular domain with a size of 6.8 km by 6.0 km was set up by rotating the domain to the NE direction so that the outer open boundary could face N40°E (Figure 4a). This was because the majority of the waves approach in the NNE and NE directions in this region, and the input wave directions forced along the open boundary could range within ±30°, which is important for the stability of the model runs. The Δx and Δy of each grid was 5 m and set uniformly in the whole domain; thus, a total of 1360 grid points were allotted in the x-direction (parallel to detached breakwater), and 1200 grids were allotted in the y-direction.

The model was run for two different cases, representing the wave conditions in BC and AC, respectively. The two cases were the same as the selected events shown in Figure 3. That is, the model was run from 05:30 on the 26 to 21:30 on the 27 April 2009 for BC and from 07:00 on the 22 to 21:00 on the 23 September 2019 for AC. The BOUSS-2D model is not stable when used with waves of periods shorter than 8 s, meaning that the two cases were run for the times during which the observed periods were longer than 8 s. As it was not practical to continuously run BOUSS-2D for the whole time of 38–40 h, which had varying wave conditions, as shown in Figure 3, the experiment was conducted by running the model at every hour by specifying the directional spectra that represented the wave conditions at the corresponding time. In the case of BC, for example, the model was run 39 times, corresponding to each hour from 05:30 on the 26 to 21:30 on the 27 April 2009, with the model being forced to be determined from the data measured at each hour. Individual runs were simulated for 3900–4000 s until a stable condition was reached with $\Delta t = 0.15$ s. These numerical experiments were designed to understand the course of wave energy reduction between BC and AC through model validation via comparisons with observational data. The results are discussed in Sections 3.2 and 4.

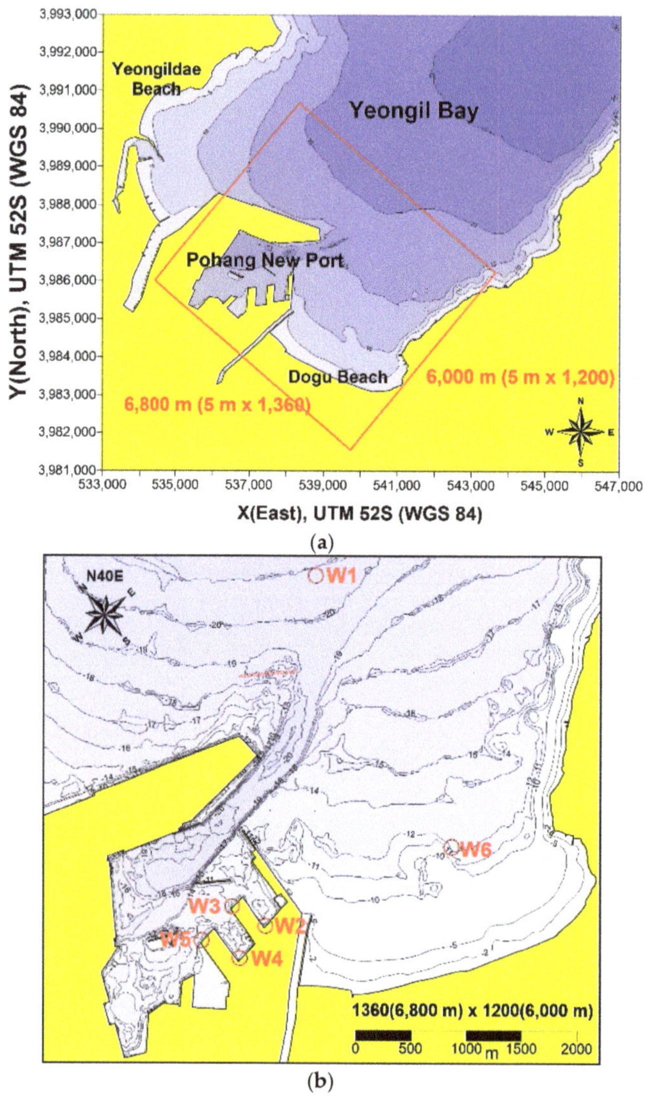

Figure 4. (**a**) Map of Yeongil Bay around Pohang New Port, in which the computational domain for the BOUSS-2D wave model is marked in a red rectangle, (**b**) model domain with the bathymetry map and the locations of wave stations. It is noted that the domain is rotated 40° in the NE direction for the outer open boundary to face N40°E.

3. Results

3.1. Observational Data

The number of wave events that consistently had wave heights greater than 1 m for at least 12 h was 38 in BC and 27 in AC, as described in Section 2. The selected events generally corresponded to the times when high waves, including storms, were present in Yeongil Bay, and H_{m0} displayed increasing and decreasing patterns during these periods, as shown in Figure 3. Therefore, there was a time when the H_{m0} value became the maximum, and the H_{m0} value at this time was selected as the representative value for each event. In Figure 5, the maximum H_{m0} values of all of the events are compared between W1 (outside

Pohang New Port) and W2, W3, W4, and W5 (inside Pohang New Port) in BC and AC. Similarly, the representative values for the wave propagation direction, D_p, were selected at the same time at which the maximum H_{m0} values for the corresponding events were observed. In Figure 6, the representative D_p value measured in W1 is compared with the H_{m0} value in W2, W3, W4, and W5.

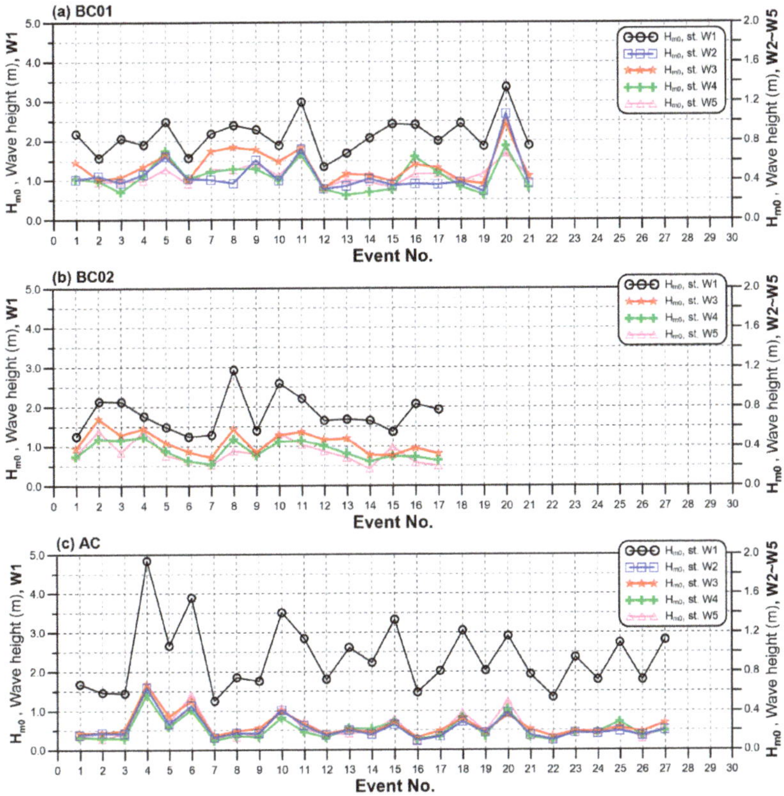

Figure 5. (**a**) Maximum values of H_{m0} observed in the 21 events before construction in W1 (black circles), W2 (blue squares), W3 (red stars), W4 (green crosses), and W5 (pink triangles) for BC01 (15 October 2008~12 July 2009); (**b**) maximum H_{m0} values for the 17 events before construction in W1 (black circles), W3 (red stars), W4 (green crosses), and W5 (pink triangles) for BC02 (21 May 2018~13 February 2019) (it is noted that the H_{m0} in W2 was not included due to poor data quality); (**c**) maximum H_{m0} values for the 27 events after the construction in W1 (black circles), W2 (blue squares), W3 (red stars), W4 (green crosses), and W5 (pink triangles). The pattern of H_{m0} variation is similar between the wave stations outside (W1) and inside (W2, W3, W3, and W4) the port.

In the case of wave height, the H_{m0} varied with the wave events and showed similar patterns at all stations, as when the H_{m0} value was high inside the port (W2, W3, W4, and W5), the H_{m0} value was generally also high outside the port (W1). It is noted that the representative H_{m0} values inside the port were mostly higher than 0.2 m, and ~50% of them were even higher than 0.4 m in BC. In contrast, in AC, the representative H_{m0} values outside the port in 23 events out of a total of 27 events were lower than 0.4 m, and ~50% of them were even lower than 0.2 m, which indicates that the breakwater effectively reduced the wave energy inside the port. In the case of the representative D_p value, its correlation with the representative H_{m0} was lower, as shown in Figure 6. In the second period of BC (Figure 6b), the representative D_p value in W1 showed a large discrepancy with the H_{m0}

measured inside the port. The D_p measured in W1 was greater than 40° in events #1–#7, whereas the D_p value became lower than 25° in events #8–#17. However, this pattern was not observed in terms of the H_{m0} value inside the port. In AC, a similar conclusion was not easily found, because there were wave events such as #4, #6, #10, and #20 when the H_{m0} value in W2, W3, W4, and W5 increased and the D_p value in W1 was also greater than 30°; thus, a correlation was observed between the wave direction and the height inside Pohang New Port.

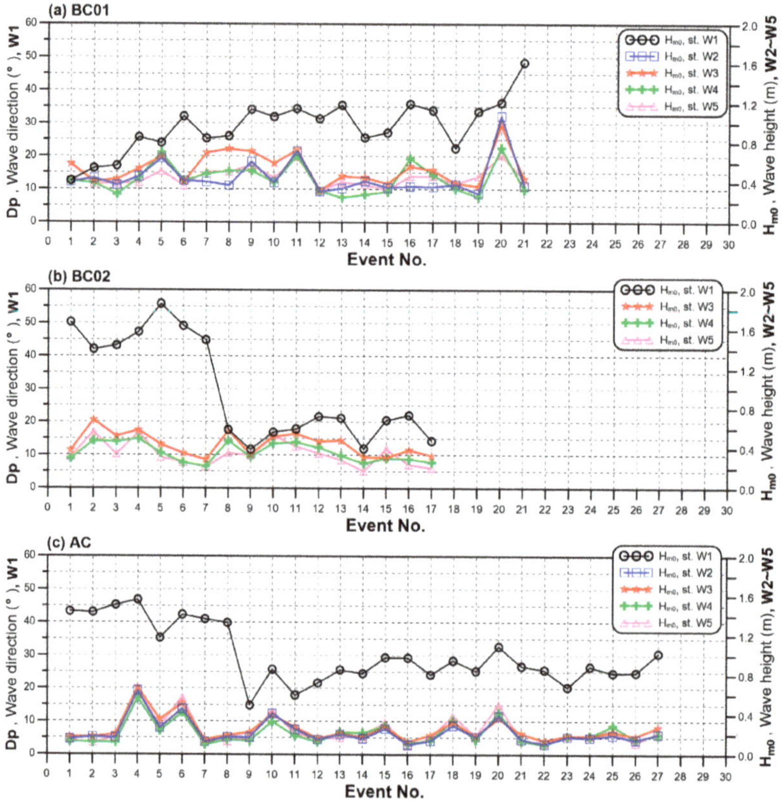

Figure 6. Same as Figure 5 but for D_p values that were observed at the same time as the H_{m0} in Figure 5 (i.e., D_p in Figure 6 was measured at the time when the H_{m0} value was maximum for each wave event). Compared to the results in Figure 5, it shows a lower correlation between D_p outside the port (W1) and H_{m0} inside the port (W2, W3, W3, and W4).

In Figure 7, H_{m0} and D_p are compared between the measurements at the two stations, W0 and W1. It is noted that the waves were not measured in W0 during BC01 (2008–2009), meaning that Figure 7 only compares the data at BC02 (2018–2019) and AC. In the case of H_{m0}, the data between the two locations show relatively high correlation with an r^2, the coefficient of determination, of 0.84. However, the r^2 for D_p was as low as 0.16 between W0 and W1, which indicates that the directions of the waves that approached Yeongil Bay were not correlated with those that approached the detached breakwater. Therefore, to carry out a detailed investigation, it was necessary to examine the impact of wave direction based on the data in W1, as proposed in Section 1, in addition to the analysis based on the data in W0, which was conducted in [17]. In the present study, therefore, the data in W1 were considered the primary wave measurements outside Pohang New Port.

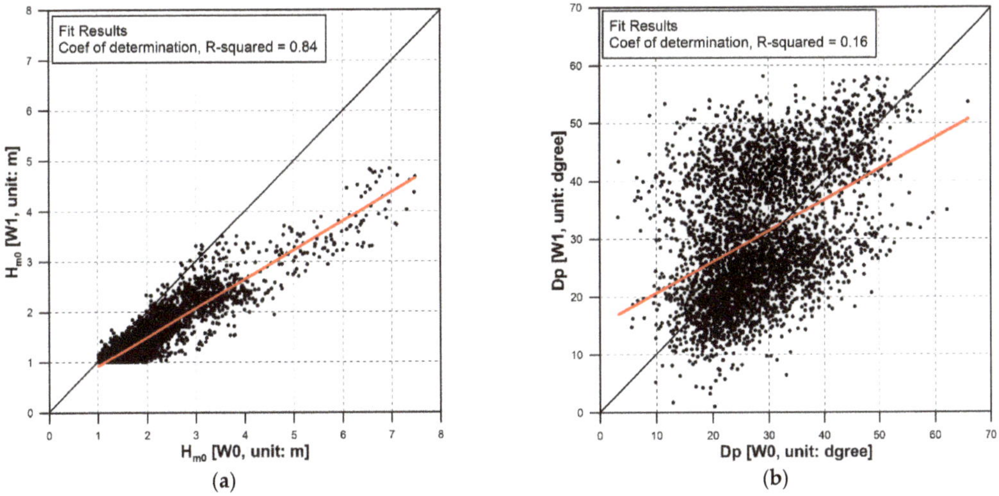

Figure 7. Comparison of the distributions of (**a**) significant wave height (H_{m0}) and (**b**) peak wave direction (D_p) between W0 and W1 for the selected events for both BC (only for the second period from 2018 to 2019) and AC. The coefficient of determination (r^2) was 0.83 for H_{m0} and 0.16 for D_p. Wave measurements show a high correlation in H_{m0} between the two wave stations of W0 and W1, whereas they show a lower correlation in the case of D_p.

The correlation between the wave parameters inside and outside Pohang New Port was examined by defining a new parameter, the wave height reduction rate ($R_{W\#}$). $R_{W\#}$ was calculated from the ratio of H_{m0} measured inside and outside the port (i.e., $R_{W2} = H_{m0}[W2]/H_{m0}[W1]$ and $R_{W3} = H_{m0}[W3]/H_{m0}[W1]$, etc.). The $R_{W\#}$ parameters were estimated from the data sets of 38 and 27 wave events selected for BC and AC, respectively, as described in Section 2.2, because the relationship between $R_{W\#}$ and D_p was not clearly observed for waves with low energy. In Figures 8 and 9, the distributions of $R_{W\#}$ versus D_p in W1 for the data selected for the wave events are compared between BC and AC. In addition to the four stations inside the port, the R_{W6} and R_{W7} values were estimated outside the port as well. Considering that the wave energy inside the port decreased with the decreasing magnitude of $R_{W\#}$, the results in Figures 8 and 9 clearly show that the wave energy was much reduced in AC compared to that in BC, confirming the effectiveness of the detached breakwater in reducing the wave energy inside the port. It is also interesting to see that the $R_{W\#}$ value measured inside the port ($R_{W2} - R_{W5}$) tended to increase with D_p, showing that the $R_{W\#}$ value increased as the wave direction in W1 changed from N to NE. In contrast, the opposite phenomenon occurred in W6, the station in front of Dogu Beach outside the port (Figure 1c), at which the R_{W6} value decreased with increasing D_p. On the other hand, the pattern in W7, the station in front of Yeongildae Beach, became similar to that inside the port, as the R_{W7} value also increased with increasing D_p. The difference in the pattern of $R_{W\#}$ with D_p in W6 and W7 might have been induced by their locations, which is discussed in Section 4. The correlation between $R_{W\#}$ and D_p in BC was repeated in AC. However, the magnitude of $R_{W\#}$ measured inside the port was much lower in AC than that in BC, as its values in AC were ~50% of those in BC, which is discussed in Section 4. In Figures 8 and 9, the red lines show linear regressions and the regression coefficients are listed in Table 3.

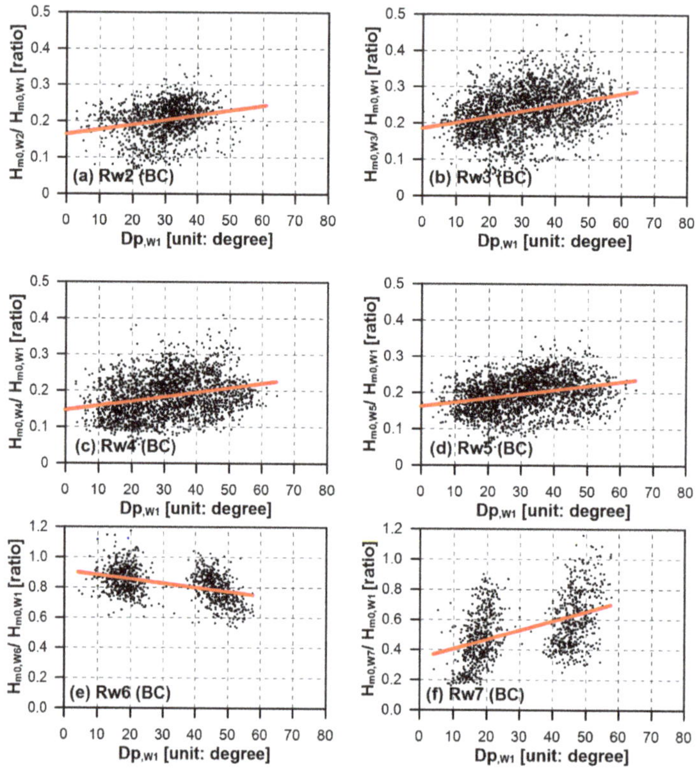

Figure 8. Scattered plots of $R_{W\#}$ $\left[R_{W2}\left(=\frac{H_{m0}[W2]}{H_{m0}[W1]}\right), R_{W3}\left(=\frac{H_{m0}[W3]}{H_{m0}[W1]}\right), R_{W4}\left(=\frac{H_{m0}[W4]}{H_{m0}[W1]}\right), R_{W5}\left(=\frac{H_{m0}[W5]}{H_{m0}[W1]}\right)\right]$ versus D_p measured in W1 for the data measured during the 38 wave events in BC. (**a**) R_{W2} vs. D_p, (**b**) R_{W3} vs. D_p, (**c**) R_{W4} vs. D_p, (**d**) R_{W5} vs. D_p, (**e**) R_{W6} vs. D_p, (**f**) R_{W7} vs. D_p. The red line in each panel shows the linear regression calculated for each data set. It is noted that R_{W2} was calculated from the BC01 data (2008–2009) due to the poor quality of the measurements in BC02 (2018–2019). In addition, BC01 data were not available in the case of R_{W6} and R_{W7}, as the waves were not measured in these stations in the 2008–2009 experimental period. $R_{W\#}$ consistently tended to increase with increasing $D_{p,\ W1}$ in the wave stations inside the port (W2, W3, W4, and W5) and that located in the west corner of Yeongil Bay (W7), whereas $R_{W\#}$ decreased with increasing $D_{p,\ W1}$ in the wave station located in the southeast corner of the bay.

Considering that the D_p value measured in W1 acted as a controlling factor on the wave reduction inside Pohang New Port, it was also necessary to examine the impact of the wave height on the wave reduction pattern. Figures 10 and 11 show the scatter plots between $R_{W\#}$ and H_{m0}, measured in W1. It is observed that $R_{W\#}$ did not show a clear correlation with H_{m0} in W1 for both BC and AC. It is noted that the time variation pattern of H_{m0} measured inside the port was similar to that of H_{m0} in W1, as shown in Figures 2 and 5. Therefore, the low correlation between $R_{W\#}$ and H_{m0} inside the port was an interesting result, indicating that the wave reduction process inside the port might not be dependent on the wave height, unlike the wave direction. The low correlation between $R_{W\#}$ and H_{m0} was not only observed inside the port but also outside the port in W6 and W7 in both BC and AC. The results in this section are important, because the effect of the breakwater was confirmed, as not only the magnitude but also the variance in $R_{W\#}$ were significantly reduced in AC compared to those in BC, regardless of the wave height. In addition, the high dependence of the wave reduction process inside the port on the propagation direction of short waves outside the port required further analysis to understand the courses of wave propagation around the breakwater and the port, which was carried out using the BOUSS-2D model, which can simulate wave transformation around these complex structures, and the results are provided in the next section.

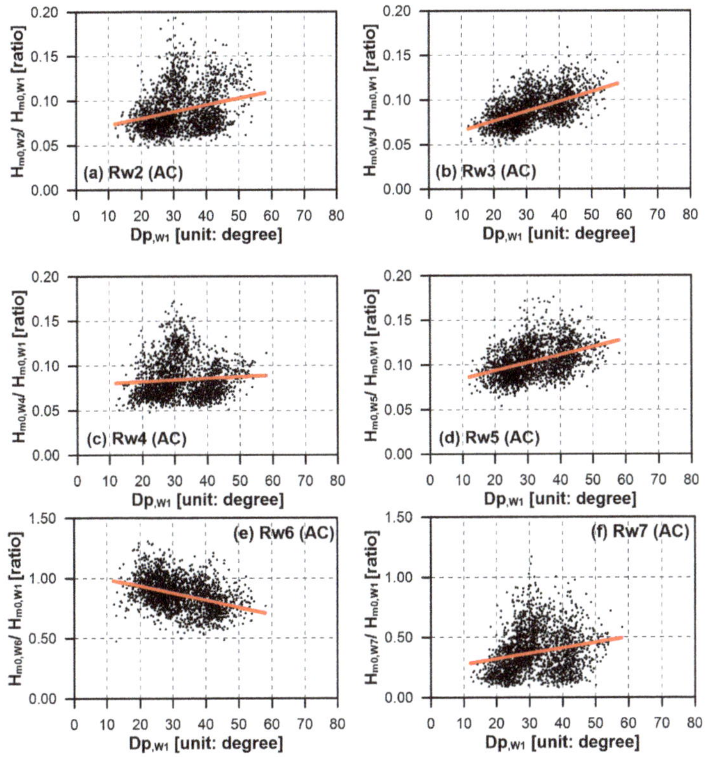

Figure 9. Same as Figure 8, but for the data measured during the 27 wave events in AC. Similar to the case of BC, it was found that $R_{W\#}$ consistently tended to increase with increasing $D_{p,W1}$ in the wave stations inside the port (W2, W3, W4, and W5) and that located in the west corner of Yeongil Bay (W7), whereas $R_{W\#}$ decreased with increasing $D_{p,W1}$ in the wave station located in the southeast corner of the bay.

Table 3. List of coefficients of determination (r^2) and linear regression coefficients (slope and intercept) for the scatter plots in Figures 8–11.

Figure 8	(a)	(b)	(c)	(d)	(e)	(f)
r^2	0.055	0.097	0.072	0.084	0.172	0.225
Linear regression	0.0013X + 0.1638	0.0016X + 0.1850	0.0012X + 0.1468	0.0011X + 0.1621	−0.0028X + 0.9121	0.0062X + 0.3428
Figure 9	(a)	(b)	(c)	(d)	(e)	(f)
r^2	0.072	0.296	0.005	0.168	0.149	0.047
Linear regression	0.0008X + 0.0654	0.0011X + 0.0546	0.0002X + 0.0787	0.0009X + 0.0760	−0.0059X + 1.0529	0.0045X + 0.2308
Figure 10	(a)	(b)	(C)	(d)	(e)	(f)
r^2	0.035	0.000	0.000	0.014	0.011	0.000
Linear regression	−0.0212X + 0.2364	−0.0031X + 0.2388	−1.2018X + 0.1810	−0.0142X+ 0.2184	0.0299X + 0.7804	0.0053X + 0.5318
Figure 11	(a)	(b)	(C)	(d)	(e)	(f)
r^2	0.104	0.032	0.016	0.011	0.002	0.034
Linear regression	0.0115X + 0.0690	0.0045X + 0.0813	0.0041X + 0.0772	0.0029X + 0.0991	0.0091X + 0.8489	0.0640X + 0.4936

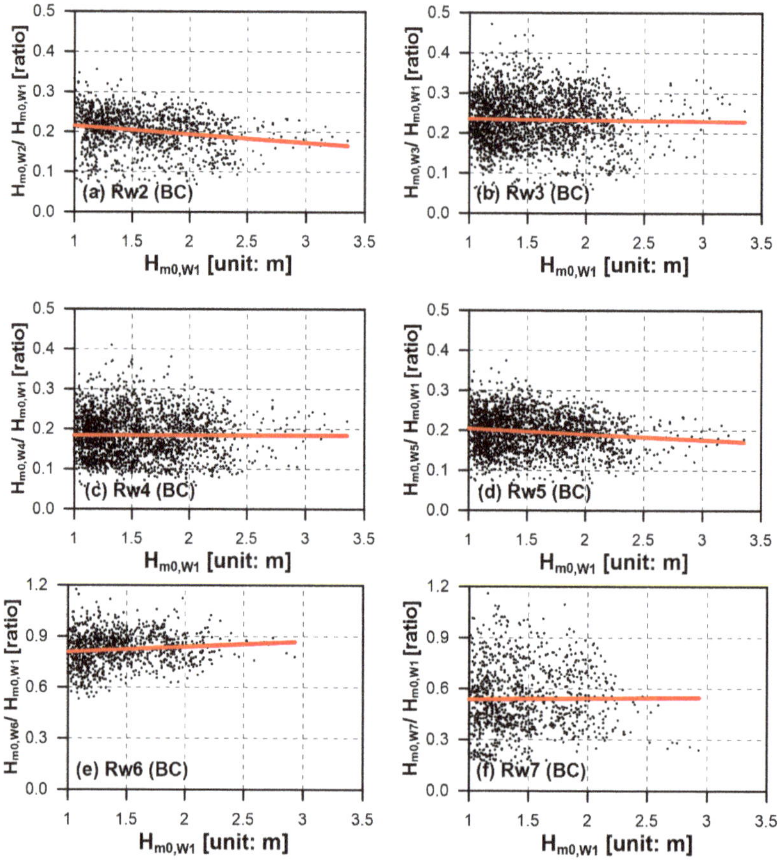

Figure 10. Same as Figure 8, but the x-axis is H_{m0} measured in W1 in BC. (**a**) R_{W2} vs. H_{m0}, (**b**) R_{W3} vs. H_{m0}, (**c**) R_{W4} vs. H_{m0}, (**d**) R_{W5} vs. H_{m0}, (**e**) R_{W6} vs. H_{m0}, (**f**) R_{W7} vs. H_{m0}.The correlation between $R_{W\#}$ and the wave propagation direction ($D_{p,W1}$) observed in Figures 8 and 9 was not found in the case of the wave height (H_{m0}).

3.2. Numerical Model Results

The BOUSS-2D wave model was set up as described in Section 2.3 for the two cases representing the wave propagation in BC and AC. The two cases were selected because they had the maximum wave heights during the period of the 38 (BC) and 27 (AC) wave events, as shown in Figure 3a,b. The events with the maximum H_{m0} were selected because the effect of the detached breakwater in reducing the wave energy inside Pohang New Port could be more clearly observed under higher wave energy conditions.

In Figure 12, the wave conditions in W1 are compared between the observed and modeled data from 26 to 27 April 2009, which were selected from the BC events. Considering the computational domain shown in Figure 4b, the location of W1 was close to the outside open boundary, and the modeled wave conditions were well matched with the observation data that were used for the boundary forcing conditions. The data show that the wave height increased to the maximum (H_{m0} ~3.4 m) at 15:30 on 26 April 2009, and then, it gradually decreased. The wave period fluctuated between 9.2 and 12.2 s, and the D_p value gradually increased from ~N12°E to ~N40°E, indicating that the waves were mainly approaching from NNE~NE in W1 during this wave event.

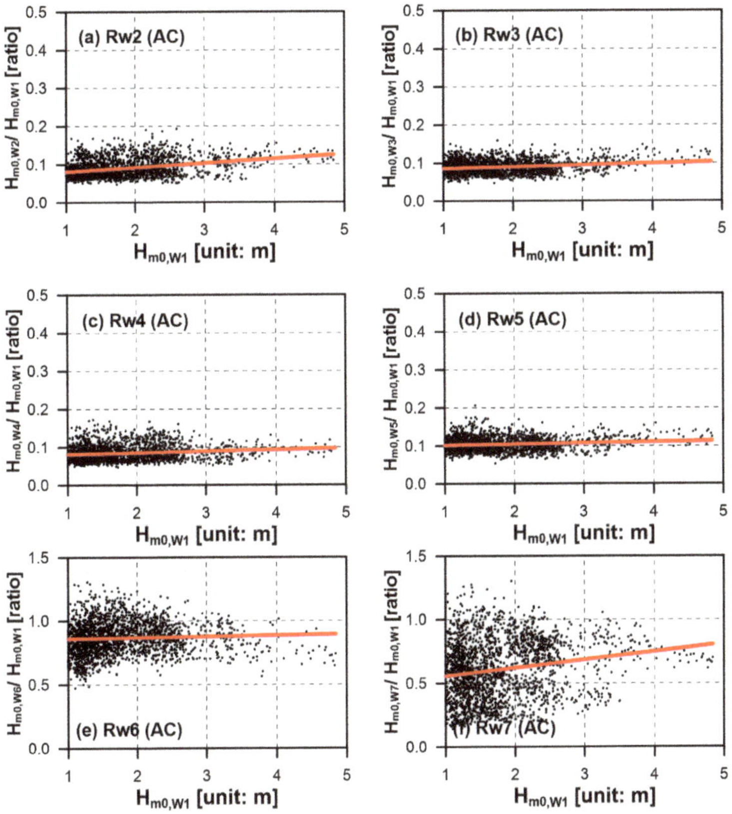

Figure 11. Same as Figure 10, but in AC. (**a**) R_{W2} vs. H_{m0}, (**b**) R_{W3} vs. H_{m0}, (**c**) R_{W4} vs. H_{m0}, (**d**) R_{W5} vs. H_{m0}, (**e**) R_{W6} vs. H_{m0}, (**f**) R_{W7} vs. H_{m0}. The correlation between $R_{W\#}$ and the wave propagation direction ($D_{p,W1}$) observed in Figures 8 and 9 was not found in the case of the wave height (H_{m0}).

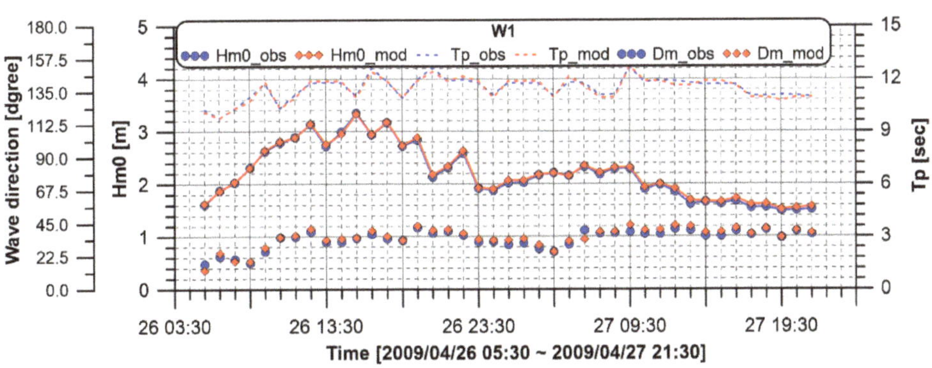

Figure 12. Comparison of wave conditions (H_{m0}, $T_{\rm p}$, and D_p) between the observational (blue) and modeled (red) data in W1 during the period of the wave event selected for BC from 05:30 on the 26th to 21:30 on 27 April 2009 (39 h). It is noted that the modeled data matched well with the observational data due to the proximity of W1 to the open boundary, as shown in the computational domain (Figure 4b).

Figure 13 shows the comparisons of the modeled wave heights with observations at the four wave stations (W2, W3, W4, and W5) inside Pohang New Port to validate the model performance. Reasonable agreement was observed among the four wave stations inside the port (marked in Figure 4b), as the overall pattern of the H_{m0} time variation generally agreed between the modeled and observation data, although the model results were underestimated, specifically in W3 and W5. The average error ($\frac{\sqrt{(H_{obs}-H_{mod})^2}}{H_{obs}} \times 100$) was calculated, and it ranged from 12% to 23% (12.9%, 18.7%, 13.6%, and 22.7% for W2, W3, W4, and W5, respectively). It is noted that the errors were greater in the stations located outside the slits (W3 and W5) than those located inside the slits (W2 and W4). Although the reason for this is not yet clearly understood, the four wave stations were located in the corners of the slits where the incoming and reflected waves were interfered with in complicated ways, which could have increased the wave height compared to that in other areas near the slits and also could have increased the modeling errors. As the pattern of the H_{m0} time variation of the model generally agreed with that shown in the observation, it could be concluded that the model was validated for the wave event, and the model results were further analyzed to understand the wave propagation pattern inside the port.

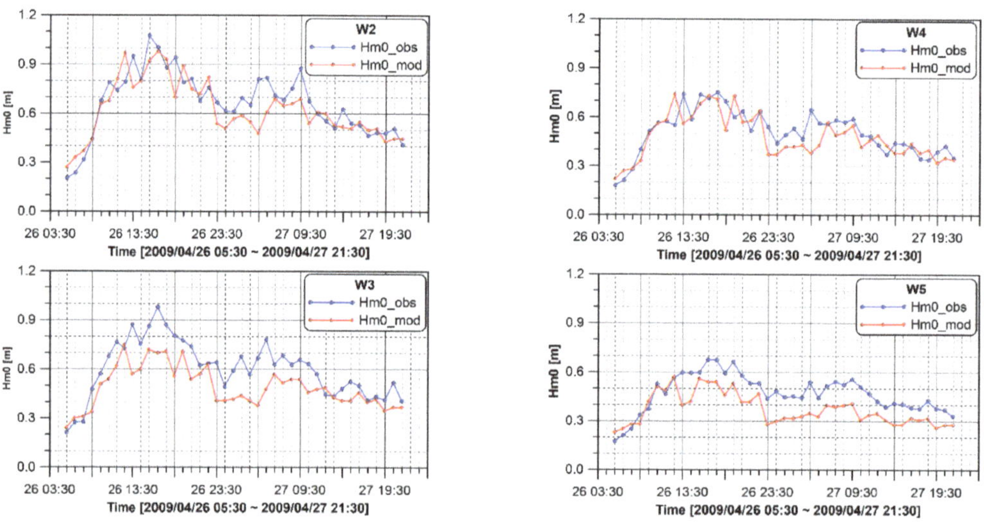

Figure 13. Comparison of wave height (H_{m0}) between the observational (blue) and modeled (red) data at the four wave stations inside Pohang New Port (W2, W3, W4, and W5) during the period of the wave event selected for BC from 05:30 on the 26 to 21:30 on 27 April 2009. Reasonable agreements are observed between the observational and modeled data, supporting the model validation.

A similar comparison between the modeled and observational data was carried out for the selected event case in AC. In Figure 14, the observed and modeled wave conditions in W1 are plotted for the period of the wave event selected for AC, from 07:00 on the 22 to 21:00 on 23 September 2019. Similar to the event case in BC, the wave conditions in W1 were nicely matched between the model and observation due to the proximity of the location of W1 to the open boundary. It was shown that the wave height increased to become the maximum (H_{m0}~4.8 m) at 19:00 on 22 September 2019 and then gradually decreased to ~2.5 m at 21:00 on the 23rd in W1. The wave period generally increased from ~8.5 s to ~11.0 s, and the D_p value fluctuated between 30° and 60°, showing that the waves generally approached from NE in W1 during the wave event.

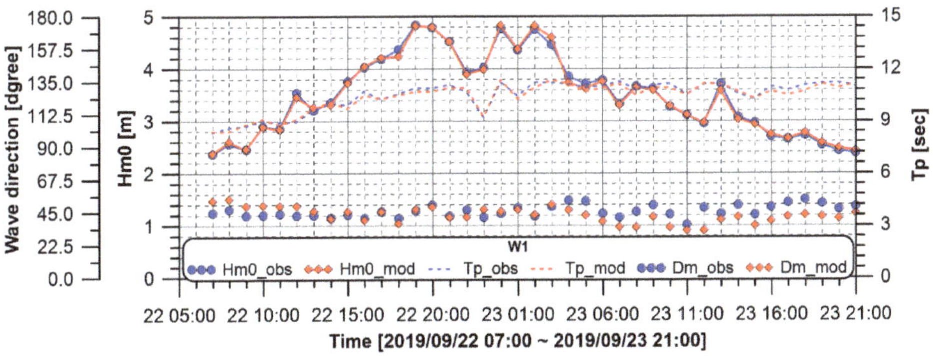

Figure 14. Comparison of wave conditions (H_{m0}, T_p, and D_p) between the observational (blue) and modeled (red) data in W1 during the period of the wave event selected for AC from 07:00 on the 22nd to 21:00 on 23 September 2019 (37 h). It is noted that the modeled data matched well with the observational data due to the proximity of W1 to the open boundary, as shown in the computational domain (Figure 4b).

Figure 15 compares the modeled wave heights with the observed data at the four wave stations (W2, W3, W4, and W5) inside Pohang New Port during the wave event selected for AC. It shows that the pattern of wave height time variation during the event was successfully generated by the model, although the errors estimated from the wave height were not improved compared to those in BC, as they were ~19.2%, 17.1%, 31.3%, and 19.2% for W2, W3, W4, and W5 respectively. The higher error range in AC was likely due to the lower magnitude of the observed H_{m0}, which would increase the error for the same deviation between the model and observation data.

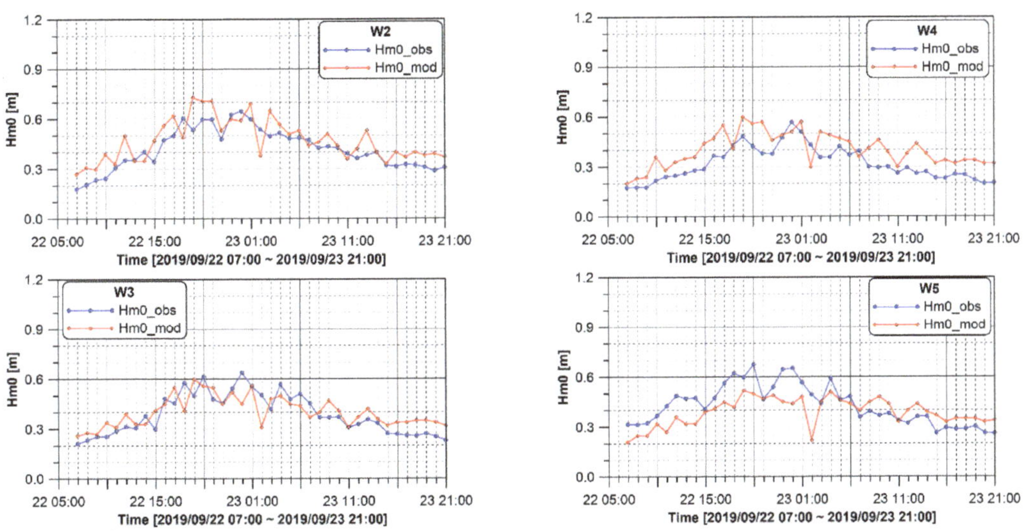

Figure 15. Comparison of wave height (H_{m0}) between the observational (blue) and modeled (red) data at the four wave stations inside Pohang New Port (W2, W3, W4, and W5) during the period of the wave event selected for AC from 05:00 on the 22 to 21:00 on 23 September 2019. Reasonable agreements are observed between the observational and modeled data, supporting the model validation.

Once the BOUSS-2D model was validated, the next step was to analyze the course of wave propagation to understand how the wave energy was reduced inside Pohang New Port through a comparison between BC and AC. In Figures 16 and 17, the contours of wave heights during the periods of the two wave events selected for BC and AC are compared. The contours are captured as a snapshot of the wave field at the time when the H_{m0} value became the maximum for each event, as shown in Figures 12 and 14. That is, in the case of BC, Figure 16 shows the field of wave heights at 15:30 on 26 April 2009 when the H_{m0} value became the maximum, 3.35 m (Figure 14). Similarly, the wave height field is contoured in Figure 17 at 19:00 on 22 September 2019, as it is the time when the H_{m0} value became the maximum (4.84 m) during the wave event selected for AC. The input wave conditions for the two cases were based on the observations at corresponding times. The wave period was similar between the two cases, as the T_p value was 11.0 s and 10.9 s for BC and AC, respectively. The wave direction was N35.0°E for BC and N46.3°E for AC, as they were propagating in the NNE–NE directions. Along the open boundary that was rotated 40° in the NE direction (Figure 4a), the incident wave direction was nearly normal to the boundary in both BC and AC (with a deviation of 5°–6°), indicating there were no significant discrepancies in terms of the input wave propagation direction between the two cases.

Both of the figures effectively compare the pattern of wave reduction inside Pohang New Port between the two cases. In BC (Figure 16), the isolines of the wave height became parallel to the port entrance where the narrow navigation channel is connected to the inside of the port. At the port entrance, the H_{m0} value was ~2.0 m (marked in a red rectangle in Figure 16), indicating that the H_{m0} value was reduced by ~40% from the input wave height at the outer open boundary (3.35 m). Once the waves entered the port, the wave height was rapidly reduced due to the combined effects of refraction and diffraction around the slits and other harbor structures, with H_{m0} varying from ~2.0 m to ~0.3 m. The wave height was observed to be 0.3 m inside the slit where the wave stations W4 and W5 were located (marked in a blue rectangle). It is noted, however, that the H_{m0} value was increased to 0.9 and 0.7 at the locations of wave stations W2 and W4, which was likely because the reflected waves were interfered with at the wall of the slits to increase the H_{m0} value to higher than that calculated off the walls.

In Figure 17, the H_{m0} isolines are contoured for the wave event selected for AC, which occurred at 19:00 on 22 September 2019. Compared to the case of BC, the detached breakwater was constructed near W1, blocking the waves. Behind the breakwater, therefore, H_{m0} was rapidly reduced, and the reduction rate was greater than that in BC. For example, the H_{m0} value became ~2.0 m near the entrance of Pohang New Port (marked in a red rectangle), and the reduction rate was 59% considering that the input wave height on the boundary was 4.84 m in AC, which was ~19% greater than that in BC. In addition, in AC (Figure 17), the red rectangle that marks the isoline position of H_{m0} = 2.0 m is located ~0.2 km further offshore compared to that in BC (compare with Figure 16), indicating that the wave heights inside the port would become lower than those in BC, assuming that the pattern of H_{m0} reduction inside the port would be same between the two cases, which was also confirmed from a comparison of the observed and modeled wave heights between BC and AC in W2, W3, W4, and W5, as shown in Figures 13 and 15.

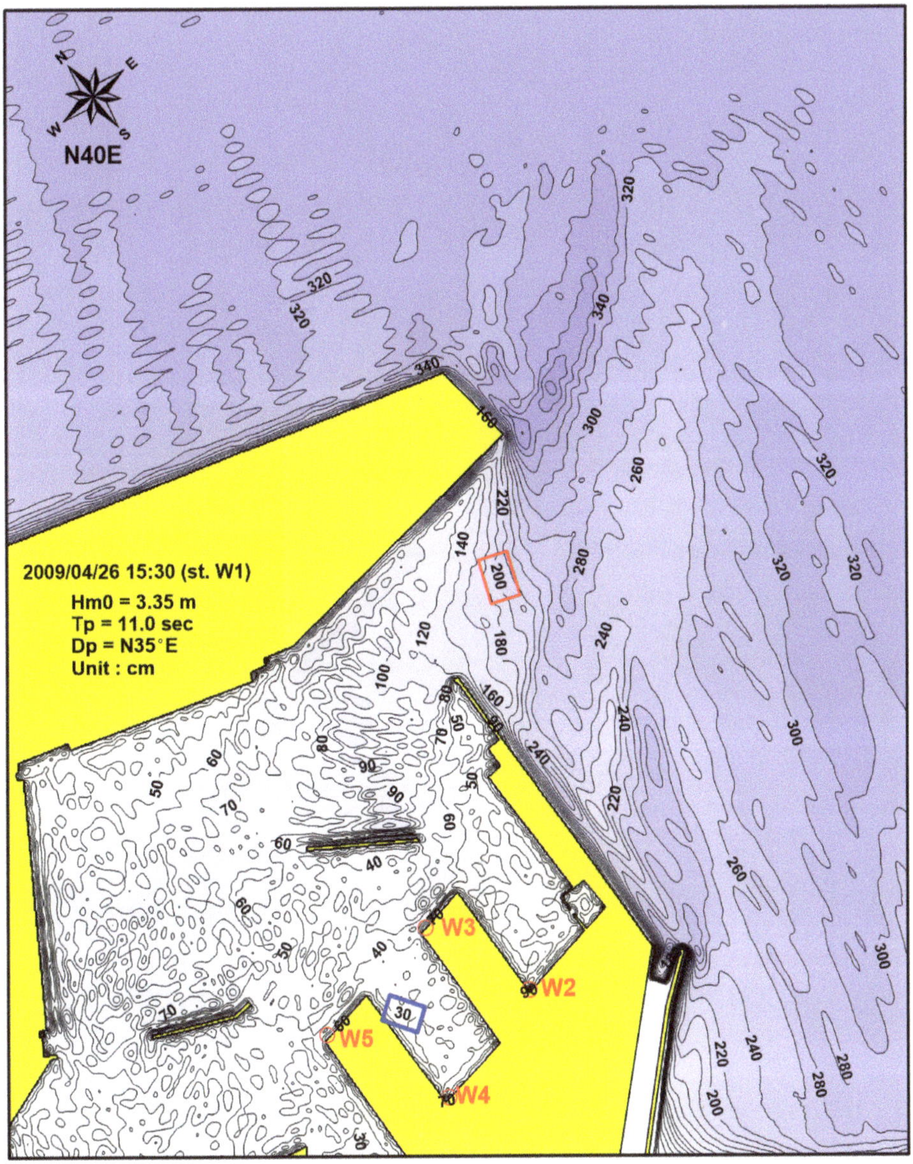

Figure 16. Contours of wave height isolines at 15:30 on 26 April 2009 when the H_{m0} value became maximum during the 39 h of the simulation time shown in Figure 12, in the case of BC. The input H_{m0} was 3.35 m at the open boundary, which was reduced to 2.0 m (40% reduction) near the entrance of Pohang New Port (marked in a red rectangle) and to 0.3 m inside the slit near W4 and W5 (marked in a blue rectangle).

Figure 17. Contours of wave height isolines at 19:00 on 22 September 2019 when H_{m0} became maximum during the 39 h of the simulation time shown in Figure 16, in the case of AC. The input H_{m0} was 4.84 m at the open boundary, which was reduced to 2.0 m (59% reduction) near the entrance of Pohang New Port (marked in a red rectangle). It is also noted that the location of the 2.0 m H_{m0} isoline was slightly outside the port entrance compared to that in BC (Figure 16).

Figures 16 and 17 show the model that simulated the wave propagation patterns using the input conditions based on the observational data, and the input wave directions on the open boundary were similar between BC and AC, as they were N35.0°E and N46.3°E, respectively. Therefore, the impact of wave directional change on the wave pattern inside Pohang New Port could not be examined using these model results, although it was observed from the observational data, as shown in Figures 8 and 9. In Figure 18, the two additional cases that were tested for BC by assuming extreme cases in the wave propagation direction are shown. It is noted that most of the measured D_p values in W1 ranged between N15.0°E and N55.0°E in the case of AC, as shown in Figure 9. In Figure 18, therefore, the results of two model runs are shown with D_p set as N15.0°E (Figure 18a) and N55.0°E (Figure 18b), as they represent the extreme wave directions for NNE and NE waves, respectively. Except for the wave direction, the other input wave conditions were the same as those given for the case of BC in Figure 16. The results of these two extreme cases in Figure 18 could help us to understand the different patterns in the wave energy reduction rate inside Pohang New Port, according to the changes in D_p in W1, as observed in Figure 8 in the case of BC. In the extreme NNE case (D_p = N15.0°E), the propagating waves outside the port mainly passed by the entrance of the port, and only the diffracted waves could enter the port, which increased the wave height reduction near the port entrance (Figure 18a). For example, the H_{m0} value was as high as 2.0 m just outside the port entrance, but it was reduced to 1.0 m just inside the port, as the distance between the two positions was only ~0.5 km (marked with two red rectangles). The high wave reduction rate near the port entrance meant that H_{m0} ranged from 0.2 to 0.7 m inside the port, and the estimated wave height reduction rate ($R_{W(model)} = \frac{H_{m0}[inside\ PNP]}{H_{m0}[\text{open boundary}]}$) was 5–20%. On the contrary, in the extreme NE case (D_p = N55.0°E) shown in Figure 18b, the wave reduction near the port entrance was less severe compared to that in Figure 18a, as shown in the two marked red rectangles (the H_{m0} value was reduced from 2.0 m to 1.4 m for ~0.5 km). Therefore, the H_{m0} value inside the port was also higher compared to those in Figure 18a, with a range of 0.4–1.2 m in the areas, providing $R_{W(model)}$ = 12–35% inside the port.

Figure 19 shows the model results of the two extreme wave directions that were tested for AC as well. That is, the other wave conditions were identical to those used in Figure 17, except that D_p was set as N15.0°E (N55.0°E), which represented an example of extreme conditions of NNE (NE) waves, as shown in Figure 19a (Figure 19b). Similar to the cases in BC, the wave propagation pattern inside Pohang New Port was affected by the different D_p input conditions at the open boundary, regardless of the existence of the detached breakwater. When D_p was N15.0°E, the H_{m0} value was significantly reduced in the lee area of the breakwater, as it was ~4.6 m offshore of the breakwater but decreased to ~2.0 m just outside the port, as shown in Figure 19a (marked with a red rectangle). The waves were then diffracted near the entrance of the port, where strong H_{m0} reduction occurred for the waves entering the port, as the H_{m0} value was reduced to ~1.0 m just inside the port (marked with a red rectangle). The range of H_{m0} was then 0.2–0.7 m inside the port, leading to $R_{W(model)}$ = 5–15% inside the port. In the case of extreme NE waves (D_p = N55.0°E in Figure 19b), the waves approached the port from the NE direction, meaning the detached breakwater became less effective in reducing the wave height. For example, the H_{m0} value was reduced from ~2.0 m to ~1.4 m (marked in the red rectangles in Figure 19b) for the similar distance marked in Figure 19a. As a result, the H_{m0} value became higher inside the port, compared to that of D_p with N15.0°E, with the H_{m0} range of 0.4–1.2 m inside the port, providing $R_{W(model)}$ of 8–25% (Figure 19b). The impact of wave direction in the case of AC is further discussed in Section 4.

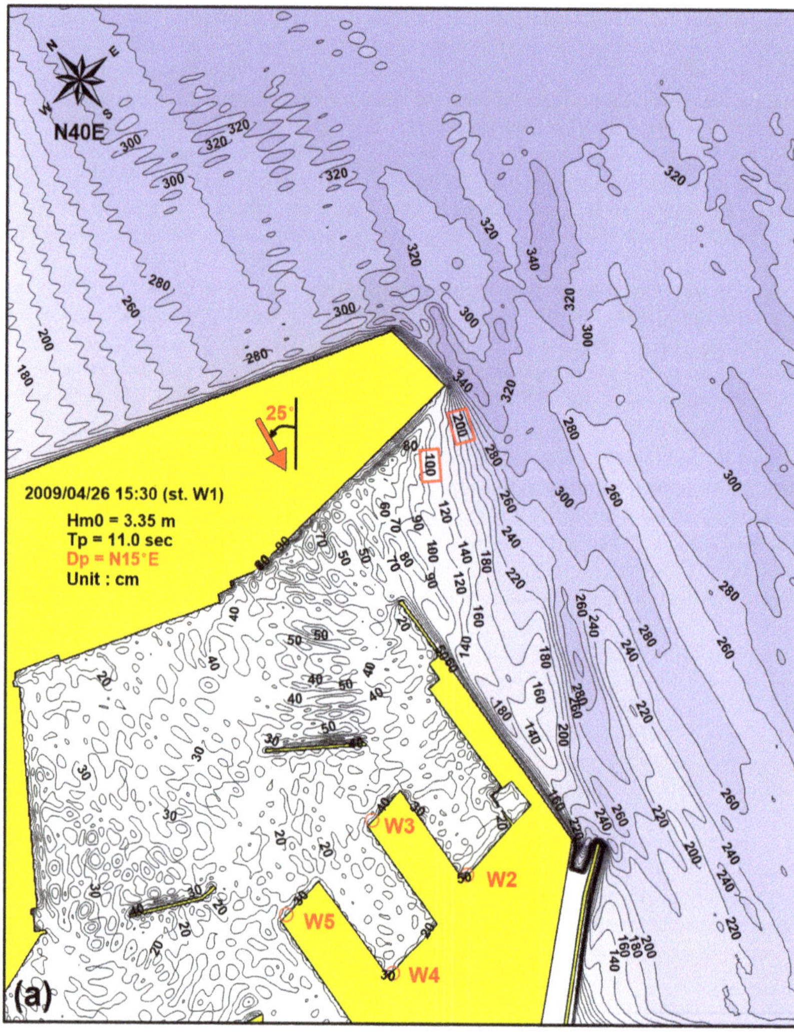

Figure 18. *Cont.*

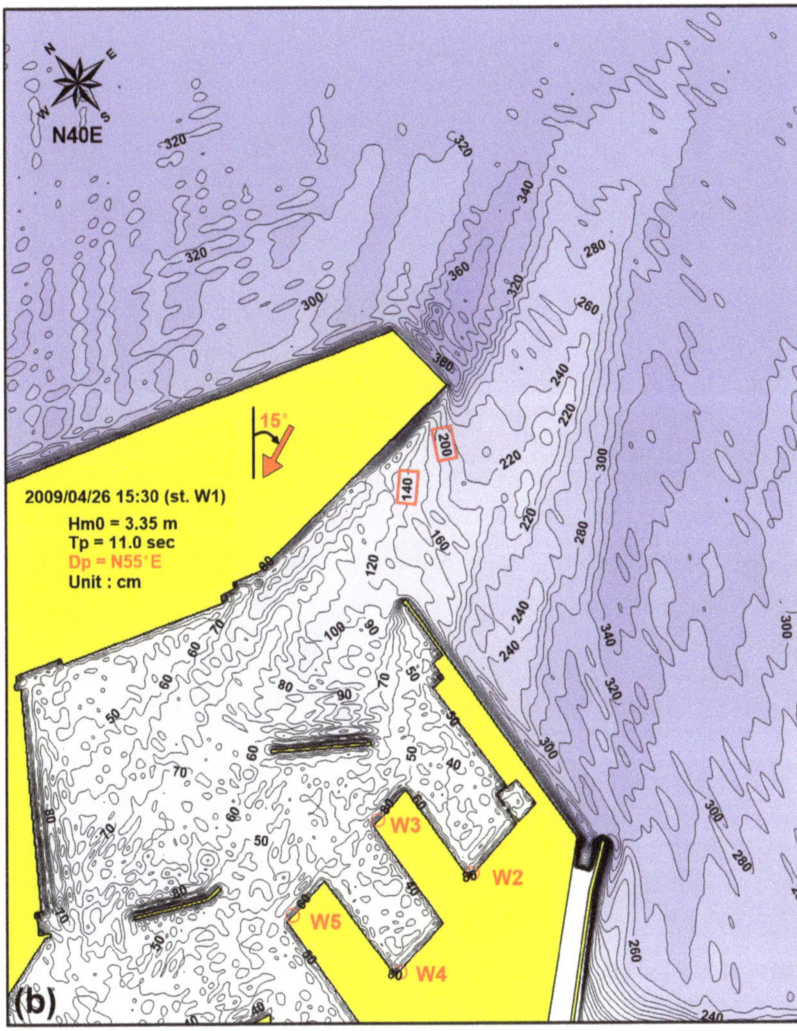

Figure 18. (**a**) Contours of wave height isolines in the case of BC. The model conditions were the same with those in Figure 16, except that D_p, the input wave direction along the outer open boundary, was set as N15.0°E as an example of extreme conditions that represented NNE waves; (**b**) the same as (**a**), except that D_p was set as N55.0°E as for an example of extreme conditions that represented NE waves. In the case of the extreme NE waves (N55.0°E), the waves could enter into the port easily compared to the case of N15.0°E.

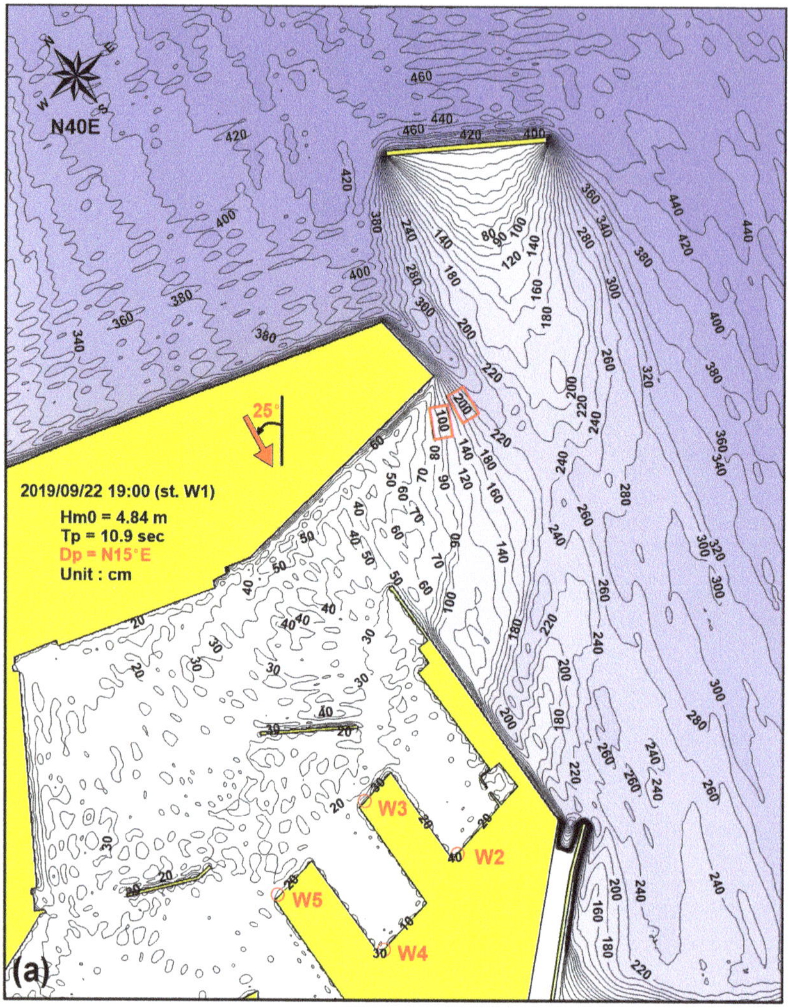

Figure 19. *Cont.*

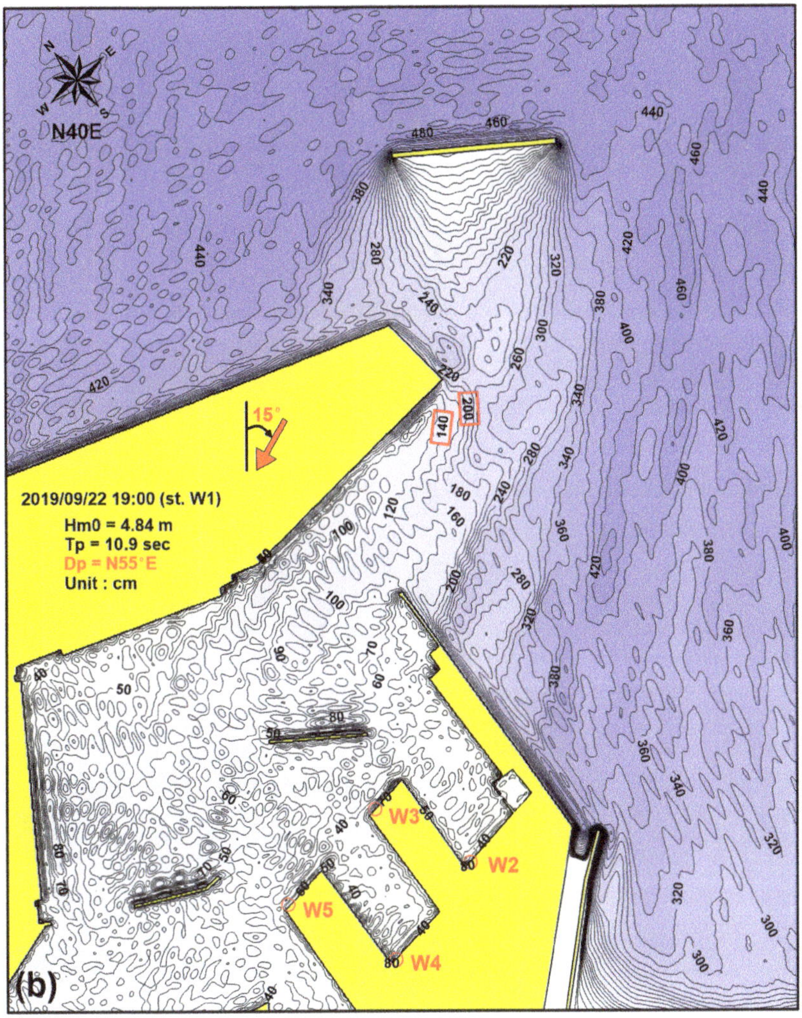

Figure 19. (**a**) Contours of wave height isolines in the case of AC. The model conditions were the same with those in Figure 17, except that D_p, the input wave direction along the outer open boundary, was set as N15.0°E as an example of extreme conditions that represented NNE waves; (**b**) the same as (**a**), except that D_p was set as N55.0°E as an example of extreme conditions that represented NE waves. In the case of the extreme NE waves (N55.0°E), the waves could enter into the port easily compared to the case of N15.0°E. When compared to the results in BC (Figure 18), wave height reduction occurred more effectively inside the port due to the detached breakwater.

4. Discussion

One of the most significant results of the present study was that the wave height reduction inside Pohang New Port was affected by the propagating wave directions that approached the detached breakwater. For example, in all of the wave stations inside the port, $R_{W\#}$ increased with increasing D_p. The close correlation between $R_{W\#}$ was observed in both BC and AC, as shown in Figures 8 and 9. However, the magnitude of $R_{W\#}$ became lower in AC. Figure 20 shows the comparison of the scatter plots of R_{W3} in terms of D_p between BC and AC. The distribution of the wave reduction rate was compared in W3 because it was located closest to the entrance of the port; thus, the impact of the wave propagation direction could be distinct in this station. The figure shows that R_{W3} increased with an increasing D_p value in both BC and AC. However, the magnitude of R_{W3} ranged from 0.12 to 0.40 in BC, but the magnitude was reduced by ~40% in AC, as R_{W3} ranged from 0.05 to 0.15. The lower magnitude of R_{W3} in AC indicates that the wave energy was effectively reduced inside the port, confirming the effect of the detached breakwater in securing the tranquility of the port.

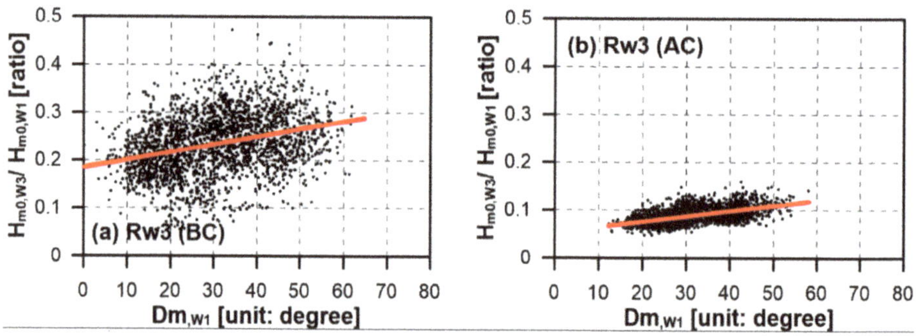

Figure 20. Comparison of the scatter plots of R_{W3} (the wave height reduction rate based on observation data) in terms of D_p in W1, (**a**) BC and (**b**) AC. The red line in each panel shows the linear regression calculated for each data set. The lower R_{W3} magnitude in AC indicates that the wave energy was effectively reduced inside the port, confirming the effect of the detached breakwater in securing the tranquility of the port.

Regardless of the observation of the significantly lower values of $R_{W\#}$ in AC, the result in Figure 20b indicates that the reduction in the wave height inside Pohang New Port became less effective even with the protection provided by the detached breakwater if the wave propagation direction increased (i.e., changed from NNE to NE). The cause for this was investigated using the BOUSS-2D model, which is displayed in Figure 19, showing that the wave heights were higher in the lee of the breakwater and also in the areas near the port entrance, as well as inside the port when the waves approached in the extreme NE direction (N55°E), which raised a question—"is the construction of a detached breakwater, as shown in Figure 1c, sufficient to reduce the wave energy in securing the tranquility inside the port for the waves approaching in all directions?" To find the answer, we needed to account for the fact that the eastern end of the breakwater was located near the western end of the navigation channel (see Figure 4b). From the engineering aspect, therefore, it was not practical to extend the breakwater further to the SE because the detached breakwater would then block the navigation channel. An alternative option would be to place another detached breakwater further SE, avoiding the navigation channel. However, this second option would not be effective either, considering that the extreme NE waves would approach in the N55.0°E direction, and thus, the new detached breakwater may be useless in blocking the waves propagating to the port entrance. Therefore, alternative options to construct additional engineering measures may not be practical, unless unavoidable.

The next step to find an answer to the question would be a statistical approach. From the results in Figure 9, it was shown that the magnitude of $R_{W\#}$ of the waves with D_p values greater than N55°E was mostly not higher than 0.15 at the wave stations located inside the port (W2–W5). In addition, as described in Section 1, the targeted maximum wave height to maintain the desired harbor tranquility inside the port was proposed as 0.3 m [13]. Therefore, considering that the maximum $R_{W\#}$ of the waves with D_p values greater than N55°E would be 0.15 for AC, the maximum H_{m0} value in W1 that would enable the tranquility inside the port to be maintained can be calculated as 0.3/0.15 = 2.0 m.

Based on the wave measurements in W1 during the whole period of AC, in total, there was only 1 h in which the observed waves had D_p values greater than N55°E and H_{m0} values greater than 2 m. As this number is too small compared to the total period of wave measurement of 278 days in AC, the criteria for the extreme NE direction could be extended to N50°E and even to N45°E. In these cases, however, the time in which the D_p value was greater than N55°E and the H_{m0} value was greater than 2 m was only 11 and 33 h, respectively, indicating that the probability of waves with heights that could be greater than 0.3 m inside the port because they approached from the extreme NE directions was ~0.0% (>N55°E), and it would only become 0.2% (>N50°E) and 0.5% (>N55°E) if the criteria for the extreme NE direction were extended. Therefore, it can be concluded that the detached breakwater located in the current position was sufficient to protect the port by maintaining its tranquility with the targeted wave height, and no further secondary engineering structures would then be required. This recommendation provided to elevate the effectiveness of the detached breakwater could be determined based on the amount of observational data measured in both periods in BC and AC. Without these wave measurements, the model results could lead to overestimation or underestimation in evaluating the effect of the breakwater. Although this result was only found in this specific case, it can also provide an insight for future studies in planning/operating ports under similar environments in which unexpected extreme cases should be carefully considered, and these extremes may need to be evaluated from the sufficient data sets observed in the field.

5. Conclusions

In this study, observational and numerical experiments were conducted to investigate the effect of the detached breakwater on reducing the wave energy inside Pohang New Port in terms of the wave parameters and specifically, the wave propagation direction. The reason for selecting the wave direction as a primary controlling factor was to examine whether the position of the detached breakwater was appropriate in reducing the wave energy inside the port for the waves that approached in all practical directions. If not, it may have been necessary to consider placing additional structure(s) to secure the tranquility inside the port. For this purpose, wave data were measured before (BC) and after (AC) the detached breakwater was constructed at multiple stations inside and outside the port. In addition, the BOUSS-2D model based on Boussinesq-type equations was employed to investigate the effect of the detached breakwater on the wave propagation pattern around the breakwater and inside the port.

The observational results show that the wave energy (the ratio of H_{m0} inside and outside Pohang New Port) significantly decreased inside the port in AC compared to that in BC, although the wave height condition was generally higher outside the port (near the detached breakwater) in AC, which confirmed the effect of the breakwater. The reduction in the wave energy was observed in all wave propagation directions measured outside the port; however, the reduction rates inside the port showed that the effectiveness of the breakwater decreased as the wave direction changed from NNE to NE. The causes for the difference in the detached breakwater's effect depending on wave directions were investigated based on the model results, which made it possible to investigate the wave propagation pattern around the breakwater. In the case of BC, the model results showed that more wave energy could be transferred inside the port without severe transformation when the waves approached in the NE direction, which could be difficult for the waves approaching in the NNE direction as the waves mostly passed by the entrance of the port. In the case of the AC model runs, the detached breakwater could effectively block the waves that approached in both the NNE and NE directions, resulting in significantly enhanced wave reduction rates inside the port. In the extreme case of NE waves (e.g., wave direction = N55.0°E) however, the wave energy increased inside the port regardless of the blocking caused by the breakwater, which might have required the consideration of additional engineering measures to secure the port tranquility. However, placing a secondary detached breakwater would not be practical because the location of the current breakwater is too close to the navigational channel, meaning that further extension of the existing detached breakwater could block the channel itself. Further investigation based on the observational data showed that the probability of the wave height inside the port exceeding 0.3 m (the target wave height to maintain the tranquility inside the port) was only 0.2–0.5% when the waves approached in the extreme NE direction, indicating that no secondary structures might be necessary and that the existing detached breakwater would be enough to secure tranquility inside Pohang New Port.

Author Contributions: Conceptualization, K.H.R., W.M.J. and Y.S.C.; methodology, K.H.R., J.-E.O., and W.-D.B.; investigation, K.H.R., W.M.J., J.-E.O. and Y.S.C.; data curation, K.H.R. and W.-D.B.; supervision, W.M.J. and Y.S.C.; visualization, K.H.R. and J.-E.O.; writing—original draft preparation, K.H.R. and Y.S.C.; writing—review and editing, W.M.J. and Y.S.C.; funding acquisition, W.M.J. All authors have read and agreed to the published version of the manuscript.

Funding: This research was funded by the Korea Institute of Ocean Science and Technology (KIOST), grant number PEA0032, and the Ministry of Oceans and Fisheries, grant number PG52810.

Institutional Review Board Statement: Not applicable.

Informed Consent Statement: Not applicable.

Conflicts of Interest: The authors declare no conflict of interest.

References

1. Oh, J.-E. *A Data-Based Downtime Forecasting Model in a Harbor: Application to Pohang New Port*; Seoul National University: Seoul, Korea, 2017.
2. PIANC. *Criteria for Movements of Moored Ships in Harbours: A Practical Guide*; Permanent International Association of Navigation Congresses, Permanent Technical Committee, Working Group (PIANC) General Secretariat: Brussels, Belgium, 1995.
3. Kankal, M.; Yüksek, Ö. Artificial neural network approach for assessing harbor tranquility: The case of Trabzon Yacht Harbor, Turkey. *Appl. Ocean. Res.* **2012**, *38*, 23–31. [CrossRef]
4. Moon, S.; Lee, J.; Kwon, S.; Song, H. Analysis of Harbor Tranquility due to Port Expansion. *J. Navig. Port Res.* **2019**, *43*, 320–327.
5. Panigrahi, J.K.; Padhy, C.; Murty, A. Inner harbour wave agitation using boussinesq wave model. *Int. J. Nav. Archit. Ocean. Eng.* **2015**, *7*, 70–86. [CrossRef]
6. Kankal, M.; Yüksek, Ö.; Kömürcü, M.; Akpınar, A.; Önsoy, H. Physical and Numerical Modeling of Harbor Oscillations: A Case Study in Ünye Harbor, Turkey. *ENE* **2010**, *7*, 8.
7. González-Marco, D.; Sierra, J.P.; de Ybarra, O.F.; Sánchez-Arcilla, A. Implications of long waves in harbor management: The Gijón port case study. *Ocean. Coast. Manag.* **2008**, *51*, 180–201. [CrossRef]
8. Bellotti, G.; Franco, L. Measurement of long waves at the harbor of Marina di Carrara, Italy. *Ocean. Dyn.* **2011**, *61*, 2051–2059. [CrossRef]
9. Kwak, M.S.; Jeong, W.M.; Kobayashi, N. A case study on harbor oscillations by infragravity waves. *Coast. Eng. Proc.* **2020**, *36*, 1–6. [CrossRef]
10. Sakakibara, S.; Kubo, M. Characteristics of low-frequency motions of ships moored inside ports and harbors on the basis of field observations. *Mar. Struct.* **2008**, *21*, 196–223. [CrossRef]
11. López, M.; Iglesias, G. Long wave effects on a vessel at berth. *Appl. Ocean. Res.* **2014**, *47*, 63–72. [CrossRef]
12. Jeong, W.-M.; Ryu, K.-H.; Baek, W.-D.; Choi, H.-J. Downtime analysis for Pohang New Harbor through long-term investigation of waves and winds. *J. Korean Soc. Coast. Ocean. Eng.* **2011**, *23*, 226–235. [CrossRef]
13. MOF. *Field Investigation Report for Swell Analysis and Improvement Measure Making of Downtime in Pohang New Harbor*; Ministry of Oceans and Fisheries: Pohang, Korea, 2010. (In Korean)
14. MOF. *Report on Pohang New Port Swell Improvement Feasibility Test for Basic Plan*; Ministry of Oceans and Fisheries: Pohang, Korea, 2012. (In Korean)
15. Ssangyong Engineering & Construction Co., L., Seoul, Korea. Personal communication, 2019. (In Korean).
16. Ssangyong Engineering & Construction Co., L., Seoul, Korea. Personal communication, 2020. (In Korean).
17. Ryu, K.H.; Jeong, W.M.; Kwon, J.; Chang, Y.S.; Baek, W.-D.; Kim, W.G. Evaluation of Harbour Tranquility Improvement in Pohang New Port by Detached Breakwater. *J. Korean Soc. Coast. Ocean. Eng.* **2020**, *32*, 230–241. [CrossRef]
18. Kumar, P. Spectral wave modeling of tsunami waves in Pohang New Harbor (South Korea) and Paradip Port (India). *Ocean. Dyn.* **2020**, *70*, 1515–1530.
19. Hanson, H.; Kraus, N.C. Shoreline response to a single transmissive detached breakwater. *Coast. Eng.* **1990**, 2034–2046.
20. Vicinanza, D.; Lauro, E.D.; Contestabile, P.; Gisonni, C.; Lara, J.L.; Losada, I.J. Review of innovative harbor breakwaters for wave-energy conversion. *J. Waterw. Port Coast. Ocean. Eng.* **2019**, *145*, 03119001. [CrossRef]

21. Elchahal, G.; Younes, R.; Lafon, P. Optimization of coastal structures: Application on detached breakwaters in ports. *Ocean. Eng.* **2013**, *63*, 35–43. [CrossRef]
22. Zhang, Y.; Zhao, X.; Geng, J.; Tao, L. A novel concept for reducing wave reflection from OWC structures with application of harbor agitation mitigation/coastal protection: Theoretical investigations. *Ocean. Eng.* **2021**, *242*, 110075. [CrossRef]
23. Oh, J.-E.; Jeong, W.-M.; Ryu, K.-H.; Park, J.-Y.; Chang, Y.-S. Monitoring of Recovery Process at Yeongildae Beach, South Korea, Using a Video System. *Appl. Sci.* **2021**, *11*, 10195. [CrossRef]
24. Marino, M.; Rabionet, I.C.; Musumeci, R.E. Measuring free surface elevation of shoaling waves with pressure transducers. *Cont. Shelf Res.* **2022**, *245*, 104803. [CrossRef]
25. Nwogu, O.G.; Demirbilek, Z. *BOUSS-2D: A Boussinesq Wave Model for Coastal Regions and Harbors*; Engineer Research and Development Center Vicksburg Ms Coastal and Hydraulicslab: Vicksburg, MS, USA, 2001.
26. Peregrine, D.M. Long waves on a beach. *J. Fluid Mech.* **1967**, *27*, 815–827. [CrossRef]
27. Nwogu, O. Alternative form of Boussinesq equations for nearshore wave propagation. *J. Waterw. Port Coast. Ocean. Eng.* **1993**, *119*, 618–638. [CrossRef]
28. Wei, G.; Kirby, J.T.; Grilli, S.T.; Subramanya, R. A fully nonlinear Boussinesq model for surface waves. Part 1. Highly nonlinear unsteady waves. *J. Fluid Mech.* **1995**, *294*, 71–92. [CrossRef]
29. Nwogu, O.G. Numerical prediction of breaking waves and currents with a Boussinesq model. *Coast. Eng.* **1996**, 4807–4820.

Journal of
Marine Science and Engineering

MDPI

Technical Note

Influence of Van Gent Parameters on the Overtopping Discharge of a Rubble Mound Breakwater

Federico Castiglione *, Martina Stagnitti, Rosaria Ester Musumeci and Enrico Foti

Department of Civil Engineering and Architecture, University of Catania, Via S. Sofia 64, 95123 Catania, Italy; martina.stagnitti@unict.it (M.S.); rosaria.musumeci@unict.it (R.E.M.); enrico.foti@unict.it (E.F.)

* Correspondence: federico.castiglione@phd.unict.it

Abstract: The choice of the values of the friction parameters may strongly influence the numerical modeling of the interaction between waves and porous media. Here, an assessment of such an influence is carried out using the OpenFOAM solver IhFoamV1 to simulate the response of the Catania harbor breakwater under extreme wave attack. The numerical model was validated by comparison with an experimental dataset, and a sensitivity analysis of the overtopping discharge estimate to van Gent parameter β was carried out testing values suggested by previous studies. A discussion on the importance of a careful estimate of such a parameter when dealing with the numerical modeling of porous coastal structures is presented. Indeed, variations in the non-dimensional overtopping discharge higher than 150% were observed as a consequence of a small variation (10–20%) in the absolute value of β.

Keywords: CFD modeling; wave-structure interaction; porous structure; sensitivity analysis

1. Introduction

The understanding of wave overtopping phenomena of new, damaged and upgraded harbor breakwaters is fundamental to assess their hydraulic performances and the resulting port operability and safety conditions, even more if the effects of climate change on coastal areas (e.g., sea level rise, intensification of extreme marine events) are considered.

Mean wave overtopping discharge and maximum wave overtopping volume are usually employed to evaluate the acceptability of the breakwater performances. Additionally, state-of-art prediction formulas of the mean wave overtopping discharge are widely used to design the crest level of the structure. However, such traditional formulas are affected by several intrinsic uncertainties, mainly due to their empirical nature. Moreover, experimental measurements in laboratory tests refer only to newly built structures, and do not consider low overtopping discharge values [1].

For instance, Stagnitti et al. [2] demonstrated that traditional formulas for the prediction of mean wave overtopping discharge may fail in representing the hydraulic response of non-conventional damaged and upgraded rubble-mound structures, with estimates up to 300 times higher and 0.9 times lower than the observed values for low and high-energy sea states, respectively. As a consequence, the use of traditional formulations for the probabilistic assessment of the performances of damaged and upgraded rubble-mound breakwaters may cause under or over-estimation of the failure probability of the structure due to excessive mean wave overtopping discharge by up to 70% [3].

Therefore, the importance of performing specific laboratory tests to investigate wave overtopping of harbor breakwaters arises clearly, above all when dealing with non conventional damaged and upgraded structures [4]. However, technical and economic limits make the implementation of extensive experimental campaigns difficult. In this context, numerical models represent a valid tool to extend experimental datasets, because after an initial calibration they allow the simulation of a huge number of forcing conditions and structure geometries with limited costs and times [5–7]. For example, starting from one or a few physical model tests used to validate the numerical model, the latter might be used to much more easily test different combinations of structure geometries, forcing conditions and seabed topography [8,9].

Reliable numerical wave overtopping results can be obtained only if properly calibrated numerical models are employed. In this regard, the setup of rubble-mound breakwaters numerical models is particularly challenging. Indeed, such a kind of structure is made up by several different porous layers, which are necessary to ensure the proper dissipation of wave energy while concurrently preventing wave transmission and reflection. The physical features of such layers have a relevant influence on the response of the breakwater to wave attack, which makes it necessary to use a modeling approach that adequately represents them [10].

Flow through porous media is often modeled using a macroscopic approach, which considers each porous layer as a single volume homogeneously defined by its macroscopic features. This approach is relatively fast and does not require a complex geometrical description of the structure, in contrast with the more accurate and much slower microscopic approach, which needs detailed information about element size distribution inside the layers [11,12]. However, the macroscopic approach can only solve the volume averaged interaction between flow and porous media, and it requires a complex procedure to evaluate some key parameters of the porous layers, such as the friction parameters [13].

The Volume-Averaged Reynolds-Averaged Navier-Stokes equations (VARANS) are a set of equations that can be efficiently used to apply the macroscopic approach [11]. The terms of the extended Forchheimer equation appear in the VARANS, where they represent the loss of energy by friction occurring inside the porous layers. A few methods have been proposed to evaluate the Forchheimer parameters, but the most extensively used in coastal engineering are the van Gent equations [13,14]. In turn, these equations contain friction parameters that, to this day, because of the lack of a more efficient method, are still calculated with a long trial and error approach based on the comparison with physical model tests results [15].

The lack of a relationship to calculate the friction parameters prevents from estimating the errors that their miscalculation could cause, and until now the relevance of their correct estimate has not been investigated yet [13]. Nonetheless, the inaccurate parameter evaluation might significantly affect the prediction of the overtopping discharge made by macroscopic numerical modeling. For example, an increase in the friction inside the armour layer could reduce the ability of the wave to penetrate inside the layer, thus enhancing run-up and overtopping.

In the present work, a damaged cross-section of the Catania harbor breakwater (Sicily, Italy) was analysed using a bidimensional macroscopic VARANS numerical model. The model was implemented in the software OpenFOAM (v1806), where the solver ihFoamV1 was used, and it was validated by comparison with the results of physical model tests carried out by Stagnitti et al. [16,17]. Nine simulations were carried out, each one with a different value of the turbulent friction parameter β assigned to the armor layer of the structure, and the overtopping discharge was calculated in each simulation.

The objective was to assess the sensitivity of VARANS numerically estimated rubble-mound breakwater overtopping discharge to the armor layer value of van Gent parameter β.

In the following paragraphs the numerical model is first described in terms of governing equations, grid resolution and modeled breakwater (Section 2). Then, the features of the carried out simulations are illustrated (Section 3). The results of the simulations are then presented (Section 4) and discussed (Section 5), and, finally, some conclusive remarks are drawn (Section 6).

2. Numerical Model

The solver ihFoamV1, implemented by Higuera et al. [11] for the software OpenFOAM (v1806), was used to solve the coupling between the RANS and the VARANS equations, with the Re-Normalisation Group (RNG) k-ϵ turbulence model [18,19]. The VARANS are obtained by adding the extended Forchheimer equation to the RANS, and then by averaging them over the volume of the porous layer [11].

The extended Forchheimer equation reads as follows:

$$I = au + bu|u| + c\frac{u}{\delta t} \tag{1}$$

where I is the hydraulic gradient, t is the time, u is the water velocity, and a, b and c are coefficients that need to be calculated. Coefficients a and b are calculated as suggested by van Gent [20]:

$$\rho a = \alpha \frac{(1-n)^3}{n^2} \frac{\mu}{D_{50}^2} \tag{2}$$

$$\rho b = \beta \left(1 + \frac{7.5}{KC}\right) \frac{1-n}{n^2} \frac{\rho}{D_{50}} \tag{3}$$

where α and β are the friction parameters, ρ is the water density, n is the layer porosity, μ is the dynamic viscosity of water, D_{50} is the mean nominal diameter of the layer elements, $KC = \frac{T_0}{n}\frac{u_m}{D_{50}}$ is the Keulegan-Carpenter number, T_0 is the period of oscillation and u_m is the maximum oscillatory velocity.

The model was set up to study the most exposed cross-section of the Catania harbor breakwater, whose armor layer is severely damaged. Figure 1 shows a sketch of the cross-section of the breakwater, where the sign of past damage is represented by the irregularities of the armour layer. The structure is composed of the following porous layers: a core made up by quarry rocks with a nominal diameter (D_n) in the range 0.57–2.24 m; a filter made up by natural rocks with D_n in the range 2.24–2.97 m; and an armour layer made up by 62 t cube-shaped concrete blocks ($D_n = 3$ m). The height of the top wall is 8.50 m above Still Water Level (SWL), and the water depth at the toe of the structure is 19.00 m [17].

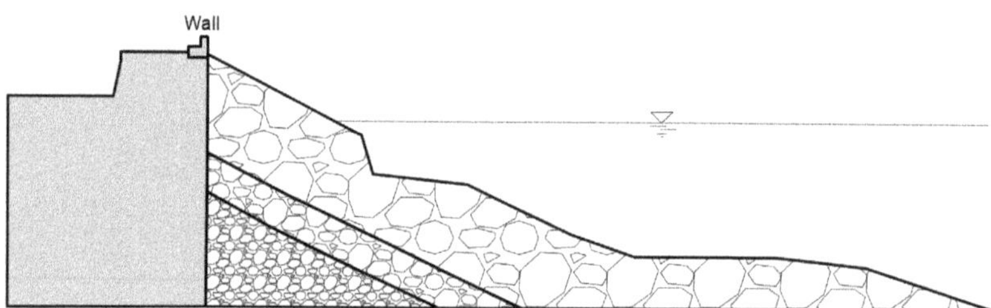

Figure 1. Sketch of the cross-section of the breakwater.

The computational domain was 310 m long and 40 m high, whereas the computational grid was made up by about 185,000 vertexes and about 95,000 cells. The dimension of each cell in the x-direction and z-direction was kept uniform, equal to 50 cm and 26 cm respectively. The solver ran in parallel on 6 Intel® Xeon® Silver 4241 2.20 GHz processors, and the ratio between the computing time and the simulated time was about 40.

The incident waves on the structure had a significant height of 6.65 m, a peak period of 11.37 s, and an average offshore wave length of about 200 m, which correspond to the 100 years return period wave climate affecting the Catania harbor breakwater [17]. Active wave absorption was used to avoid contamination of the wave signal that would otherwise be caused by the reflection of the waves at the paddle [21].

Given the long computational time, the duration of the simulations was optimized so as to balance the computational time with the accuracy of the measurement of the mean overtopping discharge. To this aim, the influence of the simulation duration on the non-dimensional mean overtopping discharge q^* was investigated. The latter is defined by the following formula:

$$q^* = \frac{q}{\sqrt{gH_s^3}} \tag{4}$$

with q being the mean overtopping discharge per unit width of structure, g the gravitational acceleration, and H_s the significant height of the incident wave.

Moreover, the percentage variation of q^* was also considered:

$$\Delta q^* = \frac{q^*(t_{i+1}) - q^*(t_i)}{q^*(t_i)} \cdot 100 \tag{5}$$

where t_i and t_{i+1} are two consecutive simulation lengths evaluated, with $t_{i+1} = t_i + 1000$ s.

Figure 2 shows the variation of the estimated mean overtopping discharge against the simulation length and against the number of overtopping waves when the value of β is 0.5 (a) and when the value of β is 4.5 (b). The corresponding percentage variations Δq^* are also shown when $\beta = 0.5$ (c) and when $\beta = 4.5$ (d).

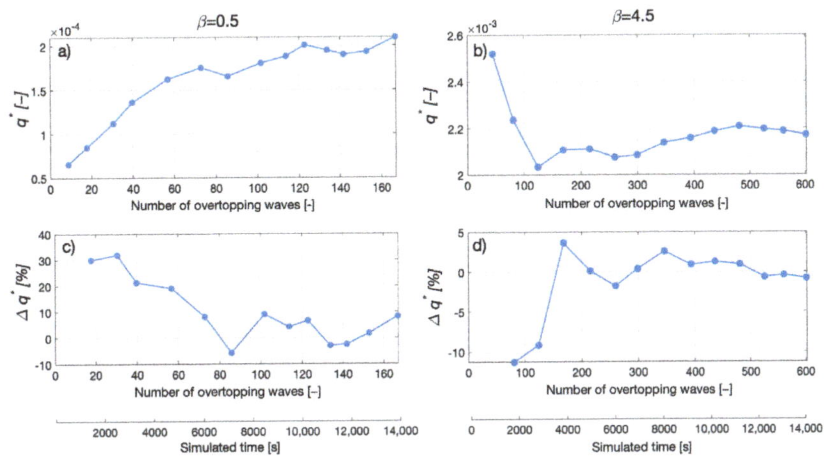

Figure 2. Estimates of q^* against the simulation duration and against the number of overtopped waves for $\beta = 0.5$ (**a**) and for $\beta = 4.5$ (**b**), and value of Δq^* against the simulation duration and against the number of overtopped waves for $\beta = 0.5$ (**c**) and for $\beta = 4.5$ (**d**).

It can be noticed that when $\beta = 4.5$ there is a much larger number of overtopping waves compared to when $\beta = 0.5$. An estimate of the sensitivity of the overtopping discharge to the value of β is the purpose of the present work, and it will be discussed in detail in Section 4. However, this phenomenon also affects the computational time. Indeed, when $\beta = 4.5$ the value of Δq^* is easily contained between $\pm 5\%$, so that the simulation of about 4000 s, corresponding to about 150 overtopped waves, can be considered enough to get a stable value of q^*. Differently, when $\beta = 0.5$, it is necessary to simulate about 13,000 s to cause the overtopping of 150 waves and to keep the value of Δq^* between $\pm 10\%$.

For the reasons described above, the simulation of 150 overtopping waves was considered enough to get a stable value of q^*. Therefore, for different values of β, the number of overtopping waves used to calculate q^* was always the same, but the simulations duration was different, leading to different values of the non-dimensional mean overtopping discharge q^*.

3. Simulations

The use of the extended Forchheimer equation (Equation (1)) and the van Gent equations (Equations (2) and (3)) implies the need to evaluate the parameters α, β and c.

Parameter α takes into account linear friction effects, being more relevant in laminar flows, whereas parameter β takes into account non-linear friction effects, being more important in the presence of turbulent flows [14]. In harbor defense structures, the armour layer is the most relevant porous layer of the structure, since it has the greatest influence on the overtopping discharge. At the same time, due to the blocks dimension, the wave-induced flow inside the armour layer can be assumed to be fully turbulent. This allows to neglect the influence that a variation in α might have on the overtopping discharge [13].

Parameter c takes into account the inertial effects due to the dynamic nature of the phenomenon, and in typical coastal engineering applications a value of 0.34 can be assumed to be adequate for all the layers [22].

For the above-mentioned reasons, the present work focused on the effect of a variation in the value of the parameter β of the armour layer on the mean overtopping discharge. Therefore, for such a layer the value of β was varied between 0.5 and 4.5 with steps of 0.5, in order to explore the range of values found in literature for similar types of structures [13]. The value of β of the remaining internal layers was instead fixed equal to 0, whereas in all the layers the values of c and α were fixed equal to 0.34 and 5000, respectively, the latter being consistent with what is used in literature for similar types of structures [13].

4. Results

The performance of the breakwater is represented by the value of the mean non-dimensional overtopping discharge, defined as in Equation (4), and it was tested for different values of the parameter β of the armour layer.

Figure 3 shows the value of q^* against the ratio between the height of the structure above the mean water level (R_c) and the significant wave height (H_s). The non-dimensional mean overtopping discharge is shown for all the values of β considered in the numerical model and for all the physical model tests carried out by Stagnitti et al. [16]. The physical model tests considered 5 different wave conditions, including the 100 years return period used in the numerical model, and each wave condition was tested three times [16].

The value of q^* predicted by the EurOtop Manual [1] and the relative confidence interval are also shown.

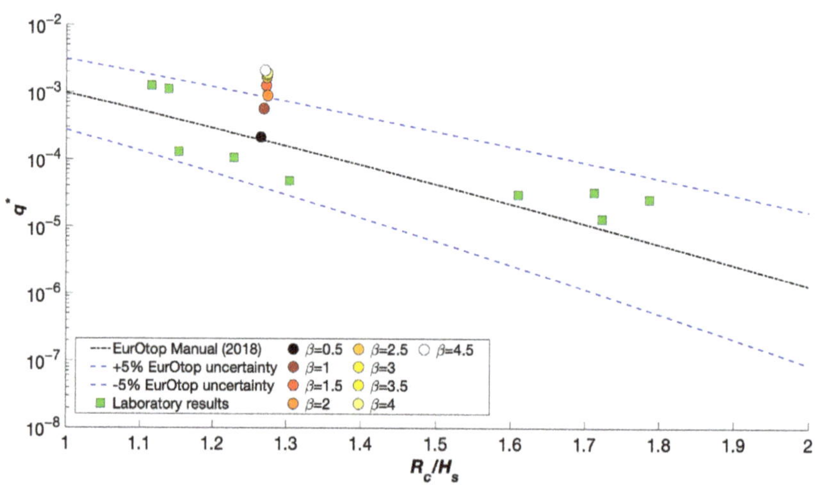

Figure 3. Comparison among the non-dimensional mean overtopping discharge q^* obtained from the numerical model for all the tested values of β, from the physical model tests and from the EurOtop formula.

The physical model results are dispersed over a wide range of R_c/H_s, and that is because of the different wave conditions tested, and also because in laboratory tests it is not possible to perfectly replicate the same conditions multiple times, so that even the three-tests groups that have the same wave conditions show some dispersion. On the contrary, numerical models allow a very stable wave conditions generation, which results in the numerical q^* values being very well aligned along a single value of R_c/H_s.

The comparison in Figure 3 shows the validation of the physical and numerical model (for different values of β), and their overall consistency with the EurOtop formula. It is possible to notice a relevant influence of the value of β on the results of the numerical model. In order to clarify such an influence, Figure 4 shows the numerical q^* against the corresponding β.

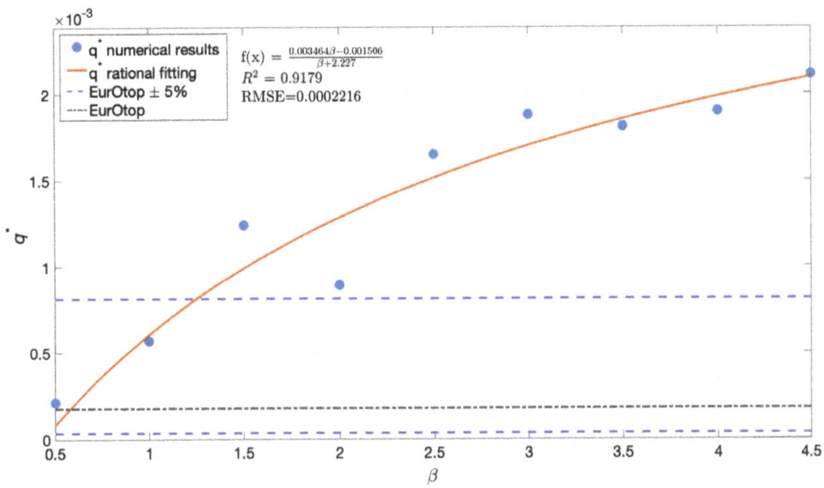

Figure 4. Non-dimensional mean overtopping discharge q^* against the variation of the friction parameter β of the armour layer. The horizontal lines represent the overtopping discharge predicted with the EurOtop formula and the relative 5% confidence interval.

The results reveal that the estimate of q^* increases with parameter β, and that the mean overtopping discharge is consistent with the one predicted by the EurOtop formula only for low values of β (0.5–1).

The fitting of a rational law to the data is shown to help visualize the trend, and it is consistent with the expected asymptotic behaviour for increasing values of β. Indeed, the increase in the overtopping discharge when β increases is physically understandable: if β increases, the friction inside the porous layer also increases, which means that the wave finds it more difficult to permeate the layer, and, as a consequence, the run-up behavior is intensified. Theoretically, if β were infinite, the armour layer could be expected to behave like a wall, since the resistance offered to permeation would be infinite (i.e., the water would not be able to enter the layer). As mentioned in Section 1, even if this qualitative result could be expected, its magnitude had not been investigated before.

To better quantify the relevance of the overtopping discharge increase, it is useful to consider the relative change of q^* as β increases. To this end, the percentage change Δq_{per} is defined as:

$$\Delta q^*_\beta = \frac{q^*_{i+1} - q^*_i}{q^*_i} \cdot 100 \tag{6}$$

where q^*_i is the value of q^* obtained with a generic value of $\beta = \beta_i$, whereas q^*_{i+1} is the value of q^* obtained with a value of $\beta_{i+1} = \beta_i + 0.5$.

Figure 5 shows the value Δq_β against the value of β. Just like in the previous figure, a parametric curve was fitted to the data to help visualize the trend. In this case a power law was used for the fitting, consistent with the x-axis being an horizontal asymptote. Indeed, for increasing values of β its effect on the overtopping discharge should decrease, and the percentage change should then become closer and closer to zero.

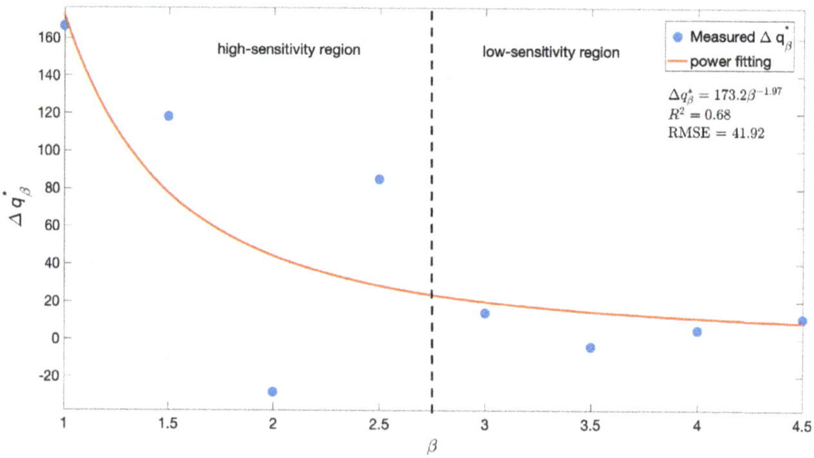

Figure 5. Relative change in the mean non-dimensional overtopping discharge q^* as a function of the parameter β of the armour layer.

A very relevant change in q^* is observed for the lower values of β (1–2.5). For $\beta > 2.5$ the results become more stable. The high sensitivity region of q^* includes the values of β that are more consistent with the EurOtop formula (see Figure 3). This proves that an inaccurate estimate of β can have a significant impact on the results. The fitting line is sometimes very far from the data, which show a sort of undulating behavior. The reason of such a behaviour is believed to be the relatively short simulation duration, which was chosen in order to have a tolerable computational time. Indeed, Figure 2 showed that variations in the overtopping discharge up to 10% are still observed for the chosen number of overtopped waves (i.e., 150). Such random variations of q^* add to the deterministic ones caused by the changing value of β, but they are not relevant enough to prevent from drawing conclusion about the sensitivity analysis. The effect described above emerges more clearly in Figure 5, but it is also present in Figure 4, where it is less evident due to the use of an absolute scale.

The results shown above prove that, for the case study under exam, the calibration of the armour layer's friction parameter β required a comparison with empirical formulas or physical model results. In particular, considering the uncertainties discussed in Section 1 about empirical formulas when dealing with damaged structures or with low overtopping discharge values, the comparison with physical model results was actually necessary. Therefore, the slow trial and error approach currently used to calibrate the friction parameters seems to be justified and necessary, considering that an estimate of their values based only on literature consultation could very easily lead to a large miscalculation of the overtopping discharge.

5. Discussion

The results of the present study greatly encourage the effort to find a faster and more efficient method to calibrate the friction parameters. Indeed, considering the growing reduction in computational times and the good results obtained when the parameters are properly calibrated, the great opportunities offered by numerical modeling lead to more and more applications [23]. A growing focus on parameter calibration can be observed in recent research works, concerning for example: uniform caisson structures and simple breakwater layouts [24], rubble-mound breakwaters [15], composite caissons having an internal rubble mound [25], and perforated breakwaters [23]. The present work contributes by quantitatively assessing the influence of the turbulent friction parameter on the modeling of the overtopping performance of a damaged rubble-mound breakwater.

Previous contributions can work as helpful guides when dealing with similar cases study, reducing the time and computational effort. However, the calibration procedure still depends on trial and errors and on the comparison with physical model results, which make the overall process very burdensome, even if methods involving machine learning are used (e.g., [15,23]), because they require a large number of validated numerical simulations with long computational times. Furthermore, the great variety of port and coastal porous structures limits the usefulness of site specific applications, and calls for a more general approach. Indeed, there is still a large uncertainty concerning the physical factors affecting the friction parameters. Lara et al. [26] suggested they depend both on flow and structure characteristics, such as Reynolds number, porosity, wave conditions, shape of the components of the structure, etc. Furthermore, the friction parameters might have a different relevance when different structure performances are considered (e.g., run-up, overtopping, reflection, pressures) [24,27]. For example, in the present work we have demonstrated that values of β larger than 2.75 have only a minor influence on the mean wave overtopping discharge. Therefore, an effort parallel to the analysis of specific cases study should be addressed toward a better physical understanding of the friction phenomenon. This is necessary to develop a more general calibration approach (e.g., physical laws tying the friction parameters to the feature of the porous structure/layers).

In this context, the results obtained in the present work suggest a future research direction to explore. Indeed, given the observed high sensitivity of the overtopping discharge to the friction parameter β of the armour layer, the mean overtopping discharge could act as guide to choose the value of β. For example, by extending the present work to a large variety of cases study, it might be possible to define a robust law expressing the non-dimensional overtopping discharge as a function of the sea state, of some key structure features, and of the friction parameters. Then, it would be possibile to choose the values of the friction parameters that allow to obtain the real overtopping discharge, calculated using empirical formulas or physical model tests. This procedure could avoid the long computational time required by the many numerical simulations necessary to calibrate the parameters' value or to train machine learning models.

6. Conclusions

Numerical modeling of port and coastal defence structures has undergone many improvements in the latest years, and it already offers many opportunities and advantages in research and design activities. Nevertheless, there are still many challenges to overcome, in particular when dealing with the modeling of wave-induced flows through porous media. Rubble mound breakwaters, being composed of many porous layers with different properties, are affected by such limitations. An approach often used in modeling is the VARANS macroscopic approach. One specific challenge of this approach is the need to calibrate the value of some parameters responsible for the energy dissipation occurring inside the porous layers due to friction phenomena. The estimate of these parameters is currently done by means of a slow procedure based on the comparison with the results of physical model tests, but their actual influence on the numerical performance of the structures had not been investigated yet.

In the present work, a bidimensional VARANS numerical model based on the solver IhFoamV1 of the software OpenFOAM (v1806) was used to study the performance of a damaged cross-section of the Catania harbor rubble-mound breakwater. The breakwater had previously been investigated also by means of physical model tests, which were here used for the numerical model validation.

Given that the most relevant friction parameter in coastal engineering is often the van Gent parameter β, in the present study the numerical model was used to estimate the overtopping discharge of the breakwater when using different values of the parameter β assigned to the armour layer. The purpose was to assess the influence of such parameter on the accurate estimate of the overtopping discharge made by macroscopic numerical modeling, as well as to verify the need to use the slow trial and error approach currently used for its validation.

It was observed, for β ranging between 0.5 and 2.75, that a variation of β of 0.5 can be enough to cause a change in the estimated value of the overtopping discharge of about 150%. Hence, a small inaccuracy (smaller than 10–20%) in the estimate of β can significantly affect the output of the model. However, values of β larger than 2.75 have only a minor influence on the assessment of the mean wave overtopping discharge.

The results of the present study highlight the need for a faster and more efficient method to calibrate the friction parameters. In particular, despite the growing interest in new approaches to parameter calibration, more research should be dedicated toward building a better physical knowledge of the friction phenomenon, so as to develop calibration approaches that can more easily be adapted to the great variety of existing port and coastal porous structures, by taking into account their specific structural features and wave conditions. Specifically, given the high sensitivity of the overtopping discharge to the friction parameter β of the armour layer, it could be useful to extend the present analysis to a large variety of cases study, in order to build a relationship between the non-dimensional overtopping discharge, the sea state, the structure features and the friction parameters. Such a relationship could allow to use a single known overtopping discharge to predict the proper values of the friction parameters without the need to carry out long calibration procedures.

Author Contributions: Conceptualization, Methodology, Writing—original draft and Writing—review & editing, F.C., M.S., R.E.M. and E.F.; Data curation, Formal analysis, Investigation, Software, Validation, Visualization, F.C. and M.S.; Funding acquisition, Project administration, Resources and Supervision, R.E.M. and E.F. All authors have read and agreed to the published version of the manuscript.

Funding: This work has been funded by: ISYPORT—Integrated System for navigation risk mitigation in PORT, funded under the PNR 2015-2020 program of the Ministero dell'Università e della Ricerca (grant number ARS01_01202); and by the projects Interazione Moto Ondoso—Strutture (IMOS) and VARIO—VAlutazione del Rischio Idraulico in sistemi cOmplessi of the University of Catania.

Institutional Review Board Statement: Not applicable.

Informed Consent Statement: Not applicable.

Data Availability Statement: The data presented in this study are available on request from the corresponding author.

Conflicts of Interest: The authors declare no conflict of interest.

References

1. Van der Meer, J.W.; Allsop, N.W.H.; Bruce, T.; De Rouck, J.; Kortenhaus, A.; Pullen, T.; Schüttrumpf, H.; Troch, P.; Zanuttigh, B. *EurOtop: Manual on Wave Overtopping of Sea Defences and Related Structures—An Overtopping Manual Largely Based on European Research, but for Worldwide Application*; HR Wallingford: Wallingford, UK, 2018.
2. Stagnitti, M.; Lara, J.L.; Musumeci, R.E.; Foti, E. Numerical modeling of wave overtopping of damaged and upgraded rubble-mound breakwaters. *Ocean. Eng.* **2023**, *280*, 114798. [CrossRef]
3. Stagnitti, M.; Lara, J.L.; Musumeci, R.E.; Foti, E. Assessment of the variation of failure probability of upgraded rubble-mound breakwaters due to climate change. *Front. Mar. Sci.* **2022**, *9*, 986993 . [CrossRef]
4. Foti, E.; Musumeci, R.E.; Stagnitti, M. Coastal defence techniques and climate change: A review. *Rend. Lincei Sci. Fis. Nat.* **2020**, *31*, 123–138. [CrossRef]
5. Vassen, J.M.; DeVincenzo, P.; Hirsch, C.; Leonard, B. Strong coupling algorithm to solve fluid-structure-interaction problems with a staggered approach. In Proceedings of the 7th European Symposium on Aerothermodynamics, Brugge, Belgium, 9–12 May 2011; Volume 692, p. 128.
6. Gerritsen, H.; Sutherland, J.; Deigaard, R.; Sumer, M.; Fortes, C.J.; Sierra, J.P.; Schmidtke, U. Composite modelling of interactions between beaches and structures. *J. Hydraul. Res.* **2011**, *49*, 2–14. [CrossRef]

7. Higuera, P.; Losada, I.J.; Lara, J.L. Three-dimensional numerical wave generation with moving boundaries. *Coast. Eng.* **2015**, *101*, 35–47. [CrossRef]
8. Gao, J.; Ma, X.; Dong, G.; Chen, H.; Liu, Q.; Zang, J. Investigation on the effects of Bragg reflection on harbor oscillations. *Coast. Eng.* **2021**, *170*, 103977. [CrossRef]
9. Gao, J.; Shi, H.; Zang, J.; Liu, Y. Mechanism analysis on the mitigation of harbor resonance by periodic undulating topography. *Ocean. Eng.* **2023**, *281*, 114923. [CrossRef]
10. del Jesus, M.; Lara, J.L.; Losada, I.J. Three-dimensional interaction of waves and porous coastal structures: Part I: Numerical model formulation. *Coast. Eng.* **2012**, *64*, 57–72. [CrossRef]
11. Higuera, P.; Lara, J.L.; Losada, I.J. Three-dimensional interaction of waves and porous coastal structures using OpenFOAM®. Part I: Formulation and validation. *Coast. Eng.* **2014**, *83*, 243–258. [CrossRef]
12. Dentale, F.; Donnarumma, G.; Carratelli, E.P. Rubble mound breakwater: Run-up, reflection and overtopping by numerical 3D simulation. In *From Sea to Shore—Meeting the Challenges of the Sea: Coasts, Marine Structures and Breakwaters 2013*; ICE Publishing: London, UK, 2014; pp. 1164–1173.
13. Losada, I.J.; Lara, J.L.; del Jesus, M. Modeling the interaction of water waves with porous coastal structures. *J. Waterw. Port Coast. Ocean. Eng.* **2016**, *142*, 03116003. [CrossRef]
14. Losada, I.J.; Lara, J.L. Three-Dimensional Modeling of Wave Interaction with Coastal Structures Using Navier–Stokes Equations. In *Handbook of Coastal and Ocean Engineering*; World Scientific: Singapore, 2018; pp. 919–943.
15. Vieira, F.; Taveira-Pinto, F.; Rosa-Santos, P. Novel time-efficient approach to calibrate VARANS-VOF models for simulation of wave interaction with porous structures using Artificial Neural Networks. *Ocean. Eng.* **2021**, *235*, 109375. [CrossRef]
16. Stagnitti, M.; Iuppa, C.; Musumeci, R.E.; Foti, E. Catania harbor breakwater: Physical modelling of the upgraded structure. *Coast. Eng. Proc.* **2020**, *36*, 2. [CrossRef]
17. Stagnitti, M.; Musumeci, R.; Foti, E. Surface roughness measurement for the assessment of damage dynamics of existing and upgraded cube-armored breakwaters. *Coast. Eng.* **2023**, *179*, 104226. [CrossRef]
18. Yakhot, V.; Orszag, S.; Thangam, S.; Gatski, T.; Speziale, C. Development of turbulence models for shear flows by a double expansion technique. *Phys. Fluids A Fluid Dyn.* **1992**, *4*, 1510–1520. [CrossRef]
19. Argyropoulos, C.D.; Markatos, N. Recent advances on the numerical modelling of turbulent flows. *Appl. Math. Model.* **2015**, *39*, 693–732. [CrossRef]
20. Van Gent, M. Porous flow through rubble-mound material. *J. Waterw. Port Coast. Ocean. Eng.* **1995**, *121*, 176–181. [CrossRef]
21. Higuera, P.; Lara, J.L.; Losada, I.J. Realistic wave generation and active wave absorption for Navier—Stokes models: Application to OpenFOAM®. *Coast. Eng.* **2013**, *71*, 102–118. [CrossRef]
22. del Jesus, M. Three-Dimensional Interaction of Water Waves with Maritime Structures. Ph.D. Thesis, University of Cantabria, Santander, Spain, 2011.
23. Dang, B.L.; Nguyen-Xuan, H.; Wahab, M.A. An effective approach for VARANS-VOF modelling interactions of wave and perforated breakwater using gradient boosting decision tree algorithm. *Ocean. Eng.* **2023**, *268*, 113398. [CrossRef]
24. Jensen, B.; Jacobsen, N.G.; Christensen, E.D. Investigations on the porous media equations and resistance coefficients for coastal structures. *Coast. Eng.* **2014**, *84*, 56–72. [CrossRef]
25. Iuppa, C.; Carlo, L.; Foti, E.; Faraci, C. Calibration of CFD Numerical Model for the Analysis of a Combined Caisson. *Water* **2021**, *13*, 2862. [CrossRef]
26. Lara, J.; Losada, I.; Guanche, R. Wave interaction with low-mound breakwaters using a RANS model. *Ocean. Eng.* **2008**, *35*, 1388–1400. [CrossRef]
27. Re, C. L.; Musumeci, R. E.; Foti, E. A shoreline boundary condition for a highly nonlinear Boussinesq model for breaking waves. *Coastal. Eng.* **2012**, *60*, 41–52.

Journal of
*Marine Science
and Engineering*

MDPI

Case Report

Climate Change Threats to Two Breakwaters

William Allsop

Principal, William Allsop Consulting Ltd., Abingdon OX12 0AR, Oxon, UK; william.allsop51@outlook.com

Abstract: This paper illustrates by practical examples how designers/owners can analyse the potential effects of climate change on coastal structures. It presents two example structures: at Brighton Marina (Sussex, England) and La Collette reclamation (St Helier, Jersey). In both instances the results of Joint Probability Analysis are used to define test conditions, noting that these will be different for different responses. Whilst the solutions for climate change challenges to each structure are different, the paper illustrates example solutions that may be relatively easily deployed.

Keywords: climate change; wave overtopping; wave forces; breakwater; coastal defences

1. Introduction

1.1. How May Climate Change Influence the Performance of Existing Structures?

The primary mechanisms by which climate change may affect breakwaters and seawalls is by elevated water levels increasing wave overtopping. The increased water depths will also reduce the influence of depth-limiting, in turn increasing incident (near structure) wave heights. Increased wave heights will not only increase overtopping but may also increase armour movement (damage) on the front face, and movement/loads to the crest and/or rear armour driven by that increased overtopping. Another (potential) failure mode that might be influenced by climate change could be toe scour, a process that can undermine an armoured slope and vertical wall sections.

Sea levels change over many time scales, some short, some much longer. Some changes are predictable, others are much more difficult to quantify. However, where we or our predecessors have built harbours or sea defences, we need those structures to continue to provide protection without attracting significant costs to resist climate change effects. Our solutions must therefore be sustainable, with an appropriately long structure life, and should be easily adaptable to resist climate change effects.

It is now appreciated that sea levels will continue to rise. Various sources are available for predicting sea level. There are however markedly fewer data on changes to wave conditions, although it is generally expected that climate changes may increase severe storm wind speeds and hence increase wave heights.

In general, the rates of most climate changes are anticipated to be slow, so changes are most likely to be significant over a typical economic life of 30 years. However, some breakwaters (including the two discussed here) defend infrastructure that requires a much longer life, perhaps longer than 100 years, and must provide protection even at the end of that duration to an annual recurrence interval equivalent to a 1:200 year return period.

When considering existing infrastructure, it is noted that any breakwater or sea defence will embody historical carbon emissions. The pursuit of 'zero-carbon' targets therefore suggest that the useful life of existing infrastructure should be extended as far as possible with minimal further carbon emissions.

1.2. Outline of This Paper

This paper describes example climate change threats (primarily increased water levels and/or wave conditions) to coastal structures in general, and then illustrates these for two sites in UK with measures recommended to resist or mitigate those threats [1,2].

The paper starts with a general introduction to the potential effects of climate change as perceived by owners of existing infrastructure and reviews guidance at the time of the analysis projects on climate change 'agents' for coastal developments in the UK. The paper then describes how climate change threats projected in 2008–2010 were simulated for structures in Brighton (Sussex) [1] then in St Helier (Jersey) [2] and identifies the mitigation measures accepted by the owners of those structures.

CAVEAT

The base studies discussed in this paper were conducted around 2006 to 2009. The guidance on climate change used in the studies reported here were those available around two to three years prior to those studies. Increases in wave conditions and water levels discussed here are therefore not necessarily those that are currently advised.

It is also noted that neither scheme required UK public funding, so the project owners were free to adopt estimates of climate change effects that they deemed appropriate for their own needs.

2. Future Climate Change Scenarios

The climate change scenarios discussed here are not the latest guidance but do summarise advice available to coastal infrastructure owners at the times of the analysis discussed in the paper. More recent discussions on sea level changes are given for the UK [3,4].

2.1. Introduction

In these projects, we distinguish between predictions, projections and precautionary practice. Predictions are typically based on measurements, analysis and extrapolation beyond the period of the data. Examples are that sea level will continue to rise in the future, based on the measurement of mean sea level rise in the recent past. Similarly, a 1:100 year return period wave condition may be extrapolated from (say) 10 or 20 years of wave measurements. Projections are typically based on measurements, coupled with assumptions (perhaps of statistical stationarity) and sometimes modelling. Examples are quantification of future temperature rise and mean sea level rise, based on validated numerical models, rerun for a range of assumed social and political scenarios, and for given relationships between physical variables. Precautionary practice may be based on a mixture of predictions and projections, the value and design life of a project, and the degree of risk to people and property that is acceptable. Examples are the choice of the return period for which to design and the future climate change allowances used, both of which are often prescribed in national best practice guidelines.

One can also distinguish between climate change allowances being built into the design and the sensitivity and future adaptability of the design based on present day conditions to changes that might occur in the future.

Both future climate change projections and guidelines for precautionary practice are updated from time to time, typically at about seven-year intervals in the UK. This section of the paper explores the different allowances for possible future changes in sea level, surge and waves that would have been appropriate under different projections and guidelines.

The Inter-Governmental Panel on Climate Change is the global scientific authority on past and projected future climate change. Below that there may be national scientific authorities, for example the UK Climate Impacts Programme. Below that there may be both country-specific and industry-specific projections and guidelines, for example adaptation to UK flood risks (https://www.ukcip.org.uk/preparing-society-for-future-weather-impacts/ (accessed on 1 November 2017)). In the context of the documents discussed in the remainder of this section, the UK Department for the Environment, Food and Rural Affairs (Defra) has the policy-making role in England and Wales and the Environment Agency the regulatory role.

2.2. Recent Observations

To put the predictions discussed below into context [3,4] provide recent data on sea level measurements around the UK.

Sea level data for five UK stations (Aberdeen, North Shields, Sheerness, Newlyn and Liverpool) have been plotted in Figure 1, (mean values in heavy blue, one standard of deviation errors in light grey) using data from Kendon et al. (2021) [3]. Sea level trends have not been simple linear increases, with the trends rising from 1.8 mm/year (1958–2018), through 2.4 mm/year to 3.6 mm/year (1993–2019) [3].

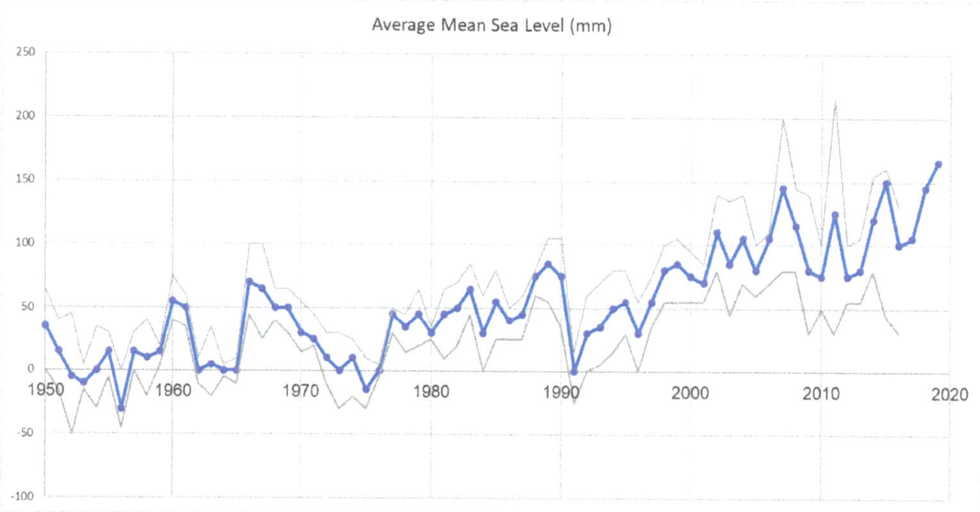

Figure 1. UK sea level indices, after Kendon et al. (2021) [3]. (Mean trend in blue, standard deviations shown in grey).

Example spatial variations of net sea level rise around UK (adjusted for land movement) [4] are summarised in Table 1. As might be expected, the ~4 mm/year rises shown in Table 1 reasonably match the recommended value of 4 mm/year recommended for use up to 2025 in Sections 2.4 and 2.5.

Table 1. Trends in sea level (linear, 1991–2020), adjusted for land movement, after [4].

Location	Central Estimate mm/year	Confidence Band mm/year
Wick	3.5	±0.6
Stornoway	5.2	±0.9
Tobermory	4.0	±0.6
Leith (Edinburgh)	3.6	±0.8
North Sheilds	3.0	±0.9
Lowestoft	3.9	±0.9
Newhaven	4.2	±0.8
Hinkley point	4.1	±0.6
Newlyn	3.6	±0.8

2.3. Guidance from IPCC (2007)

In the absence of national or local climate change projections or guidelines, appropriate future climate change allowances might be derived from the IPCC Inter-Governmental Panel's physical science basis report [5].

The global projections from IPCC might be tempered by the recent rate of mean sea level changes measured by tide gauges local to the project sites. Long term tide gauge data from southern England and northern France, show that the area would be well represented by the globally averaged rates of mean sea level change [6]. The recent rate of rise for St. Helier on Jersey is 2.30 ± 1.37 mm/year and for Newhaven (the nearest long term tide gauge to Brighton) is 2.27 ± 0.27 mm/year.

2.4. Guidance from UKCIP (2009)

The United Kingdom Climate Impacts Programme's UKCP09 marine and coastal projections [7] provided the projections of future mean sea level rise for three modelled scenarios, namely low emissions, medium emissions and high emissions, and one desk study scenario called H++, representing a plausible upper limit to future changes. Its mean sea level rise projections, averaged over the whole of the UK between 1980–1999 and 2090–2099, are summarised in Table 2 [7].

Table 2. UKCIP sea level rise projections averaged over the UK (1980–1999 and 2090–2099), after [7].

Scenario	Mean Sea Level Rise (m)		
	5th Percentile	Central Estimate	95th Percentile
High emissions	0.15	0.46	0.76
Medium emissions	0.13	0.37	0.61
Low emissions	0.12	0.30	0.48
High++		between 0.93 and 1.90	

Time-based graphs by UKCIP project an accelerating rate of rise, with about 40% of the rise summarised in Table 2, occurring in the first half of the period and 60% in the second half [7]. For the relative (that is relative to the local land level) mean sea level rise in London and on the south coast of England, UKCIP adds a further 0.075 m per century based on a current land level fall rate for SE England of 0.75 mm/year.

UKCIP also model changes in surges and waves for the low, medium and high emissions scenarios, but projected changes are minimal for the south coast of England and the English Channel (the sea area between England and France). Any allowance for change in wave conditions in coastal studies would therefore be precautionary.

2.5. Guidance from UK Government, Defra (2006)

UK Government Department for the Environment, Food and Rural Affairs (Defra) provided guidelines on future climate change allowances starting from about 2007 to 2011 [8], giving precautionary allowances for mean sea level rise for coastal defence projects seeking national government funding in England, Scotland and Wales.

The mean sea level rise allowance is given, relative to present day levels, as:

add 4.0 mm/year to 2025;
and add 8.5 mm/year from 2025 to 2055;
and add 12.0 mm/year from 2055 to 2085;
and add 15.0 mm/year from 2085 to 2115.

The wave condition allowance is given, relative to present day conditions, as:

either add 5% to offshore wave height to 2055;
or add 10% to offshore wave height from 2055 to 2115.

The Defra guidelines [8] were simple to be applied, offered users very little discretion, and assisted the comparison on an equal basis of projects competing for limited funding.

2.6. Guidance from the UK Environment Agency (2011)

The Environment Agency provides guidelines on future climate change allowances for coastal defence projects seeking national government funding in England [9]. It expects that a wider range of scenarios will be considered, demands more of users and permits them rather more discretion than Defra [8]. The four scenarios for future mean sea level rise described by Environment Agency are:

Lower end, based on the 50th percentile projection for the UKCP09 low emissions scenario;

Change factor, based on the 95th percentile projection for the UKCP09 medium emissions scenario.

Upper end, relative to present day levels:

add 4 mm/year to 2025;
and add 7 mm/year from 2026 to 2050;
and add 11 mm/year from 2051 to 2080;
and add 15 mm/year from 2081 to 2115.

High++, relative to present day values:

add 6 mm/year to 2025;
and add 12.5 mm/year from 2026 to 2050;
and add 24 mm/year from 2051 to 2080;
and add 33 mm/year from 2081 to 2115.

Possible future changes in surges and waves conclude that best estimates are for minimal changes. However, for the upper end and high++ scenarios, it recommends an additional allowance for change in surge when estimating future extreme sea levels:

either add 0.20 m to extreme sea levels by 2020s;
or add 0.35 m to extreme sea levels by 2050s;
or add 0.70 m to extreme sea levels by 2080s.

3. Brighton Marina Breakwaters, Sussex, UK

3.1. Background

The Brighton Marina to the east of Brighton, on an open coastline marked by shingle beaches backed by chalk cliffs, was first conceived around 1840. The harbour idea was then revived in 1963, and the current harbour was constructed in 1972–1977 [10].

The harbour area is protected from waves of $H_s \approx 5$ m by two breakwaters formed by circular caissons placed end-over-end. The harbour is divided into the Outer Harbour, Inner Harbour and the Tidal Harbour, see Figure 2.

Over the 30+ years since its original construction, the selected lengths of the breakwater foundation have required repair to enhance and extend toe protection. The sections of the concrete superstructure have needed repair to degraded concrete. The integrity of the West and East Breakwaters is essential to the safety and operation of infrastructure around both Inner and Outer Harbours. Whilst some damage or settlement (even the loss of a single caisson) might be tolerated from the perspective of hydraulic performance, structural damage would undermine public confidence in the safety of the harbour so could not be tolerated.

The long-term strategy study by HR Wallingford for Brighton Marina Estate Management (BMEM) assessed key hydraulic and structural responses under various present and future climate change scenarios. The study [1] was conducted in five main stages:

- Stage 0—Wave conditions and water levels.
- Stage 1—Local scour/bed condition.
- Stage 2—Wave loadings and overtopping.
- Stage 3—Option analysis using results of Stages 0–2.
- Stage 4—Physical model to measure overtopping and loading responses.

There was naturally a degree of circularity in the studies as the Stage 4 model testing did require some revisiting of results from earlier stages. In particular, the options analysis in Stage 3 was revisited during and after the physical model tests in Stage 4.

For convenience, the breakwaters were divided into five lengths, each assigned a representative orientation, toe level and approach bed slope. This paper discusses the outer part of the West Breakwater (ringed in Figure 2) where wave forces and overtopping are likely to be greatest.

Figure 2. Outer Harbour, Inner Harbour and Tidal Harbour of Brighton Marina viewed from the east (outer part of West Breakwater modelled in this study is ringed) (courtesy of Brighton Marina Estates Management).

3.2. Wave Conditions and Water Levels

The analysis of waves and water levels in Stage 0 focussed on setting appropriate water level and wave conditions to be used in the 'futures' analysis. The main outputs from Stage 0 were presented as joint probability analysis (JPA) curves derived by the method of Hawkes [11] showing wave height and sea level combinations for specified total or overall return periods (see example in Figure 3). One present day and three future scenarios were used to represent potential climate change scenarios: present day; two projections based on Defra guidance taken to 2057 or to 2107; and UKCIP medium-low projection taken to 2057. Each future scenario increased water levels and wave heights. Waves were modelled from 170° N and 210° N, directions likely to cause the most response.

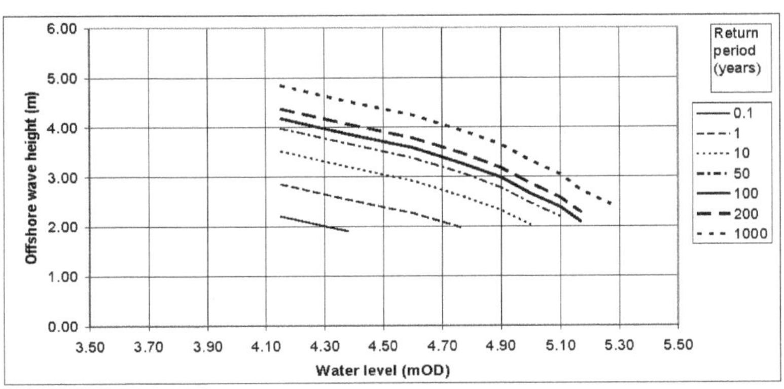

Figure 3. Joint Probability curves for 210° N, DEFRA 2107 future scenarios.

The use of the JPA curves emphasises that there is no single 'worst case' combination, so the analysis of different responses needs to explore each response over a range of wave and water level combinations. For instance, with the overtopping of the quay areas within the Inner Harbour, the main threat would be very high water levels but probably small wave heights. Contrarily, the wave overtopping of the breakwaters may be greatest under combined high water level and moderate to high wave conditions.

Bed scour may, however, reach a maximum for lower water levels and high wave conditions. The various different combinations of wave height and water levels (at the same return periods) were therefore tested to identify response maxima.

The author's experience of applying JPA curves to example structures is that wave forces generally reach their maxima when wave heights are maximised, but overtopping is low, so water levels are below maxima and coincide with none of the simple combinations above.

3.3. Scour

The analysis of scour/erosion of the chalk bed in front of the breakwaters used the original bed levels and then compared them with levels measured 30 years later. The potential for bed scour was assessed using bed shear stresses calculated by Soulsby's method [12] for 1:50 year return conditions. Historical data on bed scour at Brighton had already suggested that the chalk might erode once shear stresses exceed 20 N/m^2.

Each futures scenario envisaged increases in water levels and of wave heights and period. By reducing bed velocities, increases in water levels on their own would reduce scour potential. Increasing wave heights might partly reverse this effect but this is complicated where waves are depth-limited because higher water levels permit larger incident waves. Taken overall, the analysis suggested that scour will not increase significantly, indeed might possibly reduce, although some areas may continue to experience scour at rates equivalent to those seen over the last 30 years.

3.4. Wave Overtopping

Overtopping for a breakwater as complex in plan and elevation as these caissons (see Figure 4) cannot easily be predicted by simple empirical methods, nor were overtopping measurements available from the original modelling. The problem therefore needed to be simplified for the initial analysis but those simplifications would need to be validated (later) against the Stage 4 model test data. In analysing the overtopping, we judged that the plan-shaped indents and caisson crest/promenade features to be small in relation to wave heights and wave lengths. As with many vertical walls, a critical step was to distinguish between 'impacting' or 'pulsating' conditions using EurOtop [13,14].

Calculations using the EurOtop methods demonstrated that overtopping discharges were always greatest for the Defra 2107 condition, see example in Figure 5, when overtopping might increase by factors of 4 to 12× from present day conditions. Rerunning the overtopping simulations for the revised crest levels, however, showed that these increases might easily be countered by raising the crest by little more than 0.75 m.

3.5. Wave Loadings

The main prediction method for wave loads by Goda [15,16] can be used to give predictions of quasi-static wave loads. The Brighton caissons are in close contact with the chalk seabed, so wave uplift forces (F_u) were not included. The physical model measurements of wave pressures on selected caissons in Stage 4 were then used to adjust the hypothetical pressures predicted by this simple method.

In the physical model tests at a scale of 1:40, a 3D section of seabed bathymetry was constructed using a cement mortar overlay on compacted fill. A representative section of the West Breakwater centred on caisson W26 was reproduced using cylindrical pipe sections to represent the caissons, see Figure 6.

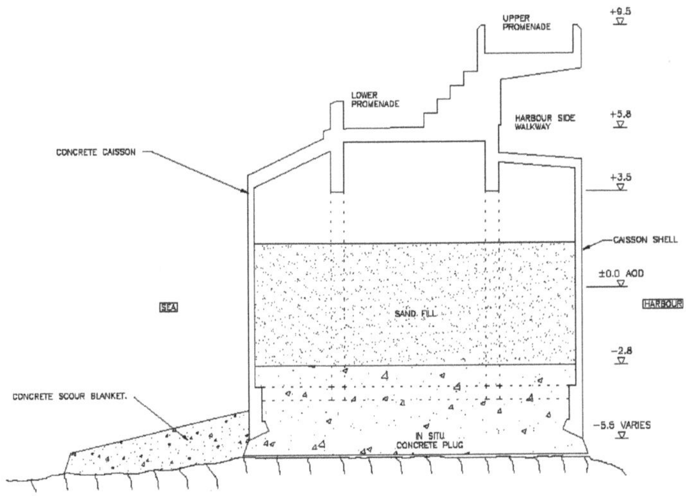

Figure 4. Example breakwater caisson.

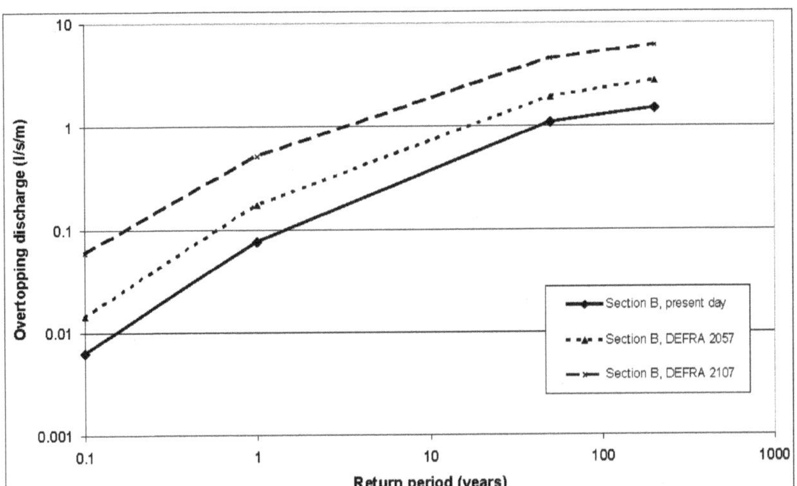

Figure 5. Wave overtopping projections.

The upper wave walls were simplified from four to two steps between lower and upper walkways. The up-stand walls were formed in plywood, with cement mortar used to form the caisson crest and to provide mass against wave forces. Waves from 170° N and 210° N were calibrated, based on the wave analysis in Stages 0–3, for the return periods of 10:1 year, 1:1 year, 1:50 year and 1:200 year.

Overtopping discharges were measured behind caisson W26 at crest level +9.5 mOD. Overtopping volumes were collected over the width of one caisson then converted to discharge rates in l/s/m width. Pressures from transducers in caisson W26 were summed to give overall forces for tests of 1000 waves logged at 790 Hz (5000 Hz model) converted to pressure (kPa) for each time step. Horizontal force (F_h) and overturning moment (M_t) were calculated at each timestep using the projected area corresponding to each transducer. Forces and moments at 1/250 exceedance ($F_{h1/250}$ and $M_{t1/250}$) were defined using the average of the top four events.

Figure 6. Model caissons made from UPVC pipe sections, plywood and mortar.

The first conclusions from the wave force measurements were that loadings under existing conditions did indeed closely match the Goda loads calculated in Stage 2. This reinforced confidence in the use of those earlier calculations.

Using those loads, the sliding analysis, however, gave Factors of Safety that fell below desired levels under present and future 1:200 year conditions (*FoS* < 1.2, for 1:50 year return conditions). This suggested that individual caissons might already have approached their stability limit at some time over the life so far.

The assumption that each caisson acts independently is, however, pessimistic, as the structural connections between caissons should allow significant load transfer. The main area of uncertainty was then geotechnical restraint against sliding. Using a simple friction relationship only, and assuming no structural attachment to the chalk, caisson stability might be under-estimated. Taken as a whole, it was, however, apparent that measures might be needed to improve restraint.

3.6. Remediation Options

Options were considered to ameliorate one or more of the key potential problems:

- Reduce scour potential, perhaps by protecting the chalk against erosion;
- Reduce wave overtopping under extreme and/or frequent conditions;
- Reduce wave loadings on the caissons, thus improving overall stability and potential for structural damage;
- Improve caisson stability by increasing resistance to loads.

Having discussed the options above with the client, possible modifications tested in the Stage 4 physical model were:

- Existing structure—baseline for the comparison of other options.
- Cusp infills to reduce local overtopping between caissons affecting boats in the harbour during frequent (>1:1 year) storms.
- A recurved wave return wall to reduce overtopping, acknowledging that this could increase wave loadings.
- Low level berm to protect against scour in front of the breakwaters, noting that overtopping and wave loads might be adversely affected.
- Armoured slopes—the most expensive option, but with the potential to eliminate scour, reduce overtopping and reduce loading on the caissons.

Since these studies, aside from repairs to the upper concrete superstructure, and local scour repairs where required, the main remedial action has been to increase caisson weight by infilling selected caissons with inert fill, increasing their weight and improving the Factor of Safety.

Taking a longer look (50 years), a possible long-term solution to both scour and stability problems could be to construct an armoured mound against the front of the existing breakwaters. An armoured mound to a moderate crest level (+4.1 mODN) provided scour protection and slightly reduced horizontal loads but did not significantly increase overtopping under extreme conditions.

A high mound (perhaps to +9.5 mODN) would give greater reductions in wave forces and overtopping for both frequent and extreme conditions, but would naturally be expensive. Large concrete armour units against the top of the caissons have the potential to impose high local bearing stresses and the armour/underlayer/caisson junction will be difficult to detail. If future developments were to be built on or behind the breakwaters, those changes of use might, however, require the lower overtopping given by this solution.

4. La Collette Breakwater, St Helier, Jersey

4.1. Background

The reclamation breakwater at La Collette, Jersey (Figure 7), was constructed during 1994–1995 for a lagoon to be filled by dry waste, and (later) to be reclaimed for port and power station users [2]. The site at St Helier is exposed to a tidal range of approximately 12 m and to wave attack along the exposed southern part of the breakwater, reaching $H_{so} \approx 8$ m at 1:200 year return. Wave heights at the breakwater are reduced by refraction and depth-limiting and (at a given return period) by the conditional probability of occurring at particular water levels.

Figure 7. La Collette reclamation, view towards North-West.

Since 1995, this breakwater has suffered damage, although perhaps not through conventional armour damage by direct wave attack or overtopping processes [2]. The users of the reclaimed area behind the breakwater now need to be able to plan their land use for the next 100 years.

This section will describe the breakwater and the damage observations. The results of a JPA of waves and water levels will be presented, including near-structure depth-limiting, then overtopping predictions for both present and future conditions, leading to the development of refurbishment/enhancement options for wave conditions and water levels some 70–100 years ahead.

Wave conditions were calculated at selected points along the breakwater/revetment for a range of water levels for each of 1:10 year, 1:200 year and 1:200 year + climate change. At each location point and water level, a check was made on depth-limiting, taking particular account of the local bed levels and slope. Using these JPA conditions projected forward, wave overtopping was simulated for present day (and future) conditions, allowing future crest levels to be derived for higher levels of protection, as discussed later in the paper. These calculations were then refined to take account of the conditional probabilities of different wave directions, and thus the effect of local wave obliquity.

The breakwater had been designed by Coode & Partners (now WSP) and modelled by HR Wallingford in 1993 for armour stability and wave overtopping. The design used core material up to +11 mCD with side slopes from 1:1.5, through 1:3, and up to 1:4 on the upper part, Figure 8. The steeper slopes at the lower elevations can be sustained as depth-limiting reduces wave heights at lower tide levels. The designers probably anticipated that the core material (50 kg–2 t) would act as a filter for the 6–10 t armour. Photographs taken on site however suggest that the core grading may actually have been wider than specified for the model testing. With hindsight, it is probable that significantly more fine material was included in the core than implied by the 50 kg lower limit.

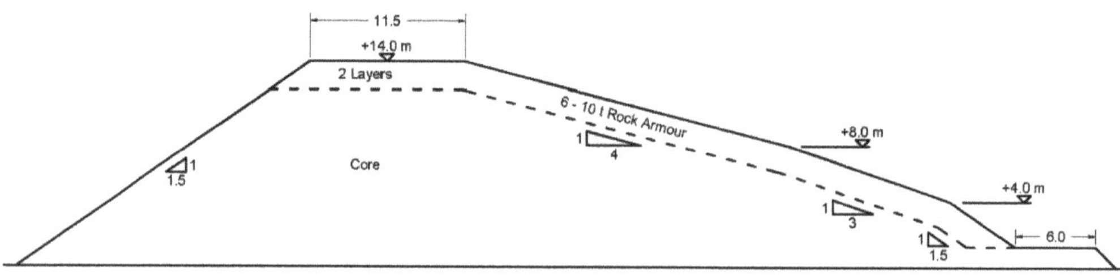

Figure 8. La Collette reclamation breakwater section, as model tested in 1993.

The filling of the lagoon started during 1995 but was slower than anticipated by the designers. By 2008, most of the lagoon had been reclaimed, and some areas had already been adopted for industrial/commercial uses. A single lagoon area remained at the time of this study, adjacent to the most wave-exposed (south-western) section of the breakwater (ringed in Figure 7).

4.2. Damage to the Breakwater

This southerly section of the breakwater is difficult to access, being backed by the tidal lagoon, accessed through a controlled tip area and having no roadway along its crest (see Figures 7 and 8). Damage to crest armour on the section behind the Dogs Nest rocks along the S/SSW section of the breakwater (Figure 9) only became apparent in aerial photos taken in 2003.

Figure 9. Wave transmission flow along core/armour interface (the author, January 2008).

The analysis of those photographs [2] suggested that the crest armour might have settled by perhaps 2–3 m along approximately 250 m where the crest width reduces from the 11.5 m on the western facing frontage to ~3.5 m on the southerly facing frontage. It was initially assumed that crest damage had been driven by wave overtopping, so initial analysis concentrated on calculating overtopping. It was, however, noted that the crest level measurements indicated that the landward part of the crest had settled the most, perhaps through the loss of supporting material. This suggested that the damage might be caused by suffusion of fine material from the core driven by intermittent flows along the core/armour interface.

This initial conclusion was supported by evidence of such flows on that interface (see Figure 9 taken by the author) and by the absence of any obvious loss of armour from the seaward face. Little or no evidence was noted of the same type or degree of damage along sections of the breakwater with the wider crest and none on those sections where the lagoon behind had been filled. This suggested that the simplest remedy to the suffusion problem would be to bring up fill material against the rear face up to crest level, removing the potential for substantial flows along the core/armour interface. The challenge then remained as to how the crest details might be configured to resist future wave events.

4.3. Water Levels and Wave Conditions

Prior to this study, a model of wave action around the island of Jersey had been set-up and calibrated using 40 years of wind data. New tide gauge data were used to refine estimates of extreme sea levels and allowed the analysis of the correlations of large waves with high water levels for various positions and wave directions to be developed in the JPA.

Separately, Jersey had previously adopted allowances for climate change to wind speed, wave height, sea level, etc., which were then used in this analysis:

- sea levels will be 0.5 m higher than present by the 2080s;
- waves will increase by 10% in terms of offshore wave height and period.

Combined wave conditions and water levels were derived around the reclamation and were then checked for local shoaling/depth-limiting. At the location of damage behind the Dog's Nest (Position C), waves track across rocks averaging +4 mCD, but bed slopes are steep at 1:10. Local depth limiting should therefore reduce wave heights for water levels below 10.5–12 mCD but less so than for shallower bed slopes.

The approach bathymetry and the relative orientation of the breakwater change significantly around section C, so it was sub-divided into sections C1, C2 and C3 (see Figure 10). Similar analysis was conducted for Sections B, D1, and D2, but are not discussed in this paper.

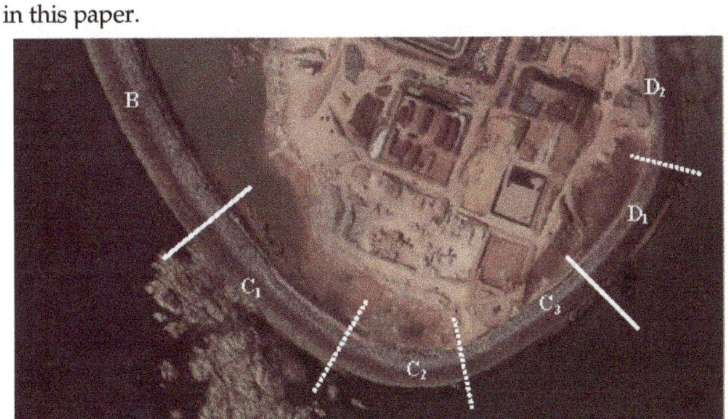

Figure 10. Locations of sub-sections C1–C3, D1 and D2.

Water level and wave conditions derived in the first overtopping calculations in Figure 11 were then refined to take increased account of local shoaling, depth-limiting and obliquity. Local wave generation was also included for easterly sectors.

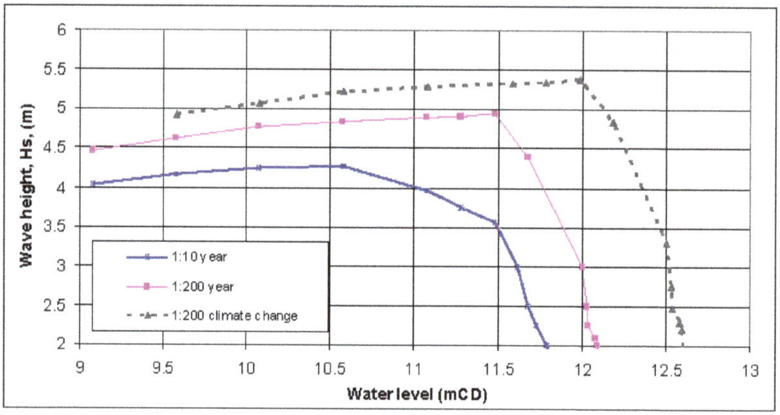

Figure 11. Initial JPA conditions for Position C.

Wave conditions and water levels for the sub-section of defence labelled C3 are shown for each wave direction in Figure 12. These follow the general form seen previously, although wave heights at C3 are reduced relative to adjoining positions. These conditions were then used as input into refined overtopping calculations.

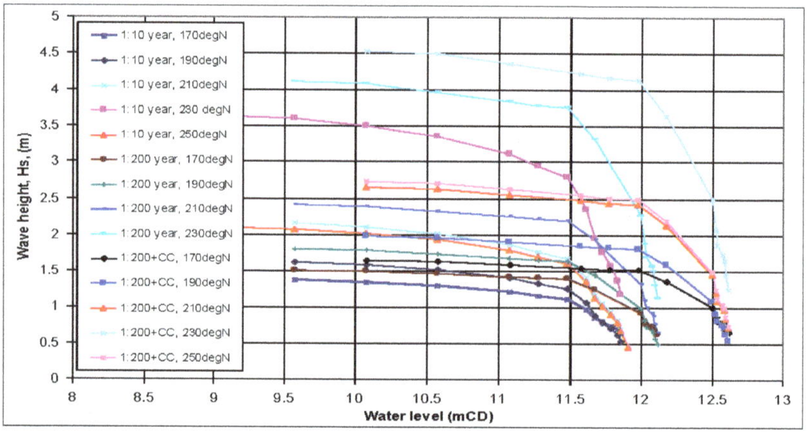

Figure 12. Refined JPA conditions for Position C3.

4.4. Wave Overtopping

It can be appreciated that it was not possible to predict beforehand which combination of water level, wave height and direction would maximise overtopping. Overtopping discharges were therefore calculated for each combination of wave direction, JPA water levels and wave conditions at 1:10, 1:200 and 1:200 year + climate change.

These calculations were then repeated using different refurbishment options and land-use scenarios:

Phase 1a—a near future conditions where land use continues as present, tested against 1:10 year combinations of water level and waves;
Phase 1b—a future condition where there would be no sensitive use close to the defence, but it would be assessed against 1:200 year and 1:200 year + climate change conditions;
Phase 2—possible future uses with industrial or similar developments close to the defences to be assessed against 1:200 year and 1:200 year + climate change conditions.

At C1, overtopping with the refined conditions was greater, but it was found that raising the crest from +14 mCD to +19 mCD would restore overtopping under the 1:200 year + climate change case to below $q \leq 1$ l/s/m, see Figure 13.

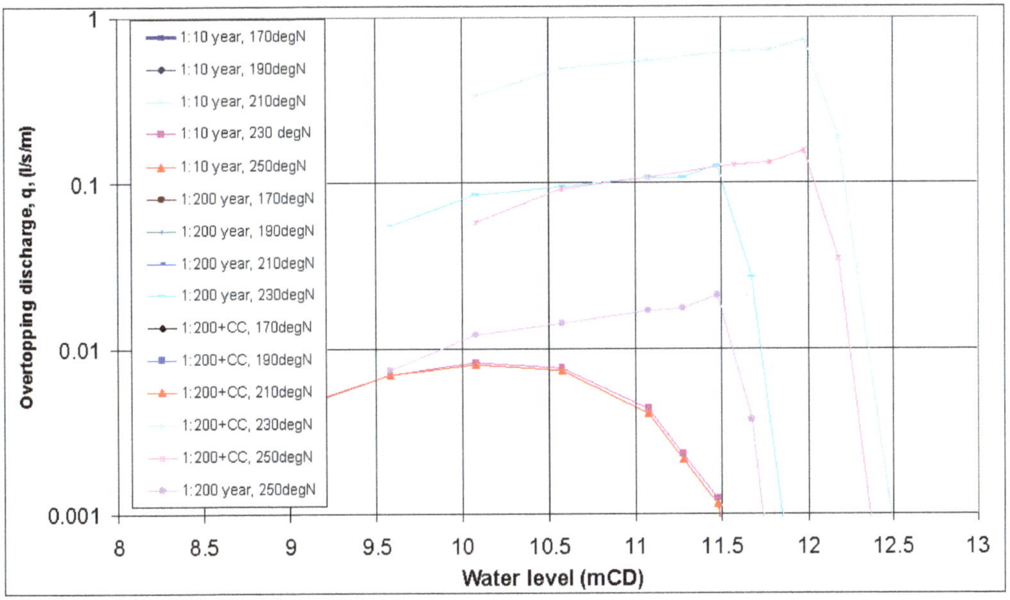

Figure 13. Overtopping at Position C1, wide crest at +19 mCD.

In contrast, at C3, refining the wave conditions as in Figure 11 had substantially reduced the calculated overtopping, so much that it would be possible to lower the (wide) crest back down to +14 mCD and still yield $q \leq 1$ l/s/m under the 1:200 year + climate change case.

The overtopping calculated in Figure 13 demonstrated that the most exposed parts of the reclamation would need increased protection before more sensitive users could safely use the reclamation land close to the revetment. The refined directional analysis however allowed the spatial extent of those enhancements to be limited, even under the full range of the 1:200 year conditions modelled.

In Phase 2, developments on reclaimed land behind the defence may need to be protected against larger threats. The defence crest might then need to be elevated. For defences at C1 and C2, the crest should be extended up to +19 mCD with an armoured slope at 1:3. This slope should end in a horizontal crest, typically three armour stones wide, backed by the access road. Example suggestions are sketched in Figure 14. If space were at a premium, it might be possible to refine these sections, perhaps using a recurved wave wall or if crest levels lower than +19 mCD were required, say for visibility.

The overtopping analysis for length C3, however, showed that the defence could be significantly reduced with a simple (wide) crest at +14 mCD. Similarly, no modifications are required for the eastern section where waves are smaller and/or generally oblique to the revetment. Along those lengths, overtopping requirements were satisfied even by a narrow crest (~2 m) under future 1:200 year conditions; indeed, the crest could (theoretically) be lowered to +13 mCD, although this was not recommended.

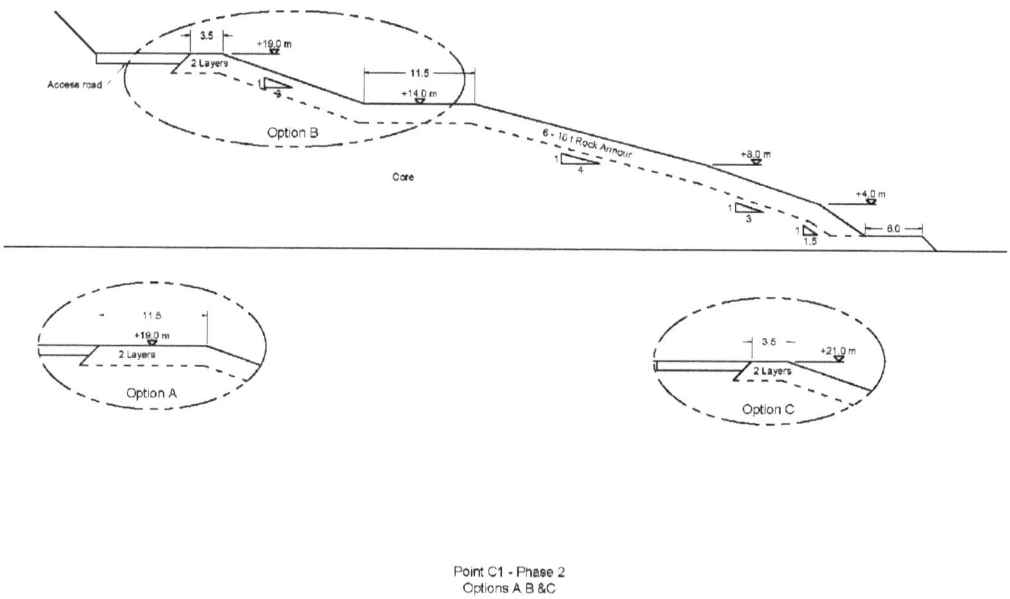

Figure 14. Suggested future defences at Positions C1 and C2.

5. Design Exercise, Rubble Revetment

The two site specific exercises discussed above might be contrasted by a hypothetical exercise by Burcharth [17], who considered that upgrading might be required for a simple rock-armoured revetment on a sandy coastline. The research under the European research project THESEUS explored the use of enhancements to the front face and/or crest and, in a few instances, the addition of new toe berms or a submerged advance rubble mound. The simple rock-armoured rubble mound chosen for the exercise faces (offshore) waves of H_{so} = 4.0 m at the 1:100 year return, and SWL = +2.3 mMWL under present conditions. Under the climate change conditions considered, the 1:100 year return wave condition increased to H_{so} = 4.2 m, and SWL = +2.87 mMWL. Unfortunately, these researchers do not seem to have conducted a joint probability analysis, so by this omission they imply that this single combination of water level and wave condition will give the 'worst cases' for each response. This is most unlikely, as is graphically illustrated in Sections 3 and 4 above.

The researchers considered possible client requirements for no change to the crest level or allow crest raising and explored the effectiveness of adding armour to the front face, extending the crest upwards by adding a new toe berm, or constructing a submerged advance mound. They also added a more challenging condition where morphological change steepened the approach slope from a 1:100 to 1:30 slope to simulate the effect of an eroding coastline. As might be expected, this change alone dramatically increased wave effects and would have substantially increased costs of upgrading.

However, for coastlines of unchanged bed slope, the upgrades required to resist climate change effects are relatively mild and the authors conclude that "the anticipated climate change does not impose a big problem as the structure can be easily upgraded at moderate cost".

6. General Conclusions

Breakwater and coastal structures are generally designed for 20–30 years of life but are often required to provide protection for much longer, e.g., for 100–120 years, when protecting dwellings or high value infrastructure.

The predictions of climate change effects, 'central estimates' or 'precautionary', have been developed by the IPCC, from which Regional Climate Models have been used to assist develop 'local guidance' by local governments or agencies. For scenarios of differing severities, these give advice on sea level increases, surges, and wave-height enhancements that should be applied to publicly funded projects or those requiring public approval. This guidance can be applied by or on behalf of any owner of coastal structures: to analyse potential threats and/or to test the efficacy of planned remediation options.

In Section 5, the study briefly reviewed various options for improving idealised (rubble mound) defence structures. This paper illustrated the development of improvement measures for a couple of (slightly) unconventional breakwaters. The key lessons are that, even for more complex structures, anticipated climate change effects can be applied over the usual 20–30 years of life of coastal structures. Climate change effects will, however, be more severe and therefore more impactful when applied to those structures expected to provide protection for extended lives for dwellings, public access or important infrastructure.

Devising appropriate measures to manage or resist increased wave overtopping is relatively routine for experienced coastal engineers. The most important aspect is always to anticipate changes in loadings and/or of usage and therefore to always anticipate the need for future adaptation.

Funding: Writing this paper was supported by William Allsop Consulting Ltd.

Institutional Review Board Statement: Not applicable.

Informed Consent Statement: Not applicable.

Data Availability Statement: Not applicable.

Acknowledgments: This paper arises from a concept originally developed by the author with Peter Hawkes at HR Wallingford and is illustrated by case studies performed by the author and colleagues at HR Wallingford. The analysis for Brighton Marina was assisted by Amanda Chapman, Katy Hooper, Alex Orbell, and Dave Baldock at Wallingford. John Scatchard (Helmsley Orrell Partnership) and John Davey (Brighton Marina Estates Management) contributed significantly to the studies [1]. The work on La Collette was again assisted by a Joint Probability Analysis by Peter Hawkes (HRW); input from Quintin Murfin, Chris Sampson, and Nigel Johnstone (Technical Services, Jersey); and colleagues from the Jersey Met Office and Jersey Harbours [2]. The writing of this paper was supported by William Allsop Consulting Ltd. without any external funding. All co-authors of the original studies have been acknowledged.

Conflicts of Interest: The author declares no conflict of interest.

References

1. Allsop, N.W.H.; Chapman, A.C.; Scatchard, J.; Davey, J. Extending the life of a coastal harbour from 30 to 150 years. In Proceedings of the International Conference on Coastal Engineering 2008, ASCE/World Scientific, Hamburg, Germany, 31 August–5 September 2008.
2. Allsop, W.; Murfin, Q.; Sampson, C. Analysis and design of refurbishment for unusual damage to a rubble mound breakwater stability of rubble mound breakwaters under wave action. In Proceedings of the Conference Coastal Structures 2011, Yokohama, Japan, 6–8 September 2011.
3. Kendon, M.; McCarthy, M.; Jevrejeva, S.; Matthews, A.; Sparks, T.; Garforth, J. State of the UK Climate 2020. *Int. J. Climatol.* **2021**, *41* (Suppl. S2), 1–76. [CrossRef]
4. Kendon, M.; McCarthy, M.; Jevrejeva, S.; Matthews, A.; Sparks, T.; Garforth, J.; Kennedy, J. State of the UK Climate 2022. *Int. J. Climatol.* **2022**, *42* (Suppl. S1), 1–80. [CrossRef]
5. IPCC. *Climate Change 2007: The Physical Science Basis. Contribution of Working Group I to the Fourth Assessment Report of the Intergovernmental Panel on Climate Change*; Solomon, S., Qin, D., Manning, M., Chen, Z., Marquis, M., Averyt, K.B., Tignor, M., Miller, H.L., Eds.; Cambridge University Press: Cambridge, UK; New York, NY, USA, 2007.
6. Haigh, I.; Nicholls, R.; Wells, N. Rising sea levels in the English Channel 1900 to 2100. In Proceedings of the Institution of Civil Engineers—Maritime Engineering 164, June 2011; pp. 81–92. Available online: https://www.icevirtuallibrary.com/doi/full/10.1680/maen.2011.164.2.81 (accessed on 7 September 2022).
7. UKCIP09. UK climate projections: Marine & coastal projections. In United Kingdom Climate Impacts Programme Report; 2009; ISBN 978-1-906360-04-7. Available online: http://cedadocs.ceda.ac.uk/1322/1/marine_and_costal_projections_full_report.pdf (accessed on 7 September 2022).

8. Defra. Flood and coastal defence appraisal guidance: FCDPAG3 Economic appraisal: Supplementary note to operating authorities: Climate change impacts, October 2006. In Defra Guidance Note; UK Department of Environment, Food and Rural Affairs: 2006. Available online: https://gat04-live-1517c8a4486c41609369c68f30c8-aa81074.divio-media.org/M4-Newport/C%20-%20Core%20Documents/17.%20Road%20Drainage%20and%20the%20Water%20Environment/17.2.20.pdf (accessed on 7 September 2022).
9. Environment Agency. Adapting to climate change: Advice for flood and coastal erosion risk management authorities. In Environment Agency Guidance Note; 2011. Available online: https://assets.publishing.service.gov.uk/government/uploads/system/uploads/attachment_data/file/902978/LIT_5707_withdrawn.pdf (accessed on 7 September 2022).
10. Terrett, F.L.; Ganly, P.; Stubbs, S.B. Harbour works at Brighton Marina: Investigations and design. *Proc. Inst. Civ. Eng.* **1979**, *66 Pt 1*, 191–208. [CrossRef]
11. Hawkes, P.J. *Use of Joint Probability Methods in Flood Management: A Guide to Best Practice, R&D Technical Report FD2308/TR2, Defra/Environment Agency Flood and Coastal Defence R&D Programme*; Environment Agency: Bristol, UK, 2005.
12. Soulsby, R.L. *Dynamics of Marine Sands*; Thomas Telford Publications, Thomas Telford: London, UK, 1997; ISBN 0-7277-2584-X.
13. EurOtop. *EurOtop: Wave Overtopping of Sea Defences and Related Structures: Assessment Manual*; Pullen, T., Allsop, N.W.H., Bruce, T., Kortenhaus, A., Schüttrumpf, H., van der Meer, J.W., Eds.; 2007. Available online: https://www.researchgate.net/publication/256197945_EurOtop_Wave_Overtopping_of_Sea_Defences_and_Related_Structures_Assessment_Manual (accessed on 1 July 2022).
14. EurOtop; van der Meer, J.W.; Allsop, N.W.H.; Bruce, T.; De Rouck, J.; Kortenhaus, A.; Pullen, T.; Schüttrumpf, H.; Troch, P.; Zanuttigh, B. Manual on Wave Overtopping of Sea Defences and Related Structures. 2018. Available online: www.overtopping-manual.com (accessed on 1 July 2022).
15. Goda, Y. *Random Seas and Design of Maritime Structures*; University of Tokyo Press: Tokyo, Japan, 1985.
16. Goda, Y. *Random Seas and Maritime Structures*, 2nd ed.; World Scientific Publishing: Singapore, 2000; ISBN 981-02-3256-X.
17. Burcharth, H.F.; Andersen, T.L.; Lara, J.L. *Upgrade of Coastal Defence Structures against Increased Loadings Caused by Climate Change: A First Methodological Approach*; Coastal Engineering 87; Elsevier: Amsterdam, The Netherlands, 2014; pp. 112–121. [CrossRef]

MDPI AG
Grosspeteranlage 5
4052 Basel
Switzerland
Tel.: +41 61 683 77 34

Journal of Marine Science and Engineering Editorial Office
E-mail: jmse@mdpi.com
www.mdpi.com/journal/jmse

www.ingramcontent.com/pod-product-compliance
Lightning Source LLC
Chambersburg PA
CBHW061440210926
51850CB00008B/1559